BRIDGING ISLANDS

Bridging Islands

Venture Companies and the Future of Japanese and American Industry

ROBERT KNELLER

OXFORD

UNIVERSITY PRESS

OXFORD

UNIVERSITY PRESS

Great Clarendon Street, Oxford OX2 6DP

Oxford University Press is a department of the University of Oxford.
It furthers the University's objective of excellence in research, scholarship,
and education by publishing worldwide in

Oxford New York

Auckland Cape Town Dar es Salaam Hong Kong Karachi
Kuala Lumpur Madrid Melbourne Mexico City Nairobi
New Delhi Shanghai Taipei Toronto

With offices in

Argentina Austria Brazil Chile Czech Republic France Greece
Guatemala Hungary Italy Japan Poland Portugal Singapore
South Korea Switzerland Thailand Turkey Ukraine Vietnam

Oxford is a registered trade mark of Oxford University Press
in the UK and in certain other countries

Published in the United States
by Oxford University Press Inc., New York

© Oxford University Press 2007

British Library Cataloguing in Publication Data

Data available

Library of Congress Cataloging in Publication Data

Data available

Typeset by SPI Publisher Services, Pondicherry, India
Printed in Great Britain
on acid-free paper by
Biddles Ltd., King's Lynn, Norfolk

ISBN 978–0–19–926880–1

1 3 5 7 9 10 8 6 4 2

This book is dedicated to the men and women in universities, technology management offices, venture businesses, venture capital companies, established corporations, and government, who are empowering scientists, engineers, and business managers to chart their professional development, expand their professional networks, and thereby to discover new outlets for their energy and creativity and to have more fulfilling and productive careers.

Preface

This book compares the role of venture companies[1] in early stage innovation[2] in new fields of science and technology in Japan and the USA. Its basic conclusion is that new companies are vital to discovery and early development in many new fields of science and technology, and thus they are vital for any industrially advanced nation whose companies seek to be competitive in new fields of science and technology. Ventures are particularly important as bridges between university discoveries and the established companies that usually undertake final commercialization of such discoveries. They are also important for the development of technologies arising in established companies that the latter, for one reason or another, do not develop.[3] The US case illustrates the importance of ventures, as well as the many preconditions for ventures to be engines of innovation.

Japan's case is interesting because at first glance it may appear that its manufacturing companies rose to the forefront in high technology fields primarily on the basis of in-house R&D, or through cooperation with affiliates within their manufacturing keiretsu.[4] Cooperation with universities was limited primarily to absorbing findings from a large number of small research projects and picking the brains of leading professors.[5] Cooperation with new, independent, domestic ventures is minimal, even today.[6]

However, in machine tools and other engineering-related fields, many companies were incorporated in the 1950s though 1970s. When they were young, some of these companies experimented with new technologies, devising new products as well as novel applications of the new underlying technologies.[7] Even after they matured, some were encouraged to innovate in areas beneficial to a main manufacturer's business. Thus innovation in new or small companies may have been a major factor in Japan's economic miracle. However, new high technology companies and small companies innovating in new fields of technology seem rare in 2006. This book explains why this was so and why, despite concerted efforts by the Japanese government to improve the environment for ventures, most ventures today are struggling and play only a peripheral role in Japanese innovation. It also shows why, in view of institutional and social factors, this situation is unlikely to change significantly in the near future.

However, the last two chapters present cases and some comprehensive data from several industries to suggest that it is dangerous for Japan, or any other industrialized country, to rely solely on established companies for innovation in new fields of technology. Although its established manufacturers may try to shore up their innovation capacity by increasing collaborations with universities, emphasizing a black-box innovation strategy,[8] or relying on spin-offs or other affiliated companies,[9] these strategies cannot substitute for the lack of new, entrepreneurial, independent companies that will take the first risky, crucial steps to develop new technologies. Without new high technology companies, Japan, and other advanced industrial countries that rely mainly on large established companies for innovation, risk being squeezed between countries that can rely on new companies to bring new technologies to proof of concept stage quickly, and countries where manufacturing can be done at lower cost with almost the same level of quality.

Confronted with this dilemma, Japan and similarly situated countries must try to improve the environment for high technology ventures, without distorting market incentives or attempting radical transforms of their innovation systems. In Japan's case, it so happens that opportunities for reform still remain, especially in the field of university–industry relations. The last pages of this book suggest reforms that would help level the playing field between large and new companies with respect to access to university discoveries. In the process they would probably also increase the quality of university scientific research.

As for the USA, another thesis of this book is that the degree to which America relies on venture companies to remain a global innovation leader is probably not generally appreciated. Moreover, as a mirror image of Japan, the ability of America to change quickly to rely more on its established companies for innovation is also constrained by deeply rooted institutional and social factors. Therefore, the economic vitality of America depends upon maintaining a supportive environment for its venture companies. This requires many factors to fall into place. There are many ways a supportive environment can be ruined. There are only a few ways to succeed. In this regard, perhaps one of the primary social and economic benefits of a strong effective patent system, of the American-style system of university–industry technology transfer, and of the so-called system of liberal market capitalism, is that they are all essential for the birth and growth of venture companies. Any change to these systems should not undermine their benefits to new high technology companies.

These are sweeping conclusions for which I tried to provide evidence in the book. However, there is much about innovation in Japan, the USA and other countries that I do not know. Working across language and geographic divides has been challenging. I have tried to find the main Japanese language information sources, but because I read Japanese more slowly than a native speaker, I may have overlooked important sources. I have a unique insider's window on how Japanese science and business is evolving. But I see only part of the whole, and I am not an insider in the sense of being privy to discussions that shape innovation policies in companies, government, or academia. Separated by distance from US companies, I often felt uncomfortably reliant on secondary sources for information on actual conditions in corporate America. Finally, because of time constraints, I have not expanded some of the analyses in this book to the degree I would like.[10]

In writing this book, I was constantly trying to achieve an appropriate balance between circumspection regarding my conclusions, and stating them clearly, sometimes provocatively. Although I have tried to point out areas of doubt as well as contradictory information, nevertheless I may have erred more often on the side of provocation than circumspection. This not because I am absolutely sure about my conclusions. Rather, by stating controversial points clearly, I hope to generate discussion about topics that are important for all countries that seek to improve their innovative capabilities and the opportunities for their scientists, engineers, and corporate managers for fulfilling, productive work. If some of my assumptions or conclusions are in error, I hope readers will come forward with information to correct them. If this book does only this, writing it will have been worthwhile.

Three additional points should be mentioned, relating to style, focus, and how I came to write this book and the persons that helped make it possible.

I have written the book so that it can be read on two levels. Reading without references to footnotes will, I hope, enable readers to understand the main points and main lines of argument easily. For this reason, I have purposely avoided almost all references in the text to the work of other researchers, detailed data, etc. My intention is that reading without footnotes will make the book interesting and accessible to a wide range of readers. However, it was essential to indicate supporting information for statements in the text, to acknowledge the contributions of others (without which this book could not have been written), and to provide relevant details and nuances. Therefore I have used footnotes liberally. I hope persons who are interested in supporting information and a greater level of detail will find this system of footnoted information satisfactory.

This book was originally conceived as a three way comparison of Japan, the USA, and Continental Europe. However, I soon realized that it would be too great a challenge to acquire information on Europe that would enable the same focused comparisons that I felt were essential to make between Japan and the USA. Nevertheless, Europe as well as China, India, and Korea were never far from mind, and I made reference to them when it seemed appropriate. The final chapter suggests that there are many similarities between Japan's innovation system and that in Korea and some countries of Continental Europe. Perhaps researchers in these countries will analyze these similarities more thoroughly.

I arrived as an Abe Fellow at the University of Tokyo in 1997 to study the Japanese system of university–industry cooperation. Previously, I had worked in cancer epidemiology and then science policy and technology transfer at the US National Institutes of Health (NIH). I was invited to join the faculty of the University of Tokyo in 1998. Some of my closest colleagues have been scientists and engineers in Research Center for Advanced Science and Technology (RCAST), an interdisciplinary research center in the University of Tokyo; a small group of RCAST scholars in intellectual property, science policy, and innovation; and staff of the University's main technology licensing organization. My initial research focused on the system of university–industry collaboration, particularly the ownership and management of university inventions. Discussion of the need to change this system had just begun when I arrived in 1997. Since then the system has changed dramatically, in form if not in actual effect. With the initial technology transfer reforms came discussions about the need for more university ventures—a topic I was already interested in from my years in NIH. By 2000, I was interviewing biomedical ventures (approximately one per month) to understand how they obtained personnel, financing, core technologies, IP rights, and customers. I was also interviewing pharmaceutical companies to understand the origins of their new drugs. This book brings together these three strands of my early research in Japan: university–industry cooperation, innovation in pharmaceutical companies, and the challenges and opportunities facing venture companies. However, in order to write this book and to compare transnationally the importance of venture companies and their environments, I had to broaden the scope beyond biomedicine and beyond the university–industry relationship.

I am extremely privileged to have had this opportunity to pursue in-depth research over a long period into a topic that combines science, medicine, law, business, public policy, economics, education, and cross-cultural studies. I am grateful to all the persons and organizations that made this possible including the Abe Fellowship Program administered by the Social Science Research Council with funds from the Japan Foundation Center for Global Partnership; the University of Tokyo and in particular RCAST which has been my institutional home for nine years; Professor Fumio Kodama who warmly received me in his RCAST laboratory as an Abe Fellow; Professor Katsuya Tamai and Professor Etsuo Niki (the director of RCAST when I arrived) who were instrumental in arranging the faculty appointment at the end of the Abe Fellowship; Professors Teruo Kishi, Yoichi Okabe, Takashi Nanya and Kazuhito Hashimoto, also former directors of RCAST who have been supportive of research; my other colleagues in the University of Tokyo who have shared with me their time, insights, and friendship; and the Ministry of Education, Culture, Sports, Science and Technology (MEXT) for providing Grants-in-aid that have greatly facilitated my research. I hope this book, in some small way, justifies the opportunity these persons and organizations granted me and returns a small portion of the benefit I have received. In this regard, I hope the perceptions and suggestions offered in this book will be regarded as the attempts of someone who cares about Japan and America and feels deeply indebted to both countries to offer constructive perceptions and suggestions for change. If I have missed the mark, the responsibility is my own, and I can only request the understanding and forbearance of the persons and organizations that made possible my research.

I have served as an adviser for several Japanese biotechnology companies, a venture capital company focused on the founding of biomedical ventures, a university technology transfer office and a company that facilitates the growth of ventures through advice to the ventures and potential investors. From all these relationships, I probably have gained more in terms of insights into the actual conditions of venture companies than I have given in the form of advice.

I am also grateful to many persons in businesses, government, and other academic institutions in Japan who have shared information and insights. I especially appreciate the corporate managers and scientists who granted formal interviews over the past seven years, some of which are presented in this book as case studies. In the case of the non-biomedical ventures, I appreciate the cooperation of the Fujitsu Research Institute (FRI) in arranging these interviews, and the willingness of FRI officials to assist me on other topics related to this book.

I owe special thanks to Professor Richard Whitley of Manchester University Business School for taking an interest in my early research, encouraging me to develop my findings into a book and providing ongoing counsel.

I am grateful to Jon Sandelin and others in US universities who have provided information and guidance. There are many others in US companies and US government institutions such as NIH and NSF who have provided helpful information. The NSF Tokyo Regional office, in particular Ms Kazuko Shinohara, has been particularly helpful and a great source of information related to science and technology in Japan and other East Asian countries.

I am very grateful to Ms Makiko Hojo who has helped me find, scan, and interpret many of the Japanese language documents this book relies on.

Finally, words cannot express the gratitude I feel towards my wife, Sachiko Shudo, and our two daughters for their forbearance and support over the four years it has taken to write this book, and also for Sachiko's advice. When I began writing, I did not realize how consuming a project this would be. Their affection and presence as I wrote and researched day by day was both a reminder of their forbearance and encouragement to press forward. As my wife is a linguist, I was additionally blessed with her insights on Japanese society and how to interpret the nuances in the various sources of information that contributed to this book.

NOTES

1. Throughout this book, I use ventures as shorthand for new independent technology-oriented companies. They may have relied on equity investments (e.g. from venture capital companies and individuals) for financing, but not necessarily.
2. By *early stage innovation*, I refer to the discovery and early stage development of new products or manufacturing processes based on new fields of science and technology, or new applications of existing fields of science and technology to discover or develop new products or processes.
3. Provided skilled employees of established companies are relatively free to leave to join new companies.
4. Keiretsu are discussed in Chapter 6. The term literally refers to *linked* companies. Bank keiretsu are linked through common reliance on a large bank for a large proportion of their loans. The importance of this linkage has declined since the late 1990s. Manufacturing keiretsu are companies linked through a large, common end product manufacturer, the archetypal examples being keiretsu associated with the major auto manufacturers. These ties are probably still important in determining the business and innovation focus of the smaller keiretsu members.
5. In other words, with the exception of some large government-funded applied, consortium research projects in which universities were major participants, the universities' role in this process was largely passive. Even in the case of the government applied research projects, there was little entrepreneurialism in the sense of universities as institutions trying to attract research funding. The degree of pro-active involvement by individual professors probably varied. On the one hand, entrepreneurial incentives such as patent royalties and industry funds to recruit more graduate students and other staff were extremely limited. However, there are examples of professors working closely with companies and contributing to innovation.
6. More precisely, as shown in Chapters 4 and 7, such collaborations take a long time to work out and must overcome high bureaucratic hurdles within the large companies.
7. In the case of machine tools, for example, applications of compact computerized numeric control devices.
8. By this I mean seeking to remain dominant in fields of manufacturing that require a high degree of non-codified, experience-based, process-specific, and sometimes delicate, art-like knowledge that is easy to keep in house and shield from rivals.

9. All of which strategies established Japanese companies are currently trying.
10. For example, I would like to expand the number of technology fields in the patent analysis in the next chapter; extend the analysis of University of Tokyo, Keio and AIST ventures through 2005 (Chapter 3, appendix); extend the analysis of high technology companies that have had IPOs to cover more years than 2000–4 (Chapter 5); include new drugs approved by the FDA in 2004 and 2005 in the analysis of the origin of new drugs (Chapter 7), and examine more infringement and unfair competition cases to really nail down the issue of whether the IP judicial system meets the needs of ventures (Chapters 4 and 7).

Contents

List of Figures   xvi
List of Tables   xvii

1. Two Worlds of Innovation   1
2. Autarkic Large Companies   19
3. Upholding the Pecking Order: Universities and Their Relations
   with Industry   42
4. Up the Rocky Road: Venture Case Studies   93
5. IPO or Bust: Venture Financing   168
6. Amoeba Innovation: The Alternative to Ventures   192
7. Innovation Across Time and Space: Advantage New Companies   232

Glossary   377
Index   386

List of Figures

1.1. Patent applications in genomics, proteomics, and related applications   2

1.2. Issued patents covering hip and knee prostheses   3

1.3. Issued patents covering video cryptography   4

1.4. Issued patents covering rewritable electromagnetic recording devices   5

1.5. Issued patents covering tomography and planar medical radiography   6

1.6. Issued patents covering irradiation devices, especially for X- or gamma ray lithography   7

1.7. Issued patents covering ion beam tubes and ion sources   8

1.8. Issued patents containing 'micromachine' in the title   9

1.9. Issued patents containing 'nano' in the title   10

3.1. Number of Japanese startups formed per year   50

3.2. New and ongoing joint research projects between private companies and national universities   51

3.3. Inventions by field reported to one university   52

3.4. Life science inventions: association with joint research and type of industry partner   53

3.5. Non-life science inventions: association with joint research and type of industry partner   54

7A4.1. Age-specific mid career departure rates for men in all manufacturing firms with $\geq$ 1,000 employees   311

7A4.2. Age-specific mid career departure rates for men in all manufacturing firms with 30–99 employees   312

7A4.3. Total male employees in manufacturing industries by firm size   313

7A4.4. Total female employees in manufacturing industries by firm size   313

List of Tables

2A.1. Summary information for the nine companies whose pipeline
drugs are analyzed   31

3.1. Leading recipients of Monbusho/MEXT grants-in-aid in 1995 and 2005   64

3.2. Leading academic recipients of commissioned research in 2004   65

3.3. Leading recipients of centers of excellence disbursements in 2006   65

3.4. Projected leading recipients of operational and administrative
subsidies for national universities, April 2004 to March 2010   66

3A. Major competitive Japanese government S&T funding programs in 2002   68

4.1. Employment in therapeutic-oriented Japanese bioventures in 2005
and US therapeutic ventures of equivalent median age in 1987 and 1999   94

4A2.1. University of Tokyo, Keio University and AIST startups, formed
1995–2003   138

4A2.2. Founding years of U Tokyo, Keio and AIST startups   141

4A2.3. Distribution by field of U Tokyo, Keio, AIST and MIT startups   141

4A2.4. Financial status of U Tokyo, Keio and AIST startups   142

5A.1. Manufacturing or R&D-focused Japanese companies with IPOs
2000–2004   178

7A1.1. Intellectual property holdings according to size and age of Utah
and New York bioscience companies   305

7A2.1. Numbers of new FDA approved drugs by type and type of
organization (where investors worked)   306

7A2.2. Numbers of new FDA approved drugs by type of drug, type of
employer institution, and location of laboratory where investors worked   307

7A2.3. Share of new FDA approved drugs from regions with few biotechs
compared with regions with many biotechs   308

7A3.1. Leading Japanese machine tool companies ranked by 2004 sales   309

1

Two Worlds of Innovation

Japan is a nation where early stage innovation, that is the discovery and early stage refinement of new products and processes, occurs primarily in large established companies. In contrast, early stage innovation in the USA is relatively evenly divided between established companies, new companies, and universities. The reasons for this difference, its persistence despite policies to improve the environment for ventures in Japan, and its implications for economic and technical progress in both countries, are the central issues of this book.

But to begin, what evidence supports this basic assertion? Others have described the challenges facing Japanese ventures.[1] However, there have been few studies that trace the development history of new products in different countries to show whether they originated in large companies, universities, ventures or other small companies. I have done this in the case of pharmaceuticals and thus I know the assertion to be true in the case of this industry. But to my knowledge, third generation mobile telecommunications technology is the only other field in which such an analysis has been conducted (see chapter 7 and note 313). One of the recurrent issues in this book is the degree to which innovation in biomedical industries differs from that in other industries.

Simply looking at available data on the number of new companies in high technology industries does not give a clear picture of sources of innovation. The rate of new company formation in Japan has been among the lowest among industrialized countries.[2] But since 1998 the numbers of Japanese university startups[3] and biomedical ventures have been increasing rapidly. Indeed on a per population basis, or comparing the numbers of Japanese startups with those in the USA *an equivalent number of years* after enactment of the laws facilitating startup formation, the Japanese numbers are quite respectable. Also, the numbers of established Japanese small and medium size enterprises (SMEs) engaged in manufacturing and even new product development are considerable.

This book addresses these inconsistencies later. But for the purpose of this introduction, a comparison of what types of organizations are obtaining US patents in a small selection of high technology fields provides preliminary

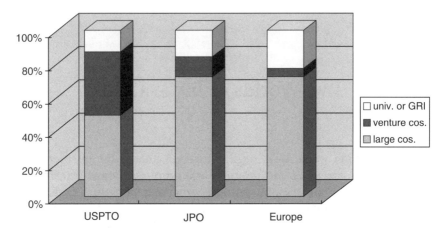

Figure 1.1. Patent applications in genomics, proteomics, and related applications (1990–97, percentages only)

Source: JPO 2002.

support for the basic assertion. Patents are not ideal indicators of innovation. Nevertheless, issued patents at least represent a subset of new discoveries that, for the applicants, merited the expenditure of funds to obtain the right to exclude others from using those discoveries.[4]

To my knowledge, the only comparative survey of the origins of patents within an entire industry[5] was conducted by the Japan Patent Office (JPO). It covered genomics, proteomics, and related patent applications in the USA, Japan, and major European countries.[6] As shown in Figure 1.1, while venture companies accounted for nearly 40 percent of US applications, they accounted for only 12 and 6 percent respectively of applications in Japan and Europe.[7] Conversely, large companies accounted for only about 50 percent of US applications but 72 percent of applications in both Japan and Europe.

Lacking similar information for other industries, I selected the International Patent Classification (IPC) codes covering six narrow nonpharmaceutical technologies that draw on new scientific or engineering knowledge.[8]

- Hip and knee prostheses,
- Video cryptography,
- Rewritable electromagnetic recordable devices,
- Tomography and planar medical radiography,
- Irradiation devices, especially for X- or gamma ray lithography, and
- Ion beam tubes and ion sources.

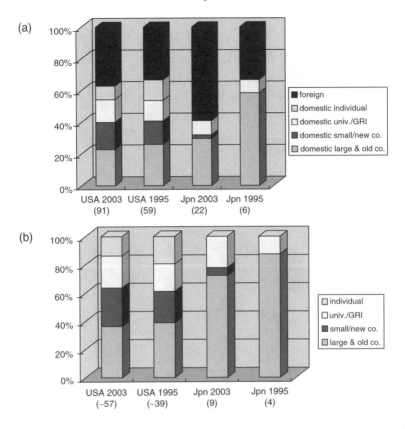

Figure 1.2. Issued patents covering hip and knee prostheses: (a) all applicants and (b) domestic applicants only

Sampling the US and Japanese patents issued in 1995 and 2003 in each of these categories and classifying the applicants according to nationality and whether they were

- individuals,
- universities or government research institutes (GRIs),
- SMEs (under 500 employees) or new companies (formed 1975 or later), or
- large companies (at least 500 employees and incorporated before 1975),

enabled me to make the following graphs (Figures 1.2–1.7). (The numbers in parentheses below each bar indicate the total number of patents for each category.)

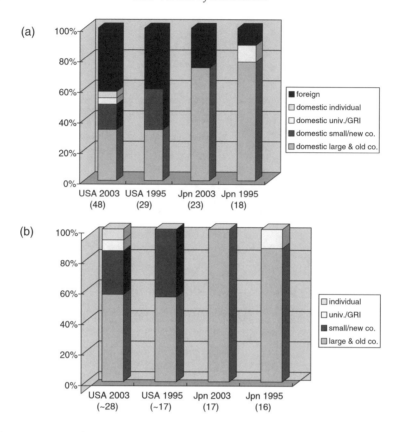

Figure 1.3. Issued patents covering video cryptography: (a) all applicants and (b) domestic applicants only

The principal difference between the US and Japanese applicants is that universities and new companies account for significant proportions of domestically originating patents in the USA, but much smaller proportions of domestically originating patents in Japan. With just a few exceptions, small or new Japanese companies do not appear as innovators in these fields, and when they do, they are usually old small companies.[9]

Of course, there is variation among technical fields. In medical tomography and radiography, innovation appears to occur almost exclusively in large companies such as General Electric. In rewritable electromagnetic recording devices such as DVDs, innovation seems confined to large foreign (mainly Japanese) companies. But in hip and knee prostheses (which often incorporate advances in materials science), video cryptography (which involves software and electrical engineering), high energy lithography (especially for integrated

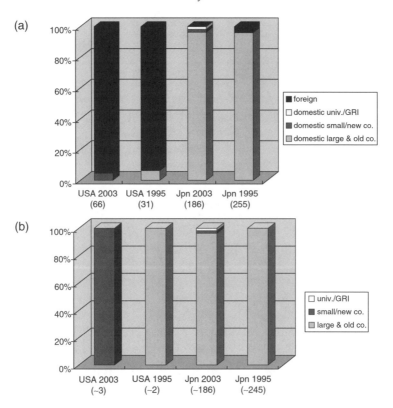

Figure 1.4. Issued patents covering rewritable electromagnetic recording devices: (a) all applicants and (b) domestic applicants only

circuit design and manufacture), and ion implantation devices (for doping various materials to improve the performance of semiconductors), US venture companies account for a significant proportion of innovative activity, but Japanese ventures very little.

Moreover, this analysis suggests that the relative contribution to innovation of US small or new companies is not diminishing. In Japan, there is no indication that this proportion is increasing. However, the share of universities and GRIs may have increased slightly between 1995 and 2003 in both countries.

This evidence is not conclusive proof of the assertion at the beginning of this chapter. There are hundreds of IPC codes and I analyzed the patents under only six. I cannot claim that these are representative of all nonbiomedical industries. Nevertheless, they do suggest differences between the two countries that may be consistent across a range of rapidly evolving scientific and engineering fields. If this is indeed the case, then there are probably many

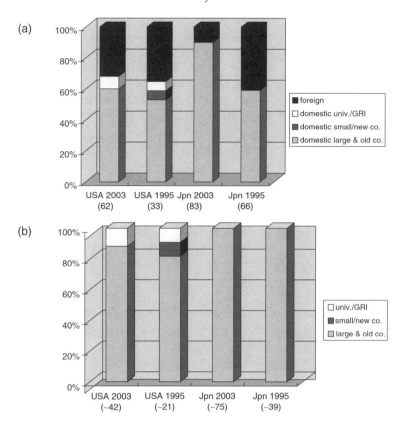

Figure 1.5. Issued patents covering tomography and planar medical radiography: (a) all applicants and (b) domestic applicants only

high technology industries in which small or new US companies are leading innovators, but few industries in which new or small Japanese companies are leading innovators.

I approached this issue in a different way by sampling the first pages of all US and Japanese patents issued in 2003 and 1995 that contained 'micromachine'[10] or 'nano'[11] as a title word or as a fragment of a title word. The inventions reflect a variety of applications of micromachine (including micro-electrical mechanical systems (MEMS)) and nanodevice or nanoparticle technologies (Figures 1.8 and 1.9). The pattern is even starker than when selecting patents according to specific IPC codes.

While both micromachines and nano patents have increased sharply since 1995, the share of small or new US companies has increased and that of large

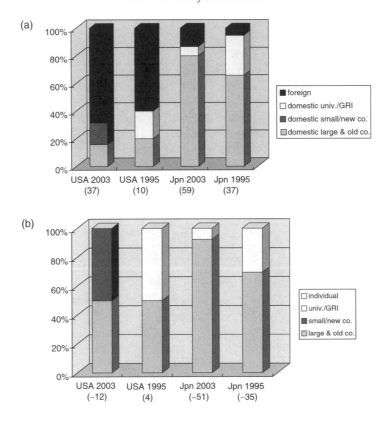

Figure 1.6. Issued patents covering irradiation devices, especially for X- or gamma ray lithography: (a) all applicants and (b) domestic applicants only

and established companies has decreased. In contrast, Japanese new or small companies are playing a negligible role.[12]

Two specific points relate to the main conclusions of this book. The first concerns patents to individual inventors. About 10 percent of the US patents issued to US applicants list no assignee. In other words, the inventors applied for the patents on their own. About one-quarter were university faculty, about 30 percent were entrepreneurs who had founded viable businesses in the field of their patents.[13] Some are prolific inventors.[14] In contrast, only one of the Japanese patents issued to Japanese applicants was unassigned, and in this case, the inventors turned out to have been University of Tokyo faculty at the time of the invention. In other words, compared to Americans, Japanese inventors rarely apply for patents on their own. Affiliation

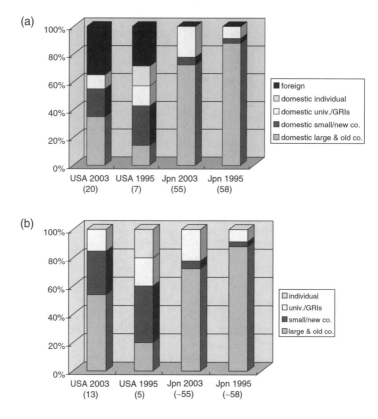

Figure 1.7. Issued patents covering ion beam tubes and ion sources: (a) all applicants and (b) domestic applicants only

with a large company seems to be necessary for inventive Japanese to realize the patenting and commercialization of their discoveries in many fields of technology.[15]

The other point concerns patenting by universities and GRIs. Overall about 13 percent of the Japanese-origin Japanese patents I surveyed were attributable, at least in part, to research in Japanese universities or GRIs, that is, they had had at least one university or GRI inventor. Over half of these patents arose under collaborative research with a Japanese company—in over 90 percent of such cases with a large, established company. In other words, these data suggest that Japanese universities and GRIs do play an important and increasing role in innovation, although probably not as great as their US counterparts, which accounted for 22 percent of the US-origin US patents in my survey.[16] University and GRI innovation frequently occurs in collaboration with large,

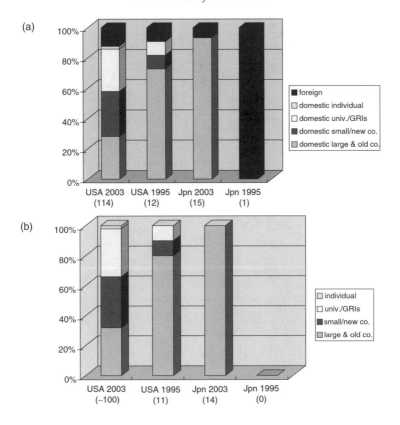

Figure 1.8. Issued patents containing 'micromachine' in the title: (a) all applicants and (b) domestic applicants only

established companies, but rarely in collaboration with small companies, and even more rarely in collaboration with new companies—at least outside of biomedicine.[17] Later chapters show that university–industry collaboration in Japan is indeed biased in favor of large companies.

New companies once flourished in Japan. The immediate postwar years saw the formation of Sony (1946), Sanyo (1947), Honda (1948), and Kyocera (1959). Sony pioneered innovations in transistor technology and their applications first to radios then to a range of other electronic products. Kyocera (short for Kyoto Ceramics) became a leader in the application of materials science to electronics and other products. Also during the 1950s and 1960s, Hayakawa Electric transformed itself from a struggling medium-size maker of radios and televisions to the world's leading pioneer of liquid crystal displays and the company we know today as Sharp.

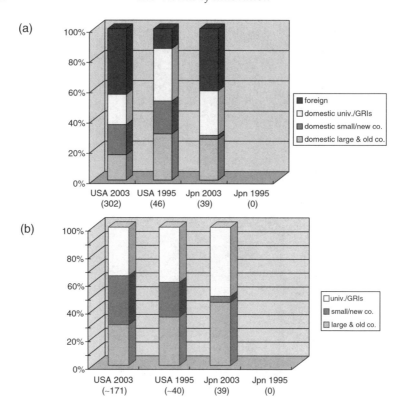

Figure 1.9. Issued patents containing 'nano' in the title: (a) all applicants and (b) domestic applicants only

SMEs still contribute significantly to the Japanese economy. In 2000, SMEs accounted for about 89 percent of employment and 57 percent of value added in Japanese manufacturing, higher levels than in 1970.[18] However, at least until recently, a majority of manufacturing SMEs probably relied on subcontracting work for most of their business, and approximately one-third relied on subcontracting from a single customer.[19] As discussed in Chapter 6, this may have limited innovation and growth opportunities for many. Recently some established high technology SMEs have been trying to diversify their customer base and to develop new products, but some remain focused on meeting the needs of a few large customers.[20] Some large companies maintain traditional relations with the SMEs that depend on them for most of their sales. Their contracts with the SMEs contain generous profit margins and they provide the SMEs with technical information so that the SMEs can manufacture state-of-the-art components. But at least in some cases, if one of these SMEs tries to recruit other customers, orders from the large company will be cut immediately.[21] In any event, the patent analysis above suggests that as of 2003, SMEs had still not become a major force in early stage innovation in new or rapidly evolving, science-based technology fields. But my limited analysis may have missed fields in which they are leading innovators,[22] or else, their contributions may yet result in new patents.

However, the focus of this book is not SMEs, in general, but rather new independent high technology companies. Chapter 4 examines the present status of these companies in Japan and the opportunities and challenges they face. It includes twenty case studies of ventures in biomedical and nonbiomedical fields. But in Japan, it has been more common for new technical opportunities to be exploited by established companies moving into new fields, and Chapter 6 examines this phenomenon and tries to explain the factors on which the success or failure of such efforts have depended. Finally, Chapter 7 addresses the larger issue of whether small/new or large/established companies are better at early stage innovation taking into account the importance of intellectual property, mobility of people, and other factors. Then, noting the current reliance of Japan on large companies and the USA on new companies for innovation, it examines prospects for change and offers some suggestions how the environment for innovation can be improved in both countries and in countries with similar innovation systems.

But to understand the environment for ventures and the challenges they face, it is necessary first to understand the role that large companies and universities[23] have played in Japan's innovation system and the degree to which they have been willing to cooperate with venture companies. Chapter 2 explores the tendency of large established Japanese companies to innovate autarkicly, that is, to rely on their own in-house R&D laboratories for new prototype products and to try to maintain control over the upstream components of a vertically integrated value chain. Chapter 3 examines the role Japanese universities have played in Japan's innovation system. It also shows how, despite policies that have improved the environment for university startups, large companies maintain preferential access to university discoveries and barriers remain to the growth of strong university startups.

For venture companies to flourish in Japan, large companies will have to come to view independent smaller companies as long-term sources of new ideas and technologies that depend on the smaller companies' ability to grow rapidly. Large companies must become less autarkic and more networked with other independent organizations, in terms of both product discovery and the flow of personnel.

However, at least with respect to building bridges between large and small companies, this process will not be easy because it will be seen as undermining the fundamental strength of Japanese manufacturing based on lifetime employment and integrated control over all steps of the process from R&D to manufacturing and marketing. Instead, large Japanese companies are cooperating more actively with universities in order to access more early stage discoveries, but they are ceding little R&D to small companies. A few large manufacturing companies actively seek alliances with independent domestic small or new companies. But this low level of engagement probably will not enable new high technology companies to become engines of innovation for Japan's industry. Furthermore, lifetime employment in large companies may always be more attractive than work in small, new companies, absent the low levels of job security in the USA. In other words, despite important changes in Japanese government policies and even some changes in corporate practices that have improved the environment for startups, Japan's innovation strength may continue for many years to rest with its large established companies, while at least for the near future that of the USA will rest to a large extent on new companies. Whether one of these systems will prove to be superior to the other is one of the main themes of Chapter 7, but ultimately time will provide the answer.

APPENDIX: METHODOLOGY FOR ANALYSIS OF PATENTS ACCORDING TO TECHNOLOGY FIELD AND TYPE OF APPLICANT

Although I used the International Patent Classification (IPC) codes to select Japanese and US patents in equivalent technology fields, I actually used the US PTO classification to select the particular technologies I would investigate.

The IPC codes are 8 character alphanumeric codes published by the World Intellectual Property Organization. They tend to be based largely on constituent materials or components, or underlying scientific processes, rather than on industrial use (at least this was the case with the seventh version codes, the latest available when I did this analysis in early 2004). Therefore, for this analysis, they are not ideal. But because US, Japanese, and most European patents are classified according to IPC codes, they can be used to compare patenting activity between countries.

The US PTO has its own unique classification system based more on the end use or overall function of an invention. US patents are classified according to both systems, but the US PTO classification cannot be used for international comparisons. (In other words, I can search Japanese patents using IPC codes but not US PTO codes.) However, I used the US classification list available at http://www.uspto.gov/

web/patents/classification/selectnumwithtitle.htm to select twenty-one three-digit classifications of possible interest, because they represented a spectrum of technologies that are rapidly evolving and, to a large extent, depend on scientific or engineering progress in several countries in a number of research centers. (In other words, I avoided fields that where progress depends on R&D in a small number of countries or laboratories, e.g. automobiles, aircraft, pulp and paper, and nuclear power.) Then I used the US PTO's concordance system at http://www.uspto.gov/ web/patents/classification/ to find equivalent IPC classifications. After reviewing the concordant IPC classifications for about half the twenty-one candidate US PTO classifications, I selected the six categories presented in this chapter, largely on the following criteria: (*a*) they would represent different types of technologies and (*b*) a small number of IPC codes would encompass a conceptually distinct and meaningful class of technologies.

Thus, I obtained lists of all US and Japanese patents issued (registered) in 1995[24] and 2003 for inventions classified under the following IPC codes:

A61F 2/32, 2/34, 2/36 & 2/38: *hip and knee (and some abdominal area) prostheses*;
H04N 7/167: *video cryptography*;
G11B 11/00: *rewritable electromagnetic recording devices*;
A61B 06/02 & 06/03: *tomography and planar medical radiography*;
G21K 05/00: *irradiation devices, especially for X or gamma ray lithography* and
H01J 27/00 & 27/02: *ion beam tubes and ion sources (for IC chip manufacturing etc.).*

I randomly sampled among each of these sets of patents to obtain about twenty patents for each year-country-IPC code category of patents. Thus, when my sample was less than the total number of patents, my numbers in parentheses represent an *estimate* of the number of *domestic applicants only* patents based on the proportion of such patents in my sample frame. (If there were fewer than twenty patents, in a category, I selected them all.) Altogether there were 1,890 patents in the 32 sampling frames (8 technology categories (including micromachine* and nano* mentioned below), two national patent offices, two years) and out of these I sampled 673 (36%).

Then I printed out at least the first page of each patent application which identifies the names and addresses of the inventors and the patent applicant(s). I assigned national origin according to the addresses of the inventors. (A few patents had coinventors from several countries and I attributed these inventions to the nationality of the majority of the inventors.)

As for the type of institution where the inventions occurred, I relied on the identity of the applicant in the case of US patents. This is reasonable because US universities and SMEs generally insist on applying for inventions by their employees, as described in subsequent chapters. However, this is not the case for Japanese universities prior to 2004. Moreover, in Japan but not the USA coinventorship involving universities/GRI and corporate researchers is common. *However, Japanese patents usually list the inventors' work addresses.* Among the small number of Japanese inventors whose affiliations were not clear from their addresses, I was able to find affiliations for almost all from public sources.

In the case of companies, I determined from public sources their years of incorporation and numbers of employees.

The vast majority of inventions were assigned by their inventors to their employers. Among those with no assignee, however, I used various public sources to determine their principal affiliation.

NOTES

1. e.g., Rowen and Toyoda (2002), Ibata-Arens (2000), Feigenbaum and Brunner (2002) Maeda (2004), Nakagawa (1999), and Suzuki (1999).
2. JSBRI (2003).
3. Throughout this book, I use the term *startup* to refer to a new, independent company whose core technology is based on university or GRI discoveries. A nearly synonymous term (more common in Japan) is university venture. As noted in the glossary at the end of this book, I distinguish startups from *spinoffs*, in that the latter are formed from technologies or personnel from existing companies.
4. Sometimes patent applications will be filed with little intention of prosecuting the application to obtain a patent. For example, in most countries, patent applications are published after eighteen months and such publication prevents rivals from patenting these discoveries. In other words, the application alone and subsequent publication can prevent patenting by rivals working in the same area. Also at least in Japan, numbers of patent applications are commonly used by employers in promotion decisions and by government agencies to evaluate the 'success' of applied research that they fund. Thus any comparative international analysis of innovation should use *issued* patents rather than patent *applications* whenever possible. Since the USA is the world's largest consumer market covered by a single, unified patent system, in many cases inventors and companies who think they have commercially valuable discoveries will try to obtain US patents. Thus issued US patents probably are appropriate to use for international comparisons of innovation. Yet these assumptions may not always hold true, especially in the case of non-US inventors and non-US companies thinking only of their domestic markets.

    However, even some issued US patents cover discoveries that the patent holders do not plan to develop. Rather, they were obtained to block competitors or to serve as ammunition or bargaining chips in case of patent disputes with other companies. Finally, even in the case of patents that are intended to protect the patent holder's discoveries related to its core businesses, it is difficult for nonspecialists to determine which patents have significant commercial value or represent significant technical achievements.

 Nevertheless, prosecution of a US patent application to issuance requires on the order of US$ 10,000. If translation fees and local attorney fees are included in the case of applications covering countries such as Japan, China, or Continental Europe, costs are substantially higher. To obtain patent protection in the world's major markets requires on the order of US$ 100,000 per patent. Thus, issued US patents represent a nontrivial investment, especially in the case of foreign applicants. Thus the discoveries they cover probably have nontrivial value for the applicants.

5. Aside from my survey of pharmaceutical patents and that of 3G mobile communication patents discussed in Chapter 7.

6. These applications were filed between 1991 and 1999 in the JPO, US PTO, and patent offices of major European countries as well as the European Patent Office. (Steps were taken to avoid duplicate counting in the case of the European applications.) Although I have just described the problems of using patent applications as a measure of innovation, I present these data here because they are the only readily available data on this subject. I hope that the analysis of pipeline drugs in Chapter 2 and the analysis of the sources of new FDA-approved drugs in the following and the last chapters will convince readers that biotechs do indeed play a major role in drug discovery in the USA but a small role in Japan and Continental Europe.

7. The JPO study used the following definition for venture company: R&D oriented, established no later than 1980, fewer than 300 employees, and less than 300 million yen invested capital (personal email communication from JPO May 17, 2006).

8. Please see the Appendix for details regarding methodology.

9. Among the small or new company US patents, over 90 percent were issued to companies incorporated in 1975 or later. In contrast, among the Japanese patents with a small or new company inventor, only about 20 percent of these inventors were from companies formed no earlier than 1975.

10. マイクロマシン in Japanese.

11. ナノ in Japanese.

12. Of the thirty-nine Japanese 2003 nano patents, one was issued to a small chemical company formed in 1951 and one was issued jointly to AIST, METI's flagship GRI, and a small Japanese pharmaceutical company established in 1955. Korean venture companies formed after 1995 accounted for five of the nano Japanese patents. None of the fifteen Japanese 2003 micromachine patents was issued to an SME or a venture company.

13. Of the 22 unassigned US patents in my sample, 7 were issued to inventors who had founded viable businesses related to the patented technology (such as Lanny Johnson, mentioned in the following note or Rameshwar Bhargava, founder of Nanocrystals Technology), 5.5 were issued to persons whose main employer was a university, and 9.5 were issued to inventors (*a*) whose affiliation I could not determine or (*b*) whose inventions seemed to be ancillary to their main

work responsibilities, i.e., two of the joint prostheses inventors are orthopedic surgeons, and one of the video cryptography inventors is a patent attorney. (The fractions are due to a patent, one of whose inventors is a university professor and the other whose affiliation I could not determine.)

14. The 22 unassigned US patents mentioned in the previous note represented the inventions of 20 inventors (2 inventors appeared twice in the patents I sampled). Ten of these had been issued at least five US patents as of May 2006, seven had been issued over ten. Some, such as Lanny Johnson the CEO of Instrument Makar and the inventor of 40 inventions mostly related to instruments for arthroscopic surgery, are famous in their fields. A few Japanese inventors, who are not employees of companies, appear frequently as co-owners of patents, but none of the 'independent' prolific inventors in my sample (who happened all to be university faculty) was ever the sole applicant. In other words, there were always coinventors from a company collaborating with the prolific inventor. In the vast majority of such cases, these collaborators are large companies (see the following note).

15. The one exception in my sample is Professor Nakayama Yoshikazu of Osaka University who is co-owner of about twenty nanotechnology-related US patents along with Daiken Chemical Co., a company founded in 1951 but with only eighty-five employees.

16. Of course, this statement is subject to the main limitation of this analysis; it cannot claim to be representative of all high technology industries. Also, I attributed Japanese patents with university and industry coinventors one-half to universities and one-half to industry (and in one case where an inventor was also from an SME, one-third each way). If I instead attributed these inventions 100 percent to universities (reasoning that they are the product of university–industry collaboration even though the university contribution may be only a fraction of the entire inventive input), then the percentage of patents attributed to US and Japanese universities/GRIs would be nearly the same. A counter argument might be that many US patents assigned solely to a company may have benefited from consultations or even joint research with university researchers, although they were not listed as inventors or they were listed but I had no way of knowing from the US patent applications that they were not company employees. Thus the attributions in my data to US universities may also be underestimated.

17. I surveyed 256 Japanese patents covering Japanese-origin inventions (i.e. the inventors had Japanese addresses). Twenty-one of these were 'pure' university or GRI inventions in that the listed inventors were only from GRIs or universities. (In fact, nineteen of these were issued to GRIs and had only GRI inventors, one was issued to a university, one (already noted above) was issued to individuals who turned out to be university inventors.) On the other hand, twenty-three were issued either (*a*) jointly to companies and universities/GRIs and had mixed inventors or (*b*) to companies alone but had one or more university (rarely GRI) inventors. Of these twenty-three, only three involved new/small companies

(only one of which was incorporated after 1975), and of these, one was issued to a large company but had coinventors from the large company, a small company and a university.

18. In 1970, they accounted for 83% of manufacturing employment and 47% of value added (JSBRI, 2003).

19. Whittaker (1997).

20. Compare, e.g., 'Corporate Japan Thrives as Subsidiaries Outshine Parents', *Nikkei Weekly*, January 17, 2005, 1 (this article describes progress by some subsidiaries to improve their technologies and to market to companies other than their parents; it could apply as well to independent SMEs that depend mainly on orders from one or two large customers) with Hotta, Takafumi, and Kame Manabu, 'Screw, Spring Makers Help Auto Industry Stay Ahead: Innovations by Basic Parts Suppliers Support Global Dominance of Carmakers', *Nikkei Weekly*, Feb. 13, 2006, 32 (describing KYB Corp. designing and manufacturing improved springs and shock absorbers for Toyota).

21. Personal communication in 2004 from the director of an SME manufacturing high quality electronic communication devices that are mostly sold to one of Japan's major telecommunications companies. The contracts with this large company are short-term which allows the company to cancel orders on short notice.

22. e.g. machine tools which I discuss in Chapter 7.

23. Henceforth, I often use the term *universities* to refer to both universities and GRIs. When necessary to distinguish between them, I refer to them separately.

24. Actually, for 1995 the JPO database contains only patents approved by JPO and then laid open for a three month of pre-grant opposition period during which other parties could challenge the patent. (Pre-grant oppositions ceased in 1996.) About 95% of laid open patents were ultimately approved (registered). Therefore, these patents are nearly equivalent to the US patents and the 2003 registered Japanese patents. By 2003, patents as actually registered were available in the JPO database.

REFERENCES

Feigenbaum, Edward A. and Brunner, David J. (2002). 'The Japanese Entrepreneur: Making the Desert Bloom', available at www.stanford-jc.or.jp/research/publication/books/cover&file/EAF_DJB.pdf

Ibata-Arens, Kathyrn C. (2000). 'The Business of Survival: Small and Medium-Sized High-Tech Enterprises in Japan,' *Asian Perspective*, 24(4): 217–42.

Japan Patent Office (JPO) (2002) 'Survey of Patent Application Trends related to Post Genome Technology' (i.e. Protein Level Analysis, including utilization of IT methods). [posuto genomu kanren gijutsu (tampaku shitsu reburu deno kaiseki to IT katsuyou) ni kansuru tokkyo shitsugan gijutsu doukou shousa] (unpublished paper by the Technology Survey Office of the General Affairs Bureau dated April 26, 2002).

Japan Small Business Research Institute (JSBRI) (2003). White Paper on Small and Medium Enterprises in Japan.

Kneller, R. W. (Summer 2003). 'University–Industry Cooperation and Technology Transfer in Japan Compared with the US: Another Reason for Japan's Economic Malaise?', *University of Pennsylvania Journal of International Economic Law*, 24(2): 329–449.

Maeda, Noboru (2004). 'Japanese Innovation System Restructuring with High-Tech Start-ups', April 19, unpublished report available from the Stanford Japan Center.

Nakagawa, Katsuhiro (Dec. 1999). 'Japanese Entrepreneurship: Can the Silicon Valley Model Be Applied to Japan?', adaptation of presentation given on May 19, 1999 at the Asia Pacific Research Center, Stanford University.

Rowen, Henry S. and Toyoda, A. Maria (2002). 'From Keiretsu to Start-ups: Japan's Push for High Tech Entrepreneurship', Asia Pacific Research Center, Stanford Unversity.

Suzuki, Hiroto (1999). *Venture Business in Japan*. Background report from the Foreign Press Center, Japan.

Whittaker, D. H. (1997). *Small Firms in the Japanese Economy*. Cambridge: Cambridge University Press.

2

Autarkic Large Companies

INTRODUCTION

Large established Japanese companies tend to innovate autarkicly. In other words, large Japanese companies have tended to rely on research within their own corporate laboratories to discover and develop new products. Also they tend to favor control over as much of the value chain as possible by performing linked steps in the R&D, manufacturing, and marketing process either in-house, in affiliated companies, or in independent companies that they regard as long-term, reliable, and loyal suppliers. In contrast, in many fields of technology, large US companies rely on venture companies for innovation leads, and will rapidly shift suppliers of intermediate products and services. This tendency toward self-reliance and control over the value creation chain creates a less favorable environment for venture companies than in the USA, because it deprives them of their natural customer (and sometimes also investment) base. Whether the large Japanese companies' tendency toward autarky extends to universities[1] is a complex issue that arises throughout this book.

What evidence is there to support these assertions? One approach would be to examine the development history of a number of products recently launched by comparable Japanese and US companies in various industries. In most industries this would require access to confidential business and technical information from various departments in large companies.

However, I have done such an analysis for the pipeline drugs of the eight largest Japanese pharmaceutical companies in 2000 and compared this to the pipeline drugs of Schering-Plough (S-P), an R&D-oriented pharmaceutical company that is comparable in size to the larger of the Japanese companies.[2] In the next few pages, I show that the major Japanese pharmaceutical companies do innovate autarkicly while major US and European companies rely more on in-licensing from biotechnology companies and, to a lesser extent, universities. Whether large Japanese companies in other industries also innovate autarkicly is the subject of the middle section of this chapter. The final section touches on collaborations between large companies and universities, a subject that Chapter 3 explores in greater depth.

AUTARKIC INNOVATION IN PHARMACEUTICALS

About one-third of the total pipeline of the eight largest Japanese pharmaceutical companies in 2000 consisted of in-licensed drugs. This is not dramatically different from S-P, half of whose year 2000 pipeline was in-licensed. However, the sources and timing of the licenses are different. About 80 percent of S-P's licenses were from biotechnology companies,[3] and S-P assumed responsibility for completing clinical trials and obtaining initial marketing approval from regulatory agencies for all its in-licensed drugs. In other words, S-P engaged in a common division of labor under which major non-Japanese pharmaceutical companies rely on biotechnology companies[4] for drug discovery and early stage development and then capitalize on their greater financial and marketing resources to complete clinical trials and to market the drugs.

In contrast, about 40 percent of the drugs in-licensed by the Japanese companies[5] had already obtained marketing approval in either the USA or Europe. Their safety and efficacy had already been demonstrated in non-Japanese populations. The large Japanese companies in general played a smaller role in the development of these drugs than did S-P with respect to its in-licensed drugs. Such late-stage in-licensing is a low-risk way for them to augment their pipelines, while sparing the licensers the trouble of obtaining regulatory approval and then marketing the drugs in Japan.[6] As for the drugs the Japanese companies in-licensed before clinical trials had been completed in any of the world's major markets,[7] over half were in-licensed from US or European pharmaceutical companies or from US biotechnology companies, and probably nearly all of these were also only for the Japanese (or Asian) markets. However, about 45 percent[8] were in-licensed from small Japanese pharmaceutical companies or from Japanese chemical or foodstuffs companies that have small pharmaceutical divisions. In other words, these small Japanese pharmaceutical operations are playing a similar role vis-à-vis the large Japanese companies as biotech companies play with respect to major European and US companies.[9] But the degree of reliance is less, 10 percent of the pipeline of the large Japanese companies compared with about 50 percent for S-P.

Prior to clinical trials Japanese pharmaceutical companies also have fewer alliances to discover drugs than do European and US pharmaceutical companies of equivalent size.[10]

It might be argued that these results are preordained because of the small number of Japanese biotechnology companies and the barriers to Japanese companies making alliances with European and US biotechs. As discussed in subsequent chapters, the phenomena are probably linked. But what about cooperation with Japanese universities? Here geographic and cultural barriers should be minimal.

Although these companies have had numerous informal consultative relationships with university professors, these have not led directly to the discovery of many new drugs or to the improvement of early stage drug candidates. According to interviews with each of the companies during which I reviewed the histories of each company's pipeline drugs, five of the eight Japanese companies discovered and improved all of their in-house-originating pipeline drugs on their own without proximate input from outside researchers.[11] Contacts with universities are important—but primarily as a way to keep abreast of scientific advances that may have implications for their internal drug discovery programs and to recruit capable young researchers. During my interviews I heard of only one example where an academic scientist was recruited to head a drug discovery project (and he left after the project was completed), whereas recruitment of senior academic scientists to head new drug discovery programs is common in US pharmaceutical companies.[12] Also recruitment of recent Ph.D. graduates or postdoctoral university researchers, who probably could bring more in-depth scientific expertise, is still rare in Japanese pharmaceutical companies, although it is standard in US companies. Finally, although I do not know details about the development history of all major Japanese-origin drugs, I do know the histories of two of drugs that represent major pioneering achievements, mevastatin, the first of the statin drugs for controlling high cholesterol, and donepezil for Alzheimer's disease. These drugs were discovered by research teams in Sankyo and Eisai, respectively. Although these scientists were familiar with relevant published academic research, they were the ones who identified the important problems to focus on and they were the ones who solved these problems.[13]

In other words, drug discovery by the large Japanese pharmaceutical companies seems to be largely an in-house process.[14] But using this autarkic innovation strategy, the in-house research teams of the Japanese companies are producing innovative drugs at rates equivalent to those of the in-house research teams in US and European pharmaceutical companies of equal or even somewhat greater size. On average, each of the Japanese companies and Schering-Plough, which is larger than any of the Japanese companies, has about four innovative drugs in its pipeline.[15] S-P's innovative pipeline drugs are no further along in development than the average for the Japanese companies.

Taking into account currently marketed as well as pipeline drugs, Japanese companies are global leaders in cholesterol, diabetes, Alzheimer's, infectious diseases, and dermatitis medications.[16] These achievements are remarkable in light of government policies that until the early 1990s encouraged developing variations of drugs pioneered by foreign companies, rather than projects to discover pioneering new drugs.[17]

However, the USA is the origin of considerably more drugs, especially innovative drugs, than Japan. This advantage is attributable entirely to drugs originating in universities and biotechnology companies, as shown in Chapter 7 which addresses the relative merits of innovation in small or new versus large and established companies.[18] In other words, on a company to company comparison, the drug discovery capabilities of the in-house research teams of the autarkic Japanese companies compare favorably with those of the in-house teams in overseas companies of equivalent size. This is also remarkable because most of the in-house Japanese pharmaceutical researchers have only MS or BS degrees, while most of their counterparts in US and European pharmaceutical companies have doctoral degrees.[19] But the overall drug discovery industry is weak in Japan compared to the USA, because of the contribution of US universities and biotechs.

Most of the Japanese companies feel they have to expand their collaborative networks in order to deal with the rapid advances taking place in many fields of biomedical science and to have strong pipelines. Since the interviews, some companies that had reputations for going it alone have reached agreements to in-license drug candidates from overseas biotechnology companies.[20] Also, many companies emphasize the need to collaborate more actively with universities.[21] Nevertheless, to my knowledge, none of the Japanese companies has initiated the sort of grant and fellowship training programs that some major non-Japanese companies have used to build long-term relationships with young university researchers—linkages that these companies value even if the grantees do not end up becoming their employees.[22]

However, the large Japanese companies hardly ever mention domestic bioventures as collaborators. Some are scornful about the new companies' technologies. Others acknowledge that some bioventures have promising technologies, but it is often difficult to convince senior managers to cooperate with these companies.[23] In some cases, such as iRNA type drugs, almost all Japanese pharmaceutical companies feel the technology is so new that the work of Japanese bioventures in this area is irrelevant to them.[24] In any case, these perspectives tend to be self-fulfilling or self-reinforcing. Lacking a natural client base among domestic pharmaceutical companies, Japanese bioventures are disadvantaged with respect to their US counterparts, which are growing faster due in part to alliances with pharmaceutical companies.[25] Because they are shunned by their natural clients, they cannot develop their technologies rapidly and thus their products are less likely to be in demand.

The rest of this chapter explores whether similar situations exist in other industries.

AUTARKIC INNOVATION IN OTHER INDUSTRIES

In early 2004 I had an opportunity to discuss relations between large and small companies with executives of major Japanese electronics manufacturers.[26] They were focused on overcoming technical bottlenecks to building faster, cheaper, smaller semiconductors, and they felt that SMEs, at least Japanese SMEs could not offer expertise in manufacturing. As for R&D, most agreed that semiconductor R&D is too capital intensive for SMEs to make contributions. Some said it is commonly perceived that the quality of SMEs' personnel and products are inferior to those of big companies. Also, it is sometimes difficult for major manufacturers to assess new technologies put forward by SMEs. Even if some engineers in the large companies are convinced of their value, it is difficult to convince others in the management hierarchy to try them. However, new US semiconductor companies have played an important role in the US industry, as discussed in Chapter 7.

Another example of the contrasting approaches between Japanese and US companies concerns the manufacture of hard drives and other PC components. As of the 1990s, Fujitsu usually either manufactured these components in house or had long-term relationships with a small number of affiliated suppliers. In contrast, IBM sourced hard drives and other components for its PCs from independent suppliers, some of whom were independent spin-offs that had made great strides in miniaturizing these components. These companies were free to innovate on their own and market to whom they please, but they were vulnerable to IBM suddenly changing suppliers or hardware components.[27]

As another example, NTT DoCoMo and J-Phone are providing wireless Internet service in an integrated manner. While they do not own every component in the value chain,

they carefully manage the interfaces with their content providers and handset manufacturers. Their interdependent [i.e, autarkic, vertically integrated] approach allows them to surmount the technical limitations of wireless data and to create user interfaces, a revenue model, and a billing infrastructure that make the customer experience as seamless as possible.

This approach has resulted in better service than the approach adopted by AT&T Wireless and other mobile Internet service providers in the USA, under which industry standards agreed on before the technology has matured have permitted various companies to provide modular services at various points along the value-added chain.[28]

In Bob Johnstone's account of how Japanese companies and engineers contributed to many of the key electronic inventions of the later twentieth century,[29] the following relatively new (or newly transformed) companies figure prominently: Sony, Sanyo, Nichia, and Casio (all formed after World War II)[30]; Canon and Fujitsu (formed in the 1930s),[31] and Sharp and Stanley Electronics (formed earlier but transformed radically in the 1960s).[32] But when these companies made their pioneering discoveries, they already had a manufacturing base, and, with some exceptions, they incorporated these discoveries into vertically integrated operations.[33] They were not brand new R&D-focused companies or even, with the exception of Nichia,[34] niche manufacturers. Finally, the large older electronics companies, Toshiba (founded as Tanaka Engineering Works in 1875), Mitsubishi Electric (1879), NEC (1899), and Hitachi (1910) figure little in Johnstone's history of breakthrough innovations, although he never remarks on this point. This is not to suggest that these older firms are not innovative, but rather that their absence from Johnstone's account may indicate that their innovative efforts are directed at areas that have been pioneered earlier.

In summary, Japan has provided fertile soil for new companies in the past. Some of these have been pioneering innovators and succeeded brilliantly. But they tend to do so on their own and to be able to incorporate their innovations into at least a partially vertically integrated manufacturing process. Furthermore, they seem to be most innovative in new fields when they are young.

The history of the software industries in Japan and the USA offers additional insights into autarky as a general characteristic of Japanese industry.[35] The US software industry evolved against the background of IBM's dominant position in mainframe and minicomputers. Other hardware manufacturers strove to make their products compatible with IBM computers. This quickly leads to the evolution of a standard hardware interface for which computer programmers could write software.

Once PC sales began to soar in the early 1980s (and many businesses and homes began to use PCs), a huge market was created for PC compatible software. This provided growth opportunities for independent software companies—even after Microsoft established its dominance in PC operating systems and some applications fields (Microsoft itself being a new independent software developer). Packaged (noncustomized) software accounted for over half of sales. Moreover, the companies developing packaged software tended to be small but also to have favorable growth prospects.[36]

In contrast, no company established a uniform hardware design standard in Japan. The computer makers themselves began to write customized applications software for clients. Major users, first steel companies, then banks and telecommunications companies began to establish their own in-house software units. Programmers from these user units would work closely with programmers from the computer companies to create high quality customized software. As this system matured, some of the computer programming units were spun off as separate companies. Independent software houses emerged but these functioned mainly as subcontractors of the computer or the user software units. Compared to the software houses of the computer manufacturers and the end users, sales per employee were low as was the educational level of the programmers. In other words a pattern of autarkic innovation also emerged in the Japanese software industry, with large companies taking the lead in innovation and new independent companies playing an auxiliary role.

But was autarky a symptom or cause of this phenomenon? The answer is probably a little of both. The fact that the hardware industry was fragmented meant that independent software makers would have a smaller market for their products—or else would have to expend considerable effort to make software that would work on NEC, Fujitsu, Hitachi, Toshiba, IBM, Oki, Sharp, and Matsushita (Panasonic) computers and peripherals. Also the Japanese PC market grew slower and was smaller than the US market. Furthermore, there was competition from US software developers, who had first-mover advantages that allowed them to gain and then hold large shares in new markets. Therefore, independent Japanese software manufacturers faced disadvantages relative to their US counterparts, and in this respect autarky on the part of computer makers and users is a symptom rather than a cause of the weak position of independent companies.

However, other factors suggest autarky was more deeply rooted. For example, stable relationships had evolved during the era of mainframe computers between computer makers and users. Users came to expect a high degree of software as well as hardware support from the computer manufacturers. The computer companies hired systems engineers from a small pool of qualified persons with computer training or experience, and these systems engineers became familiar with their employer's unique hardware. Users in turn allocated personnel to become familiar with the new hardware and software. These user personnel would specify precisely what type of software was needed and would often upgrade the software. So users came to expect a high degree of specialized service, and computer makers were able to earn sufficient revenue from the sale of hardware to provide such services. The independent software manufacturers could not compete for skilled computer science graduates. Even if they could hire such graduates, they would have had to use them to provide a high level of customized services and support rather than to develop software that could be sold to many customers. Links with universities were sparse.[37] Finally, users were willing to pay surcharges on hardware to cover the costs of customized software services.

The above analysis is based on studies of the international software industry in the early 1990s. However, after ten years the situation appears to have changed little. Microsoft OS has become the standard OS for PCs and some independent software developers focus on applications software for PCs (or for Apple computers). However, mainframe and minicomputers still run with middleware that is unique to each manufacturer and they use application software that is often custom developed by the manufacturers. Almost all the middleware sold by Toshiba is written by Toshiba employees, whereas about half the authors of middleware copyrighted by IBM are employees of software companies that have been purchased by IBM or that have alliances with IBM.[38]

Under IBM's Emerging Business Opportunities program, young companies developing applications software provide IBM access to the software and in return receive from IBM middleware, hardware, service, systems support, other technical assistances, and even direct investment. Oracle has a similar business partnering program. No major Japanese company had such a program as of 2003.[39]

According to Japanese programmers whom I have interviewed, even when the middleware is based on standard architecture such as Unix or Linux, the hardware manufacturers usually adapt it specially for their machines.[40] In addition, the hardware manufacturers rarely share their source code and interface information with independent software developers, making it difficult for them to develop compatible middleware and application software that will run on the hardware makers' computers. Occasionally hardware makers will subcontract with independent software developers, but generally the tasks they assign are circumscribed and they release their source code only on a need to know basis under strict secrecy conditions.[41]

Although the hardware manufacturers compete fiercely on price of their computers, they avoid price competition in the field of services, including software. It is difficult for an independent software developer to obtain contracts even it offers a substantially lower price. Even if it wins the confidence of persons in the software department of a user corporation, multiple levels of administrative approval are required. If the software ultimately is found not to perform as expected, bad judgment is attributed to the persons who approved the contract, whereas, if the software had been developed by a major hardware manufacturer, the latter's brand reputation would insulate them from blame. If the software instead works smoothly, the persons who approved the contract are not rewarded, even in the case an independent developer produced the software at below market price. In other words, the operative principle with respect to choosing a software developer is 'don't make mistakes'.

However, if the software company is affiliated with a major computer manufacturer, is affiliated with a prestigious university, or has been awarded contracts by a foreign company, then its chances of obtaining contracts from user corporations increase.

Discussions with programmers in US companies indicate that independent US software developers are still being formed even post-Internet bubble by university researchers and persons leaving existing companies. They spend most of their effort on developing the functionality of their products rather than customizing it to a particular computer or OS. Standardization has helped, as has the development of software modules that depend on standardization. Working with standard hardware or OS interfaces, they are usually able to demonstrate their software's capabilities to potential customers. Attracting customers is usually a straightforward matter of capabilities, cost, and backup service—but not customized design.

The example of software provides a more nuanced picture of autarkic innovation by large Japanese companies. Only a relatively small number of very large high technology Japanese companies are completely vertically integrated from research to sales. At the interface between companies, great effort is devoted to ensuring customer needs are satisfied. The interface is tight and well meshed. While this improves integration of the production process and, in many cases, product quality, it makes entry by new companies difficult— not only in software but also in other industries.

The CEO of a small engineering company founded in 1999 that has become a leading manufacturer of machines to apply thin films of light sensitive nanoparticles onto optico-electronic devices, such as the lenses of digital cameras, remarked, 'The best aspect about doing business in Japan is the high quality of manufacturing here. I manufacture my machines here for this reason, and also because it is possible to combine many quality components into a high quality final product. It is easier to do so here than in China, Taiwan or even America. Here companies coordinate well among each other and pay attention to quality.'[42] Nevertheless, as described in Chapter 4, breaking into the Japanese market was difficult and this company, like many others, relied initially on overseas customers.

The importance of Japanese manufacturers working closely with customers has been noted by others.[43] An IBM physicist, who helped initiate IBM's collaboration with Toshiba in the 1980s and 1990s to develop a manufacturing process for active matrix flat-panel displays for laptop computers, observed that, in contrast to US engineers, Japanese engineers trade off subprocess optimization in return for interface optimization. They are good at integrative work that meets the specific needs of customers. In contrast, US engineers tend to optimize individual components or subprocesses, but because they de-emphasize interface optimization, they tend to adopt whenever possible modular components to solve interface problems.[44] This provides entrées to ventures that produce products that fit into the value chain at these modular interfaces (or that themselves provide the modular interfaces), even in

industries where the technology is science-based and rapidly evolving[45]—
entrées that are less available in Japan.[46] The contrast between US modularization and subprocess optimization and Japanese interface optimization is
not absolute. The interface between companies is also important in the USA,
especially for ventures.[47] Also other reasons limit alliances between large companies and ventures in Japan—factors external to the large companies. But to
anticipate a key conclusion of this book, the fact that many successful Japanese
ventures made their initial breakthrough deals with foreign companies and
only then were able to obtain Japanese customers,[48] suggests that the problem
often lies not with the ventures' technologies and products, but the process of
acceptance by large companies.

RELATIONS WITH UNIVERSITIES

University–industry relations is the main focus of Chapter 3. However, the
fundamental outlook toward cooperation appears similar for large companies
in the two countries. In other words, outside of drug discovery discussed above
and with the following additional exceptions, large Japanese companies are
probably no more, and no less autarkic with respect to university research
than large US companies.[49]

One difference concerns researcher mobility. A review of how NSF support
for university research in three engineering fields facilitated innovation found
that both magnetic resonance imaging (MRI) and development of the Internet
benefited greatly from researcher mobility.[50] In the case of MRI, university
researchers (including tenured faculty) were recruited by large companies,
formed their own companies, and served on corporate boards of directors. In
some cases, researchers who joined industry eventually returned to academia.

Development of the Internet[51] relied even more on mobility. Unlike
computer hardware and operating systems, there was no industry leader.
Government[52] played the key organizational, funding, and infrastructure support roles during the early evolution of the Internet beginning around 1969.
As the focus of innovation shifted from government to industry, university
researchers continued to play keys role in developing new systems. The roles of
government, industry, and universities were a 'collage', and the career paths of
some of the major contributors to the Internet involved multiple job changes.

But in the case of both MRI and the Internet, the large companies that
are today's innovation leaders were slow to enter, leaving early development
largely to university researchers and smaller, usually new companies, some
of them startups founded by MRI and Internet pioneers. Large companies

completely dropped the ball with respect to the early development of the Internet.[53] As it became apparent the technology was feasible and demand was substantial, then the large companies became innovation leaders, buying many of the smaller companies, and expanding their own R&D, in the process hiring some of the key university pioneers.[54] This mobility is still rare in Japan, as is the founding of nonbiomedical startups that play an important role in innovation.

Another exception is the degree to which large US companies outsource clinical testing to universities. In the case of MRI, for example, they felt they had to do this because universities had the patients and the medical staff necessary to evaluate their prototype machines and to conduct the tests necessary to obtain marketing approval. In the process, physicians and medical researchers in the clinical sites provided considerable feedback and developmental help. 'There was much infusion from the university folks. The university researchers would propose a design, and a company would build it to try it out.'[55]

Are Japanese companies different? Interviews by a specialist in MRI electrical engineering with Japanese academic researchers and the four companies that hold the largest share of the Japanese MRI market (Toshiba, Hitachi, Siemens, and General Electric-Yodogawa Medical Systems (GE-Y)) indicated that Siemens and GE-Y sought broader collaborations with university researchers and empowered them more to pursue MRI development research.[56] Academic collaborations are important for all four companies for human testing of new MRI hardware and signal processing software. However, collaborations with Toshiba and Hitachi involved a clear division of labor, with clinical work done by physicians at research sites, and engineering work done by company employees. Typically new imaging sequences (software that controls imaging for various clinical purposes) were developed by the Japanese manufacturers and then sent to university clinical collaborators for validation. All improvements were made by the companies, in a manner reminiscent of the software customization described earlier.[57] In contrast a wide range of Siemens and GE-Y research on clinical applications is carried out in close collaboration with researchers in academic clinical centers. The academic collaborators (physicians and engineers) can be involved in coil development, pulse sequence design, and image reconstruction (all core engineering issues) in addition to clinical testing. They themselves can change the software to make sequence improvements. GE-Y and Siemens did not assign engineers to the academic centers, but they provided the infrastructure and established communication channels to allow the academic researchers to improve on their own the companies' prototype systems while pursuing groundbreaking research.[58]

In addition, Siemens was investing in training and development of long-term human networks, similar to Pfizer and SmithKline Beecham in the pharmaceutical industry. It created focus groups involving Japanese academic researchers, foreign university researchers, and its own in-house engineers. It had established fellowships to enable its Japan-based company researchers as well as Japanese academic collaborators to spend time in Siemens laboratories and collaborating institutions around the world. By promoting the professional development of outside researchers, Siemens was trying to create an environment conducive to future cooperation between Japanese academic and its in-house researchers. I do not know of similar examples of large Japanese companies promoting human resource development outside their companies.

This example recalls the earlier discussion concerning Japan's strength as an environment where many companies work together and contribute various components to a finely integrated quality product. The case of MRI development suggests that while coordination among customers and multiple suppliers may be high, technology transfer in the sense of sharing technologies or promoting training for outside persons is not common. Companies work together, but it appears that their roles are usually well defined. Decisions to share information and promote development of expertise outside the company appear to be made cautiously and infrequently.

APPENDIX: METHODOLOGY OF STUDY OF PIPELINE DRUGS

Table 2A.1 compares the sizes of the eight largest Japanese pharmaceutical companies, and Schering-Plough, which I used as the main comparator company. I wanted comparison companies that are research-oriented and whose business deals mostly with pharmaceuticals, rather than chemicals, diagnostics, etc. But most research-oriented 'pure' US and European pharmaceutical companies are much larger than any Japanese company. S-P is one of the smaller research-oriented US companies. I also considered using Bayer, Abbott, and Sanofi for comparitors, but I was not able to find complete lists of pipeline drugs for these companies. Nevertheless, based on my review of these companies' pipeline drugs that are publicly available (as well as the publicly known pipeline drugs of other non-Japanese companies), S-P's relative mix of in-licensed and in-house-originating drugs (and even, within the latter category, the relative mix of innovative vs. follow-on drugs) is probably typical of other major US and European pharmaceutical companies.

Because new drugs have to go through an internationally standardized process of testing in animals and then humans to prove they are safe and effective before they can be marketed, information about new drugs entering clinical trials has already been made available by pharmaceutical companies to regulatory agencies such as the

Table 2A.1. Summary information for the nine companies whose pipeline drugs are analyzed

Name	2000 world sales (billion US$)	R&D as % of world sales	No. pipeline drugs in 2000
Takeda	5.0	13.5	14*
Sankyo	3.2	12.1	20
Yamanouchi	2.5	13.3	7
Dai-ichi	2.1	11.8	19
Eisai	1.9	15.6	14
Shionogi	1.6	15.6	19
Fujisawa	1.5	16.9	9
Chugai	1.4	20.5	23
Schering-Plough	10.0	17.2	17

* Includes 2 drugs that I was not able to classify as to origin or innovativeness and were therefore excluded from the total of 150 that I analyzed.

Sources: Japan Pharmaceutical Manufacturers Association, UBS Warburg, HSBC.

US FDA and the Japanese MHLW. Moreover, in order to demonstrate to investors and the medical community their progress in developing new drugs, pharmaceutical companies often make public the names of their drugs in clinical trials. Thus in the pharmaceutical industry, this unique window enables identification of many new products under development and makes more manageable the next step of determining their origins.

In the case of the Japanese companies, I confirmed through interviews information from public sources and investment reports on the origins of the pipeline drugs. In the case of S-P, I relied on searching a variety of public sources to determine the origins of that company's pipeline drugs.

Altogether these companies had 150 pipeline drugs in 2000 (133 for the Japanese companies and 17 for S-P) which I classified according to whether they were in-licensed or developed mainly through in-house research. I subclassified the latter group according to their innovativeness (first in therapeutic class, second in therapeutic class, new use of a first or second in class drug from the same company, or none of the above, i.e., a follow-on or derivative drug).

More details of this study are in Kneller (2003) and note 15.

NOTES

1. In other words, whether large Japanese companies rely more than their US counterparts on collaboration with university laboratories and consultation with university scientists for key R&D needs.
2. *Pipeline drugs* are new pharmaceuticals being tested in human clinical trails in order to obtain marketing approval from regulatory agencies. In fact, as shown

in the Appendix, S-P is larger than any of the Japanese companies. Please also refer to the Appendix for information on methodology.

3. The remainder are from universities or charities sponsoring university research.

   In this book unless otherwise specified, I use *biotechnology* (or *biotech*) *company* to refer broadly to a relatively new company (formed no earlier than 1975) aiming to discover and sometimes to apply new biological technologies. (In contrast, the narrow definition encompasses only companies using genetic engineering techniques to produce naturally occurring hormones, antibodies, and other proteins that can be used as drugs, industrial catalysts, etc.) I usually use the term *bioventure* synonymously. In contrast, I use the term *pharmaceutical company* to refer to a company that has been in existence prior to 1975 and that is focused on drug commercialization (including clinical trials and often marketing) as well as drug discovery. Pharmaceutical companies traditionally focused on developing small molecule drugs rather than the larger naturally occurring proteins. The analysis in this book depends not on an increasingly artificial distinction between *biotechnology* and *pharmaceutical* companies, but on the distinction between established companies with significant downstream drug development (i.e. clinical trial and marketing) capabilities and small or new companies focusing on drug discovery research.

4. And less frequently, directly on universities.

5. Eighteen out of forty-eight pipeline drugs in-licensed by the Japanese companies.

6. Sixteen of the eighteen of these late in-licensed drugs were from large US or European companies. The other two were in-licensed from US biotechs.

7. The major markets are Europe, Japan, and the USA.

8. Thirteen in all for the eight Japanese companies, 27% of their total 48 in-licensed drugs and 10% of their total 133 pipeline drugs.

9. Also many of these licenses probably included worldwide marketing rights.

10. See Kneller (2003) which compares the numbers of such alliances for the eight largest Japanese pharmaceutical companies between 1997 and 2002 with those of S-P, Bayer, and Abbott. The average for the Japanese companies is five to eight times less than for these companies of roughly comparable (but still larger) size.

11. Of the eighty-five drugs originating from the in-house laboratories of the Japanese companies, eight had significant input from outside sources. Discovery of six drugs benefited from input from Japanese universities—five from informal collaborations with university laboratories (see Chapter 3 for more about these donation-based collaborations) and one from recruitment of a professor to set up an HIV/AIDS drug development laboratory. Two drugs benefited from collaborations with foreign universities—including one which also had input from a Japanese university. One drug was discovered in the course of collaboration with a US biotech company. Five of these eight drugs are in the pipeline of one company which is unusual in terms of benefiting from outside collaborations.

12. See Henderson, Orsenigo, and Pisano (1999) and Zucker and Darby (1996). The low levels of recruitment of academic scientists was due in large part to prohibitions on national university faculty consulting for private companies and taking leaves of absence to work in companies. These restrictions were eased beginning in 2000. However, another reason is the continuing practice of lifetime employment and internal promotions which has limited the mid-career hiring of outside scientists to at most a few per year in the companies I interviewed.

13. The discovery of mevastatin is described in Hara (2003). The discovery of donepezil was described to me by Dr Sugimoto in 2002.

14. Some smaller Japanese pharmaceutical companies are pursuing a less autarkic drug discovery strategy. I know of several such companies that have turned frequently to university researchers for help on drug discovery projects or that have important collaborations with foreign biotechnology companies.

15. I classified a pipeline drug as *innovative* if available information indicates it is the drug furthest along in development in its therapeutic class. The average for the 8 Japanese companies is 3.5, that for S-P is 4. If I broadened the definition to include second in class drugs or drugs that since launch have risen to first or second place in terms of market sales, then the average for both S-P and the Japanese companies becomes 6. Although I was not sure if I had complete lists of Bayer's and Abbott's pipeline drugs, the available lists suggested that they probably had fewer innovative in-house originating pipeline drugs than did S-P (Kneller 2003).

16. Sankyo discovered the first of the statin (HMG-CoA reductase inhibitor) drugs, mevastatin. Although development was halted during clinical trials, samples of mevastatin and data showing proof of concept in animals were supplied to Merck which went on to develop lovastatin (Mevacor®), which in 1988 became the first statin to be approved by the FDA. Pravastatin (marketed by Bristol-Myers Squibb as Pravachol®), Sankyo's follow-on to mevastatin, was approved in Japan in 1989 and by FDA in 1991 (Hara 2003: 131–49). As of 2005, statins remained the leading class of drugs to control high cholesterol. Shionogi's recently FDA-approved rosuvastatin (marketed by Astra Zeneca as Crestor®), albeit a derivative drug, probably produces better cholesterol control than any other approved statin. Sankyo pioneered the first of the thiazolidine-diones/glitazones, the current mainstay drug therapy for adult onset diabetes, with troglitazone (Resulin®). Troglitazone has been withdrawn because of safety concerns. However, Takeda's pioglitazone (Actos®) is now vying with Glaxo's rosiglitazone (Avandia®) for market leadership among diabetes drugs. In 1996, the FDA approved Eisai's donepezil (marketed by Pfizer as Aricept®), the first β-amylase inhibitor and, as of 2005, still one of the leading medicines to treat Alzheimer's disease. Donepezil was created entirely by an in-house researcher team, whose dynamic head scientist was warned repeatedly that the project would fail. Dai-ichi's levofloxacin (marketed by Ortho-McNeil as Levaquin®)

vies with Bayer's ciprofloxacin (famous following the 2001 anthrax attacks in the USA) for sales leadership among the oral quinolone antibiotics. In 2000, the FDA-approved Fujisawa's tacrolimus (Protopic®), an ointment derived from its leading immunosuppressant Prograf®. Protopic is the first new drug in decades to treat eczema and other forms of dermatitis (Brody 2001). To my knowledge, none of these drugs depended on collaborations with outside researchers for their discovery.

17. Before the mid-1970s, the Japanese pharmaceutical industry was protected against foreign competition. Patent protection was not available on the core chemical constituents of drugs. Incentives to copy foreign drugs were high and incentives to invent innovative drugs were low. Japan's national health insurance system reimbursed consumers the cost of drugs and the reimbursement price allowed prescribing physicians, wholesalers, and the manufacturers to reap healthy margins from pharmaceutical sales. However, beginning in the late 1980s, financial pressures caused the government to periodically cut reimbursement prices and thereby squeeze the pharmaceutical companies' profit margins. The Ministry of Health and Welfare (MHW) still approved new drugs that offered little or no improvement over existing drugs. It also gave only marginally higher reimbursement prices for new innovative drugs. So financial incentives still favored development of derivative drugs that were cheaper to develop but often offered little improvement over existing drugs. Also it was still difficult for foreign companies to gain approval to market drugs in Japan. So Japanese pharmaceutical companies tended to focus on the domestic market, which was relatively free from foreign competition and where they could receive satisfactory profits for minimal research effort. The government price reimbursement system and lack of foreign competition also restrained merger pressures and kept companies small (Kimura et al. 1993; Thomas 2001). Now, however, greater competition from foreign companies and stricter MHLW policies against approving drugs that offer no substantial benefit over existing medicines are driving the larger Japanese companies to invest in the discovery of innovative drugs.

18. See Chapter 7. In addition, lists of drugs in clinical trials published by Pharmaceutical Research and Manufacturers of America (PhRMA) indicate that most drugs in clinical trials to obtain FDA approval were being sponsored by biotechs (Kneller 2003).

19. However, the lack of in-depth scientific expertise among in-house researchers has had negative consequences for Japanese pharmaceutical companies. For example, when Sankyo suspended mevastatin clinical trials because of some worrisome data from animals, Merck pushed ahead with the development of lovastatin in part because Merck's top management was well versed in lipid science and thought the animal data did not indicate risk for humans. Also as Merck continued clinical trials, it clarified the biochemical mechanism of action of the statins and this gave assurance that the animal data did not imply human risk (Hara 2003: 142; Kneller 2003, for other examples).

20. See e.g., Ogiso, Yoshinori and Matsuzaki, Yusuke (2006): 'Takeda aims to be all-around player: Market leader known for developing own products now buys rights from others' (*Nikkei Weekly* March 6, 12).

21. Kneller (2003). The CEO of Fujisawa, Mr Hatsuo Aoki, also made this point in his keynote address to the International Patent Licensing Seminar 2005 (Jan. 25) in Tokyo.

22. See Lam (2002) describing Pfizer's program and Leigh (2000) describing SmithKline Beecham's.

23. Described in more detail in Kneller (2003) and chapter 7.

24. Based on conversations with industry and investment personnel in early 2006. Ironically, in the specific field of interference RNA (iRNA), i.e., short nucleotide sequences complementary to key coding regions of messenger RNA that can be used either therapeutically to block production of undesirable proteins or for research to determine the function of various genes, several Japanese companies have had contracts with overseas biotechs to use iRNA to find disease related genes, whereas Japanese biotechs focused on similar applications of iRNA have struggled to find customers.

25. See Chapter 4.

26. Semiconductor Technology Roadmap Japan (STRJ) Specialists Workshop, held March 4–5, 2004 in Tokyo, organized by the Japan Electronics and Information Technology Association (JEITA), as part of the International Technology Roadmap for Semiconductors (ITRS) initiative.

27. Chesbrough (1999). Chapter 7 summarizes the role ventures played in improving disk drives.

28. This account (including the quoted passage) is from Christensen and Raynor (2003: 137–41), who suggest that relatively mature technologies are most suitable for entry by independent specialist companies that can insert themselves at particular points along the value-added chain, provided suppliers and customers know what to specify, can verify that specifications are met, and understand how various components of the system will work together. However, the success of biotechnology companies in discovering new drugs that they hand off to pharmaceutical companies, as well as the success of semiconductor and disk drive companies described in Chapter 7, show that venture companies need not be confined to production of specialized modular components. In other words, at least in some industries, venture companies can be lead innovators even though they are active in only part of the production process. However, this seems to occur more often in the USA than Japan.

29. Johnstone (1999).

30. In 1946, 1947, 1956, and 1957, respectively.

31. In 1933 and 1935, respectively. Fujitsu developed the high electron mobility transistor that allowed satellite broadcast receiving dishes to be shrunk from meters to centimeters.

32. Sharp was known as Hayakawa Electric until the 1960s (see end of Chapter 1). Since 1920, Stanley had produced conventional automobile lights until

its research director pushed it in the late 1960s to pioneer the development of aluminum gallium arsenide light emitting diodes (LEDs) that emit much brighter light than conventional LEDs.

33. Some of the exceptions are the LEDs developed by Stanley for incorporation into automobiles and the blue diodes developed by Nichia for various industrial users. Sanyo too manufactures many intermediate products, such as batteries and solar cells, based on its pioneering inventions which it sells to other companies. But as a large diversified company it also has extensive manufacturing and product development capabilities in many fields.

34. Nichia began business in 1956 by producing fluorescent compounds to coat the insides of the tubes for fluorescent lights. It later branched into phosphors to coat the inner screens of color TV tubes. Like many of the ventures described in Chapter 4, it initially could not find Japanese customers and got its first commercial break selling to foreign companies (Sylvania and Philips). In 1993 Nichia and Shuji Nakamura stunned the world by revealing the world's first commercially feasible blue LED. Nakamura, Nichia's lead scientist in this effort, went on to invent the first commercially feasible blue laser in 1996. Blue light emitting LEDs and lasers have numerous applications including CDs that can store more information, and much longer lasting and cheaper traffic lights and outdoor display lights (Johnstone 1999).

35. The following is based on the chapters by Baba, Takia, and Mizuta (1996), Cotrell (1996), and Merges (1996) listed under References.

36. The stock of companies making prepackaged software tended to have higher price to earnings ratios. This suggested that investors considered the growth prospects of such companies to be particularly strong—although these data are from prior to the bursting of the Internet bubble.

37. Due in part to the delay in establishing computer science programs in universities, as well as to inability to compete for the limited number of university graduates with computer training.

38. This is according to discussions in 2003 with an official of IBM Japan. Middleware is software that provides the interface between hardware and applications software and includes operating system (OS), database and web server software. In the late 1990s IBM made a probably farsighted decision to abandon development of application software and concentrate on middleware, specifically middleware that would run on any IT system, thus permitting a single consistent software architecture (Foremski, Tom (2003). Big Blue lays down new foundations *Financial Times* July 12).

39. According to the official of IBM Japan mentioned in the previous note.

40. According to one of my sources, one rare exception to this pattern was NTT DoCoMo, which adopted Sun Microsystem's OS for its mobile phones without substantial customization.

41. In contrast, Sun Microsystems has made its source code available to independent software developers in order to encourage wider use of its middleware.

The programmer who told me this believes that this example has never been followed by a Japanese company.

42. Interview in March 2004. The CEO is Chinese and has technical and business contacts in mainland China, Taiwan, and the USA. He probably has the option of manufacturing in overseas.

The CEO's remark about companies coordinating well among each other is interesting in terms of comments (summarized in Chapter 7) by executives of IT companies to the effect that engineers in large companies are often not good at communicating with outside persons as well as accounts of friction between scientists from the various large companies involved in MITI's large consortia projects in the 1970s. My sense is that both perceptions are correct, although this would have to be confirmed by anthropologic studies Coordination between companies can be good, provided it has been approved at all levels of the corporate hierarchies, and specific influential persons have been designated to oversee and ensure the success of the cooperation.

43. See '(Still) made in Japan: How are Japan's manufacturers faring against low-cost competition from China?' *Economist Magazine*, April 7, 2004; and Kodama (1991) which provides examples of suppliers working closely with customers and rapidly adjusting their products to meet customers' needs (what Kodama calls *demand articulation*). Toyota's sophisticated system of parts suppliers also requires close coordination among manufacturers and attention to quality.

44. Myers (1999). Myers notes that while a US industrial laboratory is built around specialists for each subprocess, with only a few generalist engineers to oversee development, Japanese engineers tend to be generalists. In many Japanese laboratories, a small group of engineers will build a prototype from scratch learning what they need as they go along. The tendency for Japanese companies not to hire Ph.D. level scientists and engineers and to rotate them frequently within the company may reinforce these different comparative advantages. The contrasting tendency of large US manufacturers to clearly distinguish between research, on the one hand, and product development and manufacturing, on the other; to hire scientists and engineers with Ph.D.s and to station them long-term in central research laboratories, may enable their research scientists and engineers to better optimize subprocesses, but leave them more dependent on standardized methods to integrate the subprocesses (see the discussion in Chapter 7, and the articles by Westney and Sakakibara (1986) and Kuzunoki and Numagami (1998) cited under the Chapter 7 references).

45. In addition to the example of software above, see the discussions related to IC chips, hard disk drives, and nano/materials technology in Chapter 7, as well as the discussion in Chapter 1.

46. In other words, modularization in Japan is indeed limited to the cost-cutting stage of a particular technology's life cycle. Newcomers have to either have extremely innovative products, already have integrated manufacturing operations, or be able to customize their products at an early stage.

47. See Kurtzman and Rifkin (2005: especially Chapter 10). See also Eduardo Porter, 'Computing skills often best at home: work done in India lacks "Creativity," ' *International Herald Tribune (New York Times Service)*, April 29, 2004, 1 (Japan edition), reporting the decisions by some US software companies headed by expatriates from India *not* to shift development operations in India, in part because of the importance of maintaining close interactions with US customers.

48. See Chapter 4.

49. Exposition of this general statement goes beyond the focus of this book. It is based on conversations with former directors of research and IP management in IBM and General Electric as well as scientists and R&D managers in Japanese companies. A review of how NSF support for engineering research facilitated corporate innovation indicates that large US companies tend to carry out research related to their core business in their own laboratories, rather than relying on universities (SRI 1997). With the exceptions noted in the text, this appears similar to the practice of large Japanese manufacturers. Both large US and Japanese manufacturers in general rely on university research more to expand their own R&D capabilities than to develop new products (Santoro and Chakrabarti 2002; Motohashi 2005). Both tend to rely more on open sources (academic publications and conferences) to access university discoveries than on collaborative research, licenses, or consulting (Cohen et al. 2002).

50. This account is from SRI (1997). The other field is reaction injection molding for producing shaped plastic products.

51. i.e., the development of interconnected communication networks, packet switching technologies, common address and routing protocols, and switchers and routers.

52. First the Defense Department's Advanced Research Projects Agency (ARPA) and then NSF. On the European side, CERN was also involved.

53. AT&T and other common carriers did not see a market for digital communication, with AT&T being extremely skeptical of the concept of packet as opposed to circuit switched networks. Computer manufacturers such as IBM and Digital saw the potential for profit but insisted on using their own proprietary protocols (SRI 1997).

54. Several of the early pioneers of MRI during the 1950s and 1960s worked in Varian, founded in 1948 and closely tied to Stanford's physics department. Raymond Damadian of SUNY Downstate Medical Center was one of the first to realize the ability of nuclear magnetic resonance to distinguish between diseased and normal tissue. He founded FONAR in 1978 and it produced the first commercial MRI scanner two years later. John Mallard, a professor of physics at the University of Aberdeen, founded M&D Technology Ltd. which was also an early designer of MRI machines. Two other young companies, Diasonics and Technicare, were the technical leaders in MRI in the early 1980s. The only established company that played a major role in early MRI development was EMI of the UK. Its MRI operations were bought by GE as GE became the dominant manufacturer in the 1990s.

Many Internet pioneers also spent some of their career in startups, BBN founded by MIT professors in 1949 and a leading center of R&D on Internet protocols and routers from the 1960s to 1980s, being a notable example. And of course the Internet has given rise to many new companies that remain innovation leaders to this day.

55. Gary Glover, senior physicist at GE Medical Research Laboratories until 1990, when he became director of Stanford Medical School's Radiological Sciences Laboratory; quoted in SRI (1997).

56. These interviews were conducted in summer 2000, by Dr Krishna Nayak, currently assistant professor of engineering in the University of Southern California. That summer Dr Nayak was a Ph.D. candidate in electrical engineering at Stanford and also studying MRI R&D in Japan as an NSF Summer Program Fellow.

57. In cases where close collaboration with clinicians was required, Toshiba sent an engineer to work full-time on-site for periods of one to twelve months. Hitachi rarely provided on-site engineering research support.

58. As an example, a laboratory in the Bioengineering Department of Kyoto University was involved in pulse sequence design on a Siemens scanner purchased by the laboratory. Siemens provides the software plus training and ongoing support for its use. Siemens's proprietary information was protected under a confidentiality agreement with the laboratory, which also provided for relevant research results to be communicated to Siemens but not to third parties.

REFERENCES

Baba, Yasunori, Takia, Shinji, and Mizuta, Yuji (1996). 'The User-Driven Evolution of the Japanese Software Industry', in David C. Mowery (ed.), *The International Computer Software Industry*. Oxford: Oxford University Press.

Brody, Jane (2001). 'Myths Abound About Eczema: New Drugs Said to Bring Relief', *International Herald Tribune (New York Times Service)*, June 21, 10.

Chesbrough, Henry W. (1999). 'The Organizational Impact of Technological Change: A Comparative Theory of National Institutional Factors', *Industrial and Corporate Change*, 8(3): 447–85.

Christensen, Clayton M. and Raynor, Michael E. (2003). *The Innovator's Solution*. Boston, MA: Harvard Business School Press.

Cohen, Wesley M., Goto, Akira, Nagata, Akiya, Nelson, Richard R., and Walsh, John P. (2002). 'R&D Spillovers, Patents and the Incentives to Innovate in Japan and the United States', *Research Policy*, 31: 1349–67.

Cotrell, Thomas (1996). 'Standards and the Arrested Development of Japan's Microcomputer Software Industry', in David C. Mowery (ed.), *The International Computer Software Industry*. Oxford: Oxford University Press.

Hara, Takuji (2003). Innovation in the Pharmaceutical Industry: the Process of Drug Discovery and Development. Cheltenham, UK: Elgar.

Henderson, Rebecca, Orsenigo, Luigi, and Pisano, Gary P. (1999). 'The Pharmaceutical Industry and the Revolution in Molecular Biology: Interactions among Scientific, Institutional and Organizational Change', in David Mowery and Richard Nelson (eds.), *Sources of Industrial Leadership, Studies of Seven Industries.* Cambridge: Cambridge University Press, pp. 267–311.

HSBC (2000). Japan Pharmaceuticals (Investment Sector Report).

Ibata-Arens, Kathryn C. (2000). 'The Business of Survival: Small and Medium-Sized High-Tech Enterprises in Japan', *Asian Perspective*, 24(4): 217–42.

Johnstone, Bob (1999). *We Were Burning: Japanese Entrepreneurship and the Forging of the Electronic Age.* New York: Basic Books.

Japan Pharmaceutical Manufacturer's Association, Data Book 2002, p. 12.

Kimura, Bunji, Fukami, Akira, Yanagisawa, Shin-ichiro, and Sato, Kumi. (1993). 'The Current State and Problems of Japan's Pharmaceutical Market', in Daniel Okimoto and Aki Yoshikawa (eds.), *Japan's Health System: Efficiency and Effectiveness in Universal Care.* Washington, DC: Faulkner & Gray's Healthcare Information Center, pp. 171–89.

Kodama, Fumio (1991). *Emerging Patterns of Innovation: Sources of Japan's Technological Edge.* Boston, MA: Harvard Business School Press.

Kneller, Robert (2003). 'Autarkic Drug Discovery in Japanese Pharmaceutical Companies: Insights into National Differences in Industrial Innovation', *Research Policy*, 32: 1805–27.

Kurtzman, Joel and Rifkin, Glenn (2005). *Startups that Work: The Ten Critical Factors that Will Make or Break a New Company.* New York: Portfolio.

Lam, Alice (2003). 'Organisational Learning in Multinationals: R&D Networks of Japanese and US MNEs in the UK', *Journal of Management Studies*, 40: 673–703.

Leigh, Beatrice (Former Assistant Director, Worldwide Academic Liaison, for SmithKline Beecham) (2000). Presentation at the annual conference of Licensing Executive Society International (May 23, Amsterdam).

Merges, Robert (1996). 'A Comparative Look at Property Rights and the Software Industry', in David C. Mowery (ed.), *The International Computer Software Industry.* Oxford: Oxford University Press.

Motohashi, Kazuyuki (2005). 'University–Industry Collaborations in Japan: The Role of New Technology Based Firms in Transforming the National Innovation System', *Research Policy*, 34: 583–94.

Myers, Robert A. (1999). The IBM Origins of Display Technologies, Incorporated. (Publication MITJP 99-01) MIT Japan Program, p. 14.

Santoro, Michael D. and Chakrabarti, Alok K. (2002). 'Firm Size and Technology Centrality in Industry–University Interactions', *Research Policy*, 31: 1163–80.

SRI International (1997). *The Role of NSF's Support of Engineering in Enabling Technological Innovation* (Prepared for the National Science Foundation) Arlington, VA: SRI International.

Thomas, L. G. (2001). *The Japanese Pharmaceutical Industry: The New Drug Lag and the Failure of Industrial Policy*. Cheltenham, UK: Elgar.

Zucker, Lynne G. and Darby, Michael R. (1996). 'Star Scientists and Institutional Transformation: Patterns of Invention and Innovation in the Formation of the Biotechnology Industry', *Proceedings of the National Academy of Sciences*, 93: 12709–16.

3

Upholding the Pecking Order: Universities and Their Relations with Industry

INTRODUCTION

Universities are the origin of many high technology ventures. In this book I use the term *startup*, specifically to refer to ventures whose core technology at founding is directly based on university discoveries.[1] More broadly, universities are the origin of many discoveries that become the basis for innovative new technologies, whether developed by old or new companies. However, the degree to which university research is the proximate origin of innovation is difficult to determine. There is variation by country and industry, as suggested in Chapter 1. In the USA, innovation in pharmaceuticals and other fields of biomedicine draws heavily on university research. About one-quarter of all new US-origin drugs are discovered in US university laboratories.[2] The contributions of UK and Canadian universities are equivalent, but in Japan and the major Continental European countries, this proportion is much lower.[3] US patents on biomedical inventions cite academic papers more than patents in other fields—suggesting that such inventions draw on academic research more than US patents in any other field, IT being a distant second. In both the case of biomedical and IT-related patents, there is an increasing trend over time to cite academic research—suggesting that university research is becoming more important for innovation, at least in these fields.[4] Biomedicine and IT are the two main growth areas in the US economy in terms of sales and employment.[5]

The proportion of high technology ventures that are university startups is also uncertain. Biotechnology is considered to be an industry based on close university–bioventure linkages.[6] However, a survey of US biotechnology companies conducted in the late 1980s showed that just under half of the founders came from academic positions, only slightly more than those that came from other companies. Also, the trend was for founders to come increasingly from industry, so now the proportion of bioventures founded by university researchers may be considerably less than half.[7]

About three-quarters of Japanese bioventures that are oriented toward developing therapeutics are based on university discoveries. Thus in Japan, bioventures are probably at least as dependent on universities as their US counterparts.[8]

As for IT and other nonbiomedical fields, most US venture companies whose business is focused on integrated circuits or computer hard disk drives are formed by persons leaving existing companies. Few are based directly on university discoveries.[9] However, in some other technologies discussed later in this book, for example, tunable lasers for optical switching devices, gene sequencers and nanotechnology/materials science in its various applications, university startups seem to be among the innovation leaders in the USA— although not in Japan.

In summary, universities are not the fountainheads of innovation in all fields of technology. However, they may have a disproportionate influence on innovation in the most dynamic areas of the economy, and in many cases, initial commercial development of university discoveries in these areas may depend on startups.

Part I of this chapter shows how, until recently, the Japanese system of university–industry technology transfer impeded the formation of startups. Recent reforms have improved the environment for academic collaboration with large and small companies alike. Although the legal framework governing technology transfer from universities to industry is now amenable to startup formation, the system still favors transfer of university discoveries to large rather than new companies.

Part II shows how other institutional and social factors, for example career paths in academia, the system of research funding, and uncertainties related to conflicts of interest, also contribute to an environment more suitable for large companies than for startups.

PART I: UNIVERSITY–INDUSTRY TECHNOLOGY TRANSFER

Overview of Technology Transfer in Japan and the USA

Because ownership and management of intellectual property (IP) is central to how university discoveries are transferred to industry, this part begins with an overview of how the system of university IP ownership has changed. Prior to 1998, either the government or the individual inventors owned inventions made in Japanese universities, depending on the source of funding that gave rise to the inventions. In a series of legal reforms between 1998 and 2004, this system of ownership changed to a system under which universities may own all inventions made by their faculty, and indeed are encouraged to do so.

The new Japanese system of ownership is similar to the system in the USA since 1980. That year the Bayh-Dole amendments to US Patent Law[10] gave universities the right to own inventions arising under R&D funded by agencies of the US government. Prior to 1980, the government funding agencies had the right of ownership. Because US government funding accounted for about 67 percent of R&D support in US universities during the 1970s,[11] a roughly equivalent proportion of university discoveries were probably subject to government ownership. Being subject to the government's right of ownership meant that it was difficult for any of these inventions to be licensed exclusively. With a few exceptions, US government agencies did not have authority to issue exclusive licenses until 1971.[12] But even after 1971, until the passage of the Bayh-Dole amendments, the number of exclusive licenses covering university inventions issued by government agencies was small. In the case of the US Department of Health Education and Welfare (DHEW), which had pushed hardest for mechanisms to permit exclusive licensing, the number of exclusive licenses issued between 1969 and 1980 was less than twenty.[13] The more important mechanism for licensing DHEW-funded university inventions were *institutional patent agreements* (*IPAs*) that DHEW began to sign with US universities in 1968. Under a typical IPA, a university that showed it was able to manage IP and abide by applicable laws could take title to DHEW funded inventions and then license them to industry. Exclusive licenses were possible, but only for terms so limited they would probably not meet the needs of startups.[14] By 1977, DHEW had seventy IPAs in effect covering most leading US universities. NSF began to conclude IPAs in 1973. Nevertheless, by the mid-1970s, the number of exclusive licenses issued by universities under IPAs was still under 100.[15]

The pressures to grant exclusive licenses to university inventions were greatest in the case of NIH funded inventions relating to pharmaceuticals.[16] Exclusive patent rights are important for pharmaceutical development, because the process is long and expensive and even late in the human trials a candidate drug may turn out to be a failure. But after safety and effectiveness have finally been shown, the main chemical constituents of drugs are usually easy to copy. Nevertheless, some of the support for universities' authority to license exclusively under Bayh-Dole came from outside the pharmaceutical industry.[17] In addition, startups in most industries need exclusive licenses in order to be able to attract private investment necessary for growth and to have some bargaining leverage to get to the negotiating table with other companies, especially to convince larger companies to become their customers.[18] One of the earliest examples of a university granting exclusive licenses involved an electronics startup financed by one of America's first venture capital firms, that together argued the company could not be formed without protection from encroachment.[19]

Founded in 1976, Genentech is one of the first successful university bio-startups that drew heavily on university research even after its founding.[20] Part of its early business strategy involved obtaining exclusive rights to university inventions, even prior to Bayh-Dole.[21] Between the founding of Genentech and the passage of Bayh-Dole, a few other bio-startups were founded, and then in 1981 the number of new bioventures jumped to nearly threefold the number formed the previous year.[22] This suggests that the liberalization of policies governing the issuance of exclusive licenses to government funded university inventions improved the environment for new company formation and may have been a necessary condition for the rise of university startups in biomedicine and even other fields.

The same link between liberalizing government restrictions on exclusive licenses and the rise of startups is evident in the case of Japan. Nevertheless, in the years before Japan liberalized restrictions on formal exclusive licenses, its system of technology transfer differed greatly from that of the USA. It was an informal system that gave no scope to academic entrepreneurship, but which made the transfer of exclusive rights to university discoveries to established companies extremely easy. When the framework became similar to that of the USA, this was not sufficient to make new company creation a principal mechanism to develop university discoveries. Instead, the patterns of university–industry cooperation established during the postwar decades accommodated to the new legal framework and persisted. Except in biomedicine, large established companies remain the main channel for commercializing Japanese university discoveries. The very closeness of links between large companies and leading university laboratories forecloses opportunities for new companies to grow.

The Pre-1998 Japanese System[23]

Similar to postwar America, in Japan there was a presumption that inventions made in *national universities*[24] belonged to the nation and should be freely available for all to use or licensed nonexclusively by central government bureaus. However in the 1970s, just as in the USA, corporate interest in university research and in securing formal IP rights to university discoveries created pressures to make the system of IP management more flexible. The solution implemented in 1978 (two years before Bayh-Dole) was to retain government ownership over all inventions arising under project-specific funding, but to let university inventors retain ownership over inventions arising from *non-project-specific funding*. The former includes funding under *formal sponsored research agreements*[25] and *government grants-in-aid* for research. The latter includes nominal *standard research allowances*[26] available to all full-time faculty engaged in research and, most importantly, *donations* from corporations or individuals.

Project-specific funding accounted for a majority of funds available for discretionary research expenditures,[27] and thus more than half of university inventions probably should have been classified as national inventions. In fact, probably less than 10 percent of inventions were classified as national inventions, and most of these were jointly owned by the corporate sponsors of formal collaborative research.[28]

Why and how did this happen? National ownership entailed management of the patent applications by government bureaucracies and nonexclusive licensing.[29] Therefore, companies considered this designation undesirable. On the other hand, donations were attractive to faculty because they were free of many of the restrictions attached to other forms of funding. It was standard practice for large companies to distribute large numbers of small donations to many university laboratories.[30] Even today, donations remain the main source of corporate support for university research.[31]

The quid pro quo for receiving donations was that professors would inform donors of their research progress (i.e. serve as de facto consultants) and let the donors file patent applications. Also, they would encourage capable students to consider the donors as places to work after graduation. Donations were an important mechanism to sustain university–industry cooperation between the end of World War II, when formal consulting was banned and other types of formal cooperation restricted,[32] until the 1998–2004 reforms that once again opened the door to formal, transparent forms of cooperation. To keep their side of the bargain, faculty inventors also wanted to avoid the national invention classification. Also most faculty inventors thought that the government bureaux did not manage their inventions competently, and that direct transfer to collaborating companies offered the best means of development. Attribution of invention funding was easily manipulated.[33] Except for inventions arising under formal sponsored research agreements with companies,[34] almost all commercially useful inventions were attributed to donations (less frequently, to the standard research allowances)—when in fact, many benefited from project-specific government funding.[35] Thus, donations and officially tolerated misattribution of funding sources enabled the donor companies to appropriate numerous publicly funded research discoveries.

This form of technology transfer was fast and low cost. Should an invention be commercialized, companies were expected to pay only token royalties to the inventor. The system enabled large companies to keep abreast of research along wide fronts related to their interests. In the case of some breakthrough discoveries, such as titanium dioxide photocatalysts, it has resulted in a large number of companies developing a variety of products based on university discoveries in this field.[36]

But because companies received university discoveries essentially for free, incentives to develop them were low unless they were clearly outstanding or directly relevant to a company's core business. The origins of pipeline drugs discussed in Chapter 2 suggests that the numerous collaborations of the large Japanese pharmaceutical companies with university researchers usually involved basic science issues or narrowly defined research tasks and rarely led to the discovery of actual drugs or drug targets. Nevertheless, they probably involved the transfer to the pharmaceutical companies of rights to many academic discoveries. One of the most successful Japanese biostartups had to license back the founder's inventions from Japanese pharmaceutical companies that had obtained ownership under the informal technology transfer system.[37] 'Sleeping university inventions' unused by companies was a key concern of the government agencies that promoted the 1998–2004 reforms.[38] Government advisory committees that recommended adopting a US-style system reasoned that ownership would give universities incentives to manage their own inventions so as to maximize their commercial and societal value.[39]

The system was disadvantageous to small companies, especially startups. Inventions that might have provided the bases for strong startups were sometimes transferred unwittingly or automatically to companies that gave donations. Small companies could not compete in terms of the numbers of laboratories to which they could give donations. Nor, at least in the best known universities, could they compete in terms of the attractiveness of the jobs they could offer the professors' students.[40] Startups were additionally handicapped because uncertainty over invention ownership could discourage private investment. However, promotion of startups was not a main goal of the reforms, nor was there much discussion to the effect that clarity of ownership and formal technology transfer mechanisms are especially important for startups.[41]

In addition, personnel regulations prohibited consulting and holding a management position in a company. Only in 2000, when such activities were legalized, did national university professors begin to establish companies.

Finally, universities as institutions had little stake in the technology transfer process. They could not receive royalties or to hold equity in start-ups, and had only limited rights to overhead (indirect cost) payments on research grants and contracts. Their administrative staffs were MEXT bureaucrats who changed jobs every two years, sometimes moving to another institution. Today administrative staff still rotate regularly, but they usually remain within the same university. Overhead payments are higher, but they mainly are plowed back to directly support research in the laboratory or department/center that received the funding. Receiving stock in lieu of cash for license royalties is still problematic for national universities. For these reasons and others, Japanese universities *as institutions* remain less entrepreneurial than their US counterparts. Their direct financial interest in the success of their startups is still less.

Legal Convergence Masking Continuing Divergence

Four laws, enacted between 1998 and 2004, changed the legal technology transfer framework:

- The 1998 *Law to Promote the Transfer of University Technologies*[42] (the *TLO*[43] *Law*) legitimized and facilitated transparent, contractual transfers of university discoveries to industry, even though it did not change the basic ownership system. It provided a fig leaf to allow contractual licensing of inventions to industry, even though a rigorous analysis of funding sources might have revealed that inventions arose under project specific government funding. It also provided for subsidies of about US$ 180,000 annually for five years for approved TLOs.[44] Starting from five TLOs approved in 1998, the number of approved TLOs increased to thirty-nine by the end of 2005.

- The 1999 *Law of Special Measures to Revive Industry*[45] (the *Japan Bayh-Dole Law*) has the same effect as US Bayh-Dole Law, *except that it did not apply to national universities until they obtained legal status as semi-autonomous administrative entities in 2004.*[46]

- The 2000 *Law to Strengthen Industrial Technology*[47] permitted national university researchers to engage in paid outside work on behalf of corporations. Implementing regulations and university policies were progressively relaxed until about 2005, at which time a wide range of consulting and even management activities were permitted. However, permission for a national university faculty member to hold an outside management/directorship position is granted only if the outside work is directed toward the commercializing of the researcher's own university discoveries, and such high level positions require a higher level of approval within the universities.[48]

Also, the Law to Strengthen Industrial Technology streamlined the procedures for company-sponsored commissioned and joint research. It cleared away bureaucratic barriers to the flow of funds under these formal research agreements.[49] In so doing, it paved the way for using sponsored research funds to pay stipends for graduate students and post-doctoral researchers and for the increase in formal joint research agreements, the most important feature of the current technology transfer landscape.

- The *National University Incorporation Law*[50] gave national universities independent legal status when it went into effect in April 2004. Previously they were merely branches of MEXT. But by gaining status as independent legal entities, article 35 of Japan's Patent Law, which enables employers to require assignment to them of employee inventions, became applicable as did the Japan Bayh-Dole Law. MEXT has urged universities to assert ownership over commercially valuable inventions.[51]

With the last of these reforms, the legal framework of Japan's technology transfer system came to resemble closely that of the USA.

Many standard indices of technology transfer activity compare favorably to US indices. Average patent applications per TLO were higher than historical US averages.[52] Average numbers of licenses were also higher.[53] However, average royalties are probably considerably lower than historical US levels.[54]

As for startups, the numbers being formed each year are impressive and their rise coincides closely with the 1998 and 2000 reforms that facilitated exclusive licensing and consulting.

Figure 3.1 should be interpreted with caution, although the general pattern is probably accurate. It includes companies whose only connection with universities is having engaged in joint research, or having graduates or faculty as advisers, investors or managers (but not founders). It also includes limited liability companies whose operations and business scope are small, as well as companies that seem to be focused only on sales or provision of services. In order to adjust these figures to represent companies that are *based directly on university discoveries*, the totals for each year should probably be discounted by about 40 percent.[55] Also my conversations with TLO and investment personnel indicate that the leveling off in the formation rate is a real phenomenon, and outside of biomedicine, the rate of startup formation is probably decreasing.

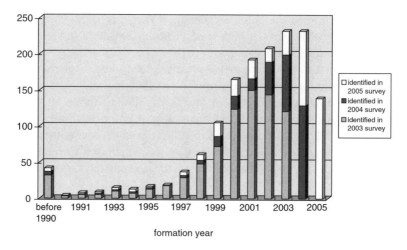

Figure 3.1. Number of Japanese startups formed per year
Source: METI (2006).

But even applying a 40 percent discount factor, the rate of startup formation is quite respectable in comparison with startup formation in the USA in the early years following Bayh-Dole. In the 1980s, rates of startup formation were probably well under 100 per year, and had only risen to around 200 per year about 14 years after Bayh-Dole.

Nevertheless, Figure 3.1 masks general weakness and difficulty to compete with established companies for access to the most important university discoveries. As discussed in Chapter 4, aside from some startups in biomedicine and a smaller number in software, most of these startups are small in terms of sales, employees, and capital, and their core technologies offer little prospect for business growth. Even in life science, the average size of the start-ups is less than half the size of US bioventures of equivalent age based on historical data.[56] Japanese bioventures (most of which are start-ups) have not been able to grow as fast as their US competitors and total sector employment is considerably less than in the USA.

The reasons for this weakness is one of the main themes of this book. However, several reasons relate specifically to the technology transfer system. As a result of the 2000 Law to Strengthen Industrial Technology, it may be too easy for professors to form startups and remain as de facto directors. Thus some startups tend to focus too much on scientific issues and not enough on business goals.[57] In a similar vein, various government programs encourage startup formation without ensuring the startups are likely to produce products for which there is market demand. For example, JST provides venture seed grants to university researchers on condition that they form a company within three years. Many recipients use these grants to pursue scientific projects, knowing it is easy to satisfy in a pro forma manner the startup requirement. Japan's corporate law permits companies to be incorporated with just 1 yen paid in capital.[58] But probably the main reason is that joint research with established companies has taken the place of donations, allowing established companies to preempt university discoveries and closing off technological niches that might otherwise have been available for entrepreneurial companies.

Joint Research and the Preemption of University Discoveries

Figure 3.2 shows that joint research has increased dramatically beginning around the start of the reforms. As already mentioned, the 2000 Law to Strengthen Industrial Technology made joint and commissioned research more attractive mechanisms for companies to collaborate with universities. Projects with large companies account for 70 percent of all projects, a proportion that has been declining only gradually since the 1990s.[59]

Incorporation of national universities in 2004 meant that the universities would own all inventions made subsequently by their employees under commissioned and joint research. Universities rarely assigned to industry partners the right to apply for patents on such inventions. Rather, like their US counterparts, they offer the partner the right to negotiate an exclusive license to

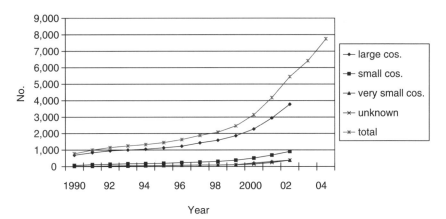

Figure 3.2. New and ongoing joint research projects between private companies and national universities

Sources and definitions: see note 60.

such inventions—or to the university's portion when there are university and industry coinventors.

However, Japan's patent law favors the industry partners in a way US patent law does not. Article 73 of the former requires the consent of all co-owners of an invention before it can be transferred to a third party, even by nonexclusive license. Thus, so long as the company is a co-owner by virtue of coinventorship or the terms of the sponsored research contract, the company can block the transfer of the university's rights to any other company. In other words, article 73 gives co-owners an automatic, de facto, nontransferable, royalty free exclusive license.[61] In order to avoid this situation, joint research contracts now usually include a clause to bypass article 73. This allows the university to give a third party a nonexclusive license to its use rights, unless the co-owning company negotiates an exclusive license to the university's rights. However in practice, few third parties are interested in nonexclusive licenses if that would put them in potential competition with a large company.[62] In addition, large companies sometimes insist that the bypass clauses be stricken from joint research contracts. The universities, often at the urging of the professor who wants to keep good relations with the company, usually agree. In such cases, the joint research sponsor usually pays most of the patent application and maintenance costs, but has no obligation to develop the invention or to pay royalties unless it licenses the invention to a third party. Under such joint research agreements, control over inventions is just like it was under the donation system.[63]

I have been fortunate to have access to the invention reports submitted by university inventors to the TLO of a major national university. As shown in Figure 3.3, over a six month period in 2005, 46 percent of the inventions were in engineering or IT hardware, 32 percent pertained to life science, 13 percent to materials or chemistry, and 9 percent to software.[64]

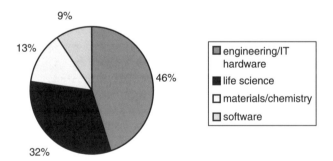

Figure 3.3. Inventions by field reported to one university

Thirty-one percent of the inventions were attributed to joint research projects with private companies, *although such projects account for less than 6 percent of activity-specific research funding in this university.*[65] In almost all these cases, a company researcher was also listed as a coinventor. If companies expect interactions between researchers that might result in inventions, they usually conclude a joint research contract in advance. Similarly, companies seem to expect that if a joint research agreement is in effect and an invention arises, at least one of their researchers will be a coinventor.

Only 18 percent of the life science inventions arose under joint research, and of these only one-third arose under joint research with large companies—the remainder arose under joint research with university startups or other small companies.[66] In other words, in life science fields, joint research accounts for only a small proportion of total inventive activity, and large companies are not using joint research as a means to appropriate a large proportion of university research results. The TLO is free to license most life science inventions to the companies it determines are most willing and able to develop them, including to new startups if the right combination of entrepreneurship, funding, and market opportunities exists.

However, in the case of non-life science inventions, nearly 40 percent were joint inventions, and over 80 percent of these were with large companies. Thus, the TLO has management authority over a smaller proportion of these inventions. Figures 3.4 and 3.5 show this graphically: a small proportion of life science inventions are attributed to joint research and of this small proportion, the joint research partner is often a small or new company. But joint research accounts for a much larger proportion of engineering, chemical, and software inventions and the joint research partner is almost always a large company.

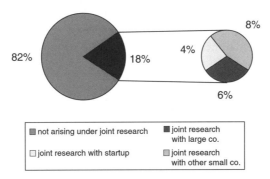

Figure 3.4. Life science inventions: association with joint research and type of industry partner

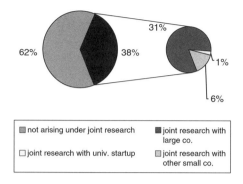

Figure 3.5. Non-life science inventions: association with joint research and type of industry partner

I have continued to monitor invention reports and as of the end of 2006 this general distribution has not changed.

This analysis deals with invention reports, not patent applications. This TLO files Japanese patent applications on roughly 30–40 percent of the reported inventions overall, but the application rate for joint inventions is 60–70 percent.[67] Thus, in terms of inventions on which applications are filed, joint research inventions probably account for about half of the total, and a majority of non-life science inventions. Considering joint patent applications and licenses as the two main mechanisms for transferring university inventions to industry, in 2005 over 60 percent of such transfers by this TLO were the former type. In other words over 60 percent of transferred inventions were transferred via joint patent applications by the university and the joint research partner (giving the joint research partner a *de facto* exclusive license to the inventions in most cases) while less than 40 percent of the transferred inventions were owned entirely by the university and transferred as licenses (with the TLO playing a major role in deciding who should be the licensee). This confirms the dominant role of joint research and joint patent applications as the means of transferring this university's inventions to industry.

As for other universities, anecdotal reports from colleagues in other TLOs indicate that rates of joint inventions are probably *higher* in most other major universities.

These findings are not necessarily negative. Economic pressures are forcing many large Japanese companies to rely more on collaborative research with universities than on basic research in their own laboratories.[68] From my vantage point in a major university, the numbers of industry researchers on campus is noticeably greater than eight years ago, an impression supported by nationwide data.[69] Conversations with various university laboratories indicate that interaction between industry and academic researchers engaged in joint research usually is quite close. Also anecdotal assessments by industry executives suggest that industry is coming to regard joint research with universities as relevant to its business goals, that is, more favorably than five to ten years ago.[70]

Industry sponsored inventions probably constitute less than 10 percent of the total among US universities, and only a small fraction of these have industry coinventors.[71] Thus, institutional barriers to cooperation between universities and established companies may be higher in the USA, and Japanese companies and professors may seek out collaborations with each other more readily than their US counterparts.[72] It has been suggested that US universities have focused too much on ownership of inventions and license revenue, whereas they ought to place more emphasis on the support that industry can provide for ongoing research.[73] Recent moves toward open collaborations in which companies support US university research in return for any resulting IP being freely available for academic and commercial use reflect this perspective.[74] But the big difference between open collaborations in US universities and joint research in Japanese universities is that, in the Japanese case, companies usually obtain exclusive IP rights.

Many Japanese universities have good researchers but weak TLOs. In these universities, joint and commissioned research is the only effective mechanism of technology transfer, if startup formation is not feasible. Also, well-known professors often engage in joint research with several companies, even from within the same industry.[75] So while preemption by established companies as a group may be of concern, preemption by individual companies probably is less so. Finally, in the university whose inventions are analyzed above, the TLO is handling the overall technology transfer process quite well, consulting closely with inventors, making timely decisions whether to file patent applications, and in the case of non-joint research inventions, licensing to a wide range of large and small companies in Japan and overseas. Its licensees include one of the strongest groups of startups in the country. In other words, this university has shown that, despite preemption by joint research of a large proportion of discoveries, promising opportunities for licensing and startup formation remain, but mainly in biomedical fields.[76]

On the other hand, having so many inventions flow automatically to established companies takes entrepreneurial initiative away from TLOs and faculty members. There is little that TLOs or inventors need to do (or can do) to influence how these discoveries will be developed. Furthermore, the prevalence of joint research raises questions about a shift in focus from fundamental to applied research in universities, and the undermining of core academic values. For example, the university of Tokyo's standard joint research contract (art. 30) obligates faculty to abide by the sponsoring company's reasonable requests to delete or change manuscripts related to the joint research. Are too many talented researchers settling too easily into a routine of doing applied research for industry while ignoring fundamental issues that hold the keys of the next generation of new products? Or conversely, does close interaction with industry lead more quickly to deeper scientific understanding and breakthroughs? Finally, the prevalence of joint research, while helping established companies to develop competence in new fields, has decreased the niches available for new companies to exploit.[77] Even when an entrepreneurial professor manages to avoid initial preemption and starts a company based on innovative and commercially promising discoveries, the presence in his or her laboratory of large companies engaged in joint research often diminishes opportunities for the startup to grow, as described in Chapter 7.[78]

Conclusions Pertaining to Technology Transfer

Two years after completing the transformation of the legal framework governing technology transfer, the system has gotten off to a credible start, with standard performance indices that are quite respectable in comparison with the US system about a decade after enactment of the Bayh-Dole Law. The high points of the Japanese system include a few TLOs that have demonstrated good competence and somewhere on the order of fifty biomedical startups and a smaller number of software startups that are making significant progress. But despite the legal framework being nearly identical to that in the USA, the Japanese system continues to favor transfer of university discoveries to large established companies. Weak nonentrepreneurial administrations coupled with the long-standing practice of faculty passing their discoveries directly to established companies have allowed joint research to take the place of donations in continuing a system of direct transfer of university discoveries from academic researchers to industry.[79] This benefits established companies but has hindered the formation of startups with strong growth prospects. Whether this contrast with the US system is beneficial for Japan's future depends on whether established companies are better at early stage innovation than new companies—whether Japan can dispense with new companies and rely on its established companies to carry forward early stage innovation in new fields of technology. These are complex questions that are deferred to Chapters 6 and 7.

At least the new Japanese technology transfer system opens alternative routes for university inventors to try to ensure their discoveries are developed. No longer do they have to rely on the large companies that provide them donations and hire their students—although, as noted in Chapter 7, it is often hard for inventors to exploit these alternatives (founding startups or licensing to outside companies) when working under the gaze of companies who have sent researchers into their laboratories.

The Japanese experience raises questions about the US system. Why did the US system evolve so that there was more separation between university researchers and established companies than in Japan? Why are faculty–company relations more at arm's length than in Japan? Turn the clock back to the 1920s and 1930s and it might seem that conditions were ripe for university–industry relations in the USA to evolve as they did in Japan. Consulting between faculty and companies was common, at least in MIT.[80] Also MIT attempted to entice industry to fund much of its R&D activities. These efforts were generally unsuccessful, although industry did welcome interactions, mainly as a means to recruit graduates. These circumstances are similar to those in postwar and probably also prewar Japanese universities.

I suspect that part of the explanation lies in US universities being stronger institutional entities than their Japanese counterparts. MIT grappled with issues related to cooperation with industry in the 1920s and 1930s, and after back and forth consultations with its faculty developed policies on faculty consulting and on IP.[81] It took steps to establish an office to handle IP, and, most importantly, it set up an office to handle research contracts. Today, the contract office serves as a gatekeeper to MIT research. Corporate research sponsors do not have free rein to suggest to professors what percentage of inventorship on joint research inventions ought to be attributed to company researchers, as is currently common practice in Japanese universities. Instead, these offices scrutinize claims of joint inventorship closely and give the sponsors a limited period during which they can negotiate licenses.[82] Another factor was the large amount of government contract and grant research support in the postwar years, much of which was either defense or health related. This enabled universities such as Stanford and MIT to build world-leading research capabilities. Government funding also brightened employment prospects in academia for graduate students. Universities could resist attempts by industry to condition sponsored research funding on free and often exclusive access to wide swaths of university discoveries.

If these hypotheses are correct, they suggest both hope and caution for the entrepreneurial prospects for Japanese universities. The incorporation of national universities is a first step to enable them to build the institutional competence to manage their discoveries, not to let them pass, as through a sieve, to companies engaged in joint research. However, these competencies are being developed slowly.[83] Government funding has increased substantially over the past ten years, and much of this increase has been in the form of large contract research projects.[84] However, as discussed in Chapter 7, such funding often involves large company collaborators. Rather than helping universities to become independent institutions that can manage their resources, large-scale government funding has often perpetuated the pass through of university discoveries to established companies. Is this beneficial to Japan? Conversely, is the more formalized, arm's-length system of university–industry cooperation beneficial to America? The answer depends on the importance of vibrant, independent startups for early stage innovation.

Part II examines other institutional and social aspects of the Japanese university environment that influence faculty and student entrepreneurialism, particularly factors related to funding, career paths, and lack of clarity regarding the appropriate scope of academic entrepreneurship.

PART II: INSTITUTIONAL AND SOCIAL FACTORS AFFECTING ACADEMIC ENTREPRENEURSHIP

If the door has been open to academic entrepreneurship, why are few academic researchers walking through it with the aim of creating strong, rapidly growing companies based on new technologies, or new applications of old technologies?[85] One reason relates to academic laboratories being under the gaze of joint research partners. Other reasons, to be discussed in subsequent chapters, concern difficulties ventures face in recruiting personnel. However, the remainder of this chapter discusses institutional and social factors related to universities.

Uncertainty Regarding Conflicts of Interest

Faculty involvement in startups usually raises more issues regarding financial conflicts of interest, and often also conflicts of commitment of time and energy, than involvement in joint, commissioned, or donation-sponsored research. Startup formation usually requires that the faculty inventor be actively involved as an adviser. The normal mechanism to encourage and compensate this involvement is for the inventor(s) to hold a substantial proportion of the startup's initial stock. He or she may, in addition, be the principal member of the startup's board of scientific advisers, and in Japan, may even hold a line management position, such as CEO or chief scientific officer. In any case, in addition to being concerned about the scientific progress of the startup, she or he often will represent the startup to investors, the scientific community and the media. If the startup's R&D and business progress favorably, the increasing likelihood of an IPO or buyout raise the prospect of substantial financial benefit for the inventor. Although cooperation with established companies through donations or contract research also offer benefits in terms of increased funding for equipment, graduate students, and so forth, and the prospect of eventual commercialization of one's discoveries, these usually do not compare to the complexity and public visibility of the conflict of interest and conflict of commitment issues that arise related to startups.

MEXT has permitted holding of pre-IPO stock in startups for several years. However, there has been little open debate about appropriate limitations or cautions related to stock ownership or other entrepreneurial activities. Moreover, conversations with journalists, university officials, researchers, and attorneys suggest that the concept, *conflicts of interest*, has a more negative connotation in Japan than in the USA. Rather than being regarded as an inevitable accompaniment to university entrepreneurship that requires management to avoid significant harm,[86] *conflicts of interest* appears widely regarded as a label of reprehension that ought to be avoided. Also, public perception is still strong that university faculty are public servants who ought not to be concerned about financial gain. Beginning about 2005, major universities began to require fairly comprehensive annual reporting of outside financial and business interests.[87] How this information will be used is not clear. The low threshold for public condemnation and the lack of open discussion on how to balance the conflicting objectives of promoting commercialization of discoveries and giving full priority to education and research create uncertainty about permissible limits of entrepreneurial behavior. The degree to which this discourages entrepreneurship and prompts university researchers to opt for collaboration with existing companies instead of forming their own startups is difficult to assess, but conversations with university researchers suggest it has an overall dampening effect.

Conflict of interest concerns are most acute in clinical trials of new drugs or medical procedures owned by a company in which a key university researcher has a financial interest. In such cases, patients' lives can be at stake.[88] In early 2006, a study committee funded by MEXT released unofficial guidelines that recommend comprehensive disclosure but leave the development of policies up to individual universities[89,90]. Several major universities have enacted official guidelines along similar lines.[91] Thus the institutional infrastructure is being developed to manage conflict of interest issues. But there is little open discussion about appropriate limits on faculty entrepreneurs participating in clinical trials, or specific procedures to protect patients and ensure the objectivity of publications. Anecdotal accounts indicate that some universities are adopting a strict approach that discourages faculty involvement in startups focused on clinical therapies, while others are adopting a liberal approach. In any case, debate on specific cases (if it occurs at all) appears to be limited to within small committees, and basic principles supported by specific, yet appropriately flexible guidelines and management procedures are slow to be articulated. Avoiding open discussion misses an opportunity to increase awareness among all parties (researchers, university administrators, businesses and the public) about the importance of conflicts of interest and how they can be managed responsibly.[92]

Demographics

In relation to overall population, there probably are more researchers in US universities who can make discoveries in new fields that might be suitable for startups or who might become startup managers. Japanese companies still tend to hire bachelors or masters degree graduates for their R&D laboratories, and this limits the number of doctoral candidates in Japanese universities relative to what they would be in the US.[93] Nevertheless, on a per-population basis, the number of Japanese S&E doctoral graduates is approaching that of the USA.[94] The number of postdoctoral researchers is, however, much larger in the USA than in Japan.[95] The extent to which US high technology ventures recruit from the ranks of postdoctoral researchers merits further inquiry. America's advantage in S&E immigrants is discussed in the final chapter. Gender imbalances are considerably greater in Japan than the USA,[96] also suggesting lost opportunities to develop scientific and entrepreneurial talent.

Kouzas

The basic organizational unit in Japanese universities is the *kouza*, modeled on the *professor chair* system in early twentieth-century German universities. A kouza typically consists of one full professor, the laboratory head, an assistant professor, who is usually the lead candidate to inherit laboratory leadership when the professor retires, and one assistant (*joshu*).[97] There is usually one laboratory per kouza. Thus laboratory facilities are under the kouza head. Applications for research funding from junior kouza members usually include the kouza head as a coapplicant, and of course must be coordinated with him or her.

In contrast, a new 30–40-year-old assistant professor in the USA is usually provided with his or her own laboratory, the startup costs for which may approach half a million dollars. The flip side of these benefits is that she or he is expected to obtain within two years competitive grants to cover not only laboratory costs but also a substantial proportion of his or her salary and the stipends for graduate students and postdoctoral researchers. Also within six years, his or her publications will have to pass muster before outside experts and a university committee that will decide whether she or he receives tenure. But young US researchers have more independence than their Japanese counterparts, who are usually constrained to follow the research leads of the laboratory head for a longer time. Consequently, young Japanese researchers probably are less likely to pursue unorthodox research directions.

Government Funding and Peer Review

The Government has placed priority on increasing research funding opportunities for young researchers. But having to rely on the professor for laboratory space, key equipment, supplemental funding, and support staff, means that even recipients of such awards still must coordinate their research with the kouza head.

Some major funding programs involve the distribution of large funds to a senior principal investigator who then distributes the funds to other collaborating kouza heads in other departments or universities.[98] More generally, over one-third of competitive funds available for universities come from programs that tend to fund large projects involving multiple laboratories,[99] a higher proportion than in the case of the US National Institutes of Health (NIH) and National Science Foundation (NSF), the two US agencies that fund the most US university research.[100] The kouza (laboratory) head is usually responsible for cooperation with companies and other university laboratories. Thus, young researchers who want to participate in these multi-laboratory projects must do so as part of the larger kouza.

Many of the government funding programs appear to have an applied emphasis.[101] Some, such as JST's CREST, ERATO, and PRESTO, seek to bring scientific talent to bear on issues that well-known, senior scientists have identified as deserving further study, and the research focus sometimes is dominated by the views of those scientists.[102] In the case of programs such as those funded by METI's New Energy Development Organization (NEDO), the aim is more explicitly to achieve advances with direct applications for industry. As described in Chapter 7, projects in the most cutting-edge fields that involve universities also tend to involve large companies, largely foreclosing opportunities for startups to develop the discoveries from these projects. Whether government projects with an applied emphasis and industry participation produce good science, or even result in substantial benefits for the large company participants, is an unresolved question.[103]

Of all the funding programs, MEXT's Grants-in-aid for Basic Research and Grants-in-aid to Support Young Researchers are the most oriented toward supporting large numbers of individual basic research projects. Together they account for about 37 percent of total competitive research funding for universities, larger than any other program.[104] However, the peer review process that these programs share is not optimal for detecting and supporting novel but well-conceived research proposals. Although review committees usually consist of at least ten academic experts, only three or six members review an application. Because some fields are broad, reviewers must often review proposals in areas where they have no expertise. Reviews occur just once a year and each reviewer may have to review over 150 applications in a limited time period. All their scores are submitted electronically. There are no discussions, no need for the reviewers to explain in detail their ratings unless they rate a particular application extremely high or low, and no feedback to the applicant beyond the overall score, in contrast to the peer review systems of NSF and especially NIH.[105]

Japan has made considerable progress in improving the quality, representativeness, and transparency of the peer review process.[106] Whether the additional benefits of an NIH-type system are justified by the large requirement of reviewers' time and administrative resources is an unanswered question. Nevertheless, in order to encourage applications from lesser known and younger researchers to explore the frontiers of science and to be able to evaluate such applications effectively, a peer review system that brings together a large number of experts and encourages them to debate the merits of competing applications may be preferable to any current Japanese system. It would help to counter the combined influences of senior professors, collaborating companies, and (in many cases) the funding agencies themselves, that tend to channel the energy and creativity of young researchers toward subjects that are not new. It is a necessary part of any successful effort to encourage capable young researchers to establish early in their careers an independent research base.

In addition, discoveries in new niche areas probably are more amenable to development by startups and less likely to be preempted by large companies, if made by independent younger researchers.[107]

Academic Recruitment and Promotion

Academic careers still depend more on patronage than a record of individual achievement.[108] Well into the 1990s, it was common for vacancies to be filled from within the kouza. The kouza represented a narrow career ladder where vacancies were usually filled by the person next below in the hierarchy, and the professor essentially picked his second generation successor when he selected a new joshu. Now internal promotions to the assistant professor level are discouraged, and joshus usually find their first assistant professorship in a different kouza, sometimes in a different university. However, appointment of joshus is still, for practical purposes, entirely a matter for kouza heads to decide. Selection of lead candidates for vacancies at the assistant and full professor levels depends on small internal committees in which a single professor often has a dominant voice. The committees' selection of a lead candidate is rarely questioned by the larger departmental faculty and university. Open debate is rare and solicitation of outside opinions even more so.

Open recruitment, in the sense of widely soliciting applications to fill vacancies and a commitment to select among applicants on the basis of merit, is still rare.[109] Rarer still is soliciting in-depth, objective evaluations of candidates' achievements from *outside* experts and giving considerable weight to these outside evaluations.[110]

The kouza system is becoming more flexible. In a few departments, formal kouza affiliations have been abandoned and professors make real collective recruitment decisions based on individual merit and the needs of the department, not on applicants' past affiliations with members of the department or the closeness of their research interests to those of particular senior professors. Even in such departments, however, there is usually no objective outside input into the process.

Unequal Funding

Exacerbating the importance of patronage for academic careers is the attraction of the Tokyo and Kansai metropolitan areas and the overwhelming prestige and access to funding enjoyed by a few elite universities. It is trite

Table 3.1. Leading recipients of Monbusho/MEXT grants-in-aid (all types, new and continuing projects)

Rank	1995			2005		
	University	Amount (10^8 yen)	% of total	University	Amount (10^8 yen)	% of total
1	U of Tokyo	125.5	13.6	U of Tokyo	201.2	11.7
2	Kyoto U	72.7	7.9	Kyoto U	131.1	7.6
3	Osaka U	61.3	6.6	Tohoku U	94.8	5.5
4	Tohoku U	41.6	4.5	Osaka U	89.8	5.2
5	Nagoya U	34.9	3.8	Nagoya U	64.6	3.8
6	Kyushu U	30.0	3.3	Kyushu U	56.8	3.3
7	Tokyo Inst. Tech	30.0	3.2	Hokkaido U	56.1	3.3
8	Hokkaido U	28.5	3.1	Tokyo Inst. Tech	45.4	2.7
9	U of Tsukuba	22.2	2.4	U of Tsukuba	30.2	1.8
10	Hiroshima U	13.2	1.4	Riken	26.3	1.5
11	Okayama U	9.5	1.0	Keio U	24.9	1.5
12	Keio U	9.1	0.9	Kobe U	24.7	1.4
Total		924.0	100.0		1714.4	100.0

Sources: **1995 data**: For individual universities: Matsuo 1997. For total: www.jsps.go.jp
2005 data: www.jsps.go.jp

but nevertheless true that most academically inclined high school students (or at least their parents) dream of entering the University of Tokyo or Kyoto University, and most academics dream of ending their careers there. As for funding, Tables 3.1–3.3 show only minor variations in the rankings of the top recipients of the three largest categories of competitive funding[111]: MEXT grants-in-aid, commissioned research,[112] and Centers of Excellence (COE) awards. Also rankings vary little over time.

The COE Program was announced in 2001 with the goals of developing up to thirty world class academic centers and differentiating research-oriented from merely education-oriented universities.[113] To no surprise, the awards have been heavily weighted in favor of universities already receiving the lion's share of other funding. Awards are usually to individual departments or new university-wide programs and usually last five years. No new awards were made in 2005, but when the program is resumed (probably in 2007), its funding will be even more restricted to a small number of institutions.[114]

The same recipients of competitive research funding are also favored recipients of operational and administrative subsidies from MEXT (Table 3.4). These are the source of salaries for full-time faculty and other general expenses in national universities, and they account for just under half of all financing attributable to research in Japanese universities.[115] These subsidies are

Table 3.2. Leading academic recipients of commissioned research in 2004

Rank	Institution	Amount (10^8 yen)	% of total
1	U of Tokyo	177.6	17.5
2	Kyoto U	81.4	8.0
3	Osaka U	77.4	7.7
4	Waseda U	44.1	4.4
5	Tohoku U	42.2	4.2
6	Kyushu U	38.9	3.8
7	Keio U	38.2	3.8
8	Hokkaido U	34.9	3.4
9	Tokyo Inst Tech	29.9	3.0
10	Nagoya U	21.1	2.1
11	Natl Insts of Natural Sci	19.1	1.9
12	Tsukuba U	13.0	1.3
Total		1012.3	100.0

Source: MEXT (2005).

gradually being reduced, but COE funding is expected to make up for the reductions in elite universities.

In contrast, 199 US universities are classified as research universities, and of these 96 are classified as highly research intensive.[116] Funding is distributed more evenly among these than among Japanese universities.[117]

Table 3.3. Leading recipients of centers of excellence disbursements in 2006 (for awards announced 2002–4)

Rank	University	10^8 yen	%
1	U of Tokyo	44.24	12.7
2	Kyoto U	33.35	9.6
3	Osaka U	24.14	6.9
4	Tohoku U	20.06	5.8
5	Keio U	17.69	5.1
6	Hokkaido U	17.39	5.0
7	Tokyo Inst. of Technology	17.21	4.9
8	Nagoya U	17.07	4.9
9	Kyushu U	12.15	3.5
10	Waseda U	10.19	2.9
11	Kobe U	8.51	2.4
12	Tokyo Medical & Dental U	5.02	1.4
Total		307.59	100.0

Source: MEXT. http://www.mext.go.jp/b_menu/houdou/18/04/06041308/003.htm

Table 3.4. Projected leading recipients of operational and administrative subsidies for national universities, April 2004 to March 2010

Rank	Institution	Amount over 6 yrs (10^8 yen)	Average per year (10^8 yen)	Approx. % of total
1	U of Tokyo	5,364	894	7.3
2	Kyoto U	3,676	613	5.0
3	Tohoku U	3,122	520	4.2
4	Osaka U	3,008	501	4.1
5	Kyushu U	2,819	470	3.8
6	Hokkaido U	2,541	424	3.4
7	Nagoya U	2,066	344	2.8
Approx. total		73,900	12,317	100.0

Sources: For seven universities: Uekusa and Takaoka (2005). For six year overall total: estimate based on overall totals for FY 2004 and 2005 at www.mext.go.jp applying the same rate of decrease over the entire period, as between FY2004 to 2005, i.e. 98×10^8 yen.

Note: These amounts represent *total* operational and administrative subsidies, not only those to support research.

Because of this concentration of resources and regional preferences in Japan, the system of recruitment in a few elite universities influences academic career strategies throughout the nation. The most elite universities manage to recruit creative and capable persons, because they attract interest from bright young researchers throughout the country. But because they generally maintain the traditional recruitment system, the need for patronage probably makes young researchers less likely to pursue unorthodox themes or research approaches. The price of initial failure is not simply losing an opportunity to work in a prestigious university. It may mean spending one's career in a university with scant research resources.

Moreover, I suspect that concentrating funding in a few institutions reduces different approaches to various scientific and technical problems, and the likelihood that less orthodox approaches will be recognized. This not only diminishes the number of discoveries that might be developed by startups, but it probably has a negative effect on Japanese science and industry as a whole.

Restraints on Communication

It has been said that Japanese society is vertically organized in that the most important relationships are subordinate–superior relationships, and that the clarity, stability, and sometimes also exclusivity of these relationships is valued because they preserve cohesion within the group and self-sufficiency.[118] As a corollary, horizontal communication is encouraged only to the extent that it undermines neither group cohesiveness nor the hierarchical relationships that define and provide the basic structure of the overall group and its subgroups.

However, the hierarchical relationships are not only about command and control but also mutual obligation, including the obligation of superiors to look out for the welfare of subordinates.[119] Also within a group, a lot of time is spent on communication, which tends to create consensus and help the group function smoothly, as well as to develop an intra-group culture of mutual obligation.[120] Close communication within shop floor factory teams, as well as between research and product development/manufacturing divisions, has given Japanese manufacturers an edge over many overseas competitors in terms of product quality and innovativeness.[121] Also Japanese society has changed from the time these observations were made over a third of a century ago. For example, students no longer tend to rely on their professors to find jobs, but instead rely mainly on their own efforts. Attendance at various group events is becoming more flexible.[122]

Nevertheless, having lived in Japan for nine years, I believe that the basic assessments about the primacy of hierarchical relationships, group identity, and group cohesion are still valid. In particular, extra-group communications seem more hesitant and restrained than in the USA, Europe, or China.[123] Of course extra-group communication occurs frequently. But it occurs smoothly only once the communication has been sanctioned by higher levels in each party's group.[124] Freelancing, even for the benefit of the group, seems to be discouraged more than in most other countries, and it is definitely discouraged if it is perceived to be for personal benefit.

An anthropological analysis of social relationships and how they might affect ventures is beyond the scope of this book. However, there are unique barriers to inter-group communication in Japanese society that pose problems for new companies that rely on rapid exchange of information and rapid decision-making to grow—communication not only between the new companies and potential collaborators and customers, but also communication within potential collaborators regarding how to cooperate with the new company.[125]

The emphasis on group cohesion, stability, and self-reliance also manifests itself in the tendency for established companies to innovate autarkicly.

Table 3A. Major competitive Japanese government S&T funding programs in 2002 (includes funding to private companies and GRIs as well as universities. Programs with annual budgets under 1 billion yen (~10 M US$) not listed)

Program name‡	2002 budget (10^8 yen)	No. new projects in 2001**	Per project annual funding range (10^8 yen) and duration
Ministry of Public Management (includes former Ministry of Posts and Telecommunications)			
Strategic Communications R&D*△	14	? (mostly large electronics cos.)	0.1–0.5, 3–5 yrs
Japan Key Technology Center (Jtech): Promotion of corporate research in basic technologies*△	107	11	unlimited
MEXT			
University, industry, and government cooperation for innovative enterprise creation*△	71	28	0.1–0.5, 3–5 yrs
Grants-in-Aid	1,703	21,000	
JSPS: Basic research	812	9,466	<1.0, 1–5 yrs
JSPS: Exploratory research	40	1,074	<0.05, 1–3 yrs
Support for young researchers (<37 y/o)	134	4,170	<0.3, 2–3 yrs
Specially promoted research	127	13	<5.0, 3–5 yrs
Priority area research	386	3,394	0.2–6.0, 3–6 yrs
Disseminating research results	34	780	Varies, 1–5 yrs
Special coordination funds	177	~150	
Ind-univ-gov't results-oriented joint res.*	28	35	~0.27, 3 yrs
Strategic human research resources development	40	2	<10.0, 5 yrs
Research support for researchers <35 yrs old	15	66	0.05–0.15, ≤5 yrs
Pioneering research in new fields△	66	24	0.5–2.0, 5 yrs
Training for emerging fields	19	7	<2.0, 5 yrs
JST Strategically Promoted Creative (Basic) Research Programs	427	~370**	
CREST†△	289	173**	avg. 0.83, ≤5 yrs
PRESTO†△	64	184**	avg. 0.17, 1–5 yrs

ERATO†△	62	4**	avg. 3.2, 5 yrs
International Cooperative Research	16	2**	?, 5 yrs
Centers of Excellence (2004 data)††	308	113**	0.10–5.0, 5 yrs
JSPS Research for the Future	90	Last project started in 1999.	?, 5 yrs
Total MEXT	2,776		
MHLW			
Grants in Aid for Health Research△	393	1,251	0.01–10.0, 1–3 yrs
Basic Research in Health and Medicine	98	10	0.5–1.0, ≤5 yrs
METI			
NEDO: Industrial Technology Research*	53	93	0.30–0.40, 2–3 yrs
MAFF			
Research to Apply Advanced Agricultural Technologies*△	18	?	0.1–1.0, ≤3 yrs
Basic Research to Create New Technologies△	42	13	≤1.0, ≤5 yrs
New Enterprise Creation R&D*△	17	6	≤0.6, ≤5 yrs
Environment Agency			
General Environmental Research△	29	13 (+7 smaller for young)	0.02–1.0, ≤3 yrs
Grants in aid for Research into Fields such as Environmental Disruptors of Biological Pathways△	10	30	0.01–1.0, ≤3 yrs
Total for all Agencies and Programs (including those not listed)	3,593		

Source: Prime Minister's Cabinet Office, www8.cao.go.jp/cstp/project/compe/haihu02/siryo1.pdf

‡ Preceded by name of semi-independent funding corporation, if applicable.

* Program generally has applied research focus or aims to develop competence in particular technical areas.

** Numbers of projects for JST programs are for 2002. The same is true for the COE program, which was launched in 2002 with 113 projects. 2003 saw the initiation of an additional 133 COE projects.

△ Program open to applicants or co-applicants from private industry.

† Program described in Kneller (2007). Budget data from www.jst.go.jp

†† COE data are for the beginning of 2004 (source www.mext.go.jp/a_menu/koutou/coe/). COE is not officially classified as 'competitive research support' but because of its importance as a major new source of S&T research support I have included it in this table. The total COE funding allocation for 2004 is 30.8 billion yen.

NOTES

1. In practical terms, this usually means the startup either receives licenses from the university covering its core technology, or is founded by faculty members or students and its core technology is closely related to the founders' academic work.

2. Details about this analysis are in Chapter 7. About 10% of the university discovered drugs were licensed directly to large pharmaceutical companies. The remainder are licensed to biotechs.

3. See the analysis of the origins of the 169 new drugs approved by the US FDA from 1998 to 2003 in Chapter 7.

4. Branstetter and Ogura (2005). This analysis was carried out using citations to papers from major California universities only, so strictly speaking, the findings apply only to California universities. The trend toward increasing citations of academic publications resembles a step function in the case of IT, with a sharp increase in the late 1980s.

5. See discussion in Chapter 7. It would be interesting to know whether new or old companies tend to cite university publications more, and also the frequency with which patents issued to universities (as opposed to companies) become highly cited patents and/or are licensed to successful startups. Branstetter's and Ogura's analysis showed that IT patents issued to universities tend to cite academic publication much more frequently than patents issued to firms. This is not surprising, academic inventors would be expected to cite academic literature. But if the university IT patents tend to be licensed to startups, which then go on to attract funding and to have sales, this would suggest that, in IT, startups are one of the main vehicles for developing inventions that incorporate a great deal of new scientific knowledge.

6. See, e.g. McKelvey (1996), Murray (2004), Powell, Korput and Smith-Doerr (1996), and Zucker and Darby (1996).

7. Dibner (1988: 90).

8. In Chapter 4, I compare numbers in Japanese biotechnology companies with a confirmed therapeutic focus with those of US therapeutic biotechs of equivalent age, and in the process, I obtained information on the percentage of Japanese companies based on university discoveries. Dibner's data (see text accompanying previous note) covers all US biotechnology companies (therapeutic and non) and focuses on the background of the founders. Thus the two proportions are not exactly comparable, and only suggestive that Japanese biotechs rely more than US counterparts on university R&D.

9. See discussion in Chapter 7.

10. Public Law 96-517, codified as 35 USC sections 200-212. Implementing regulations issued in 1987 are at 37 CFR, §401.

11. NSB (2006: A5-3).

12. This authority was granted in a statement of Government Patent Policy, issued by President Nixon, and published at 36 Federal Register 16,886 (1971). NASA already had statutory authority to do so. Even without explicit legal authority, the DHEW had assumed the right to issue exclusive licenses in 1969. See Bayh-Dole25 (2006). Prior to Bayh-Dole, invention management practices differed among agencies. In the case of university inventions, probably the dominant policy logic was that they should be dedicated to public, in other words, either not patented or, if patented, then the patents would not be enforced or else licensed nonexclusively. However, in the 1970s, there were calls for flexibility on the issue of licensing, particularly in response to concerns about the need for exclusivity in the case of some biomedical inventions. (See the accounts of the history of Bayh-Dole in Eisenberg 1996, and Bayh-Dole25, 2006.)

13. Latker (1977) states that between 1969 and May 1977, DHEW issued 19 exclusive licenses and 90 nonexclusive licenses covering IP in its portfolio of approximately 400 patents and patent applications. Most of these dealt with inventions from laboratories under DHEW, such as the NIH *intramural* laboratories. Only a minority covered *university* inventions funded by NIH, or other agencies within DHEW.

    As the Patent Counsel for DHEW in the 1970s, Mr Latker was responsible for managing DHEW inventions in the 1970s. He was one of the main proponents for authority to license government-funded university inventions exclusively, and for delegating licensing authority to universities.

14. Up to three years from the date of first commercial sale or eight years from the date of agreement, whichever came first. (Personal communication, May 2006, from Howard Bremer, Emeritus Patent Counsel, Wisconsin Alumni Research Foundation.)

15. In 1974 (the latest year for which Latker, 1977, provides data) universities were managing 329 DHEW funded inventions under IPAs and had issued 78 exclusive and 44 nonexclusive licenses. These numbers were trending up.

16. Latker (1977), Eisenberg (1996), and Bayh-Dole25 (2006). Most of the pressure to facilitate exclusive licenses came from established pharmaceutical companies rather than biotechs, of which there were few before 1980 (communication from Mr Bremer, see note 14 above).

17. See, e.g. the statement of W. Novis Smith, Director of Research and Development, Thiokol Corp. in The Role of the Federal Laboratories in Domestic Technology Transfer: Hearings Before the Subcommittee on Science, Research and Technology of the US House of Representatives Committee on Science and Technology, 96th Congress (1979), at n. 125, pp. 621–22; referenced in Eisenberg (1996: 1699).

18. Regarding the former reason, see Shane (2004, especially: 69–7, 173, 232, and 260–1). Regarding the latter reason, see Barnett (2003).

19. High Voltage, Inc. was established in 1946 with the financial support of the venture capital firm American Research and Development to commercialize

Van de Graaff generators, Van de Graaff having conceived of these generators while a student at Oxford, then Princeton, and later a researcher at MIT. MIT managed the patents and, after considerable discussion about the propriety of exclusive licenses to academic inventions, decided that exclusive licenses to High Voltage were justified in light of 'the essential business requirements for bringing an invention into use by the public'. Of special note, some of the exclusively licensed inventions were funded by government contracts. (Eisenberg (1996) notes that the Defense Department which funded the MIT research, usually let contractors retain rights to inventions it funded.) Also, MIT had earlier licensed Van de Graaff's inventions to either General Electric or Westinghouse, but these companies had failed to develop the technology. By 1955, High Voltage was a 'reasonably successful company'. (Account and quotations from Etzkowitz 2002.)

20. Cetus Corp., founded in 1971 by Nobel Physics Laureate Donald Glaser of the University of California at Berkeley and two nonacademic colleagues, may be able to claim the distinction of being the first university biostartup that achieved a notable degree of success, developing genetically engineered interleukin-2 and beta interferon and, most notably the polymerase chain reaction (PCR) method for gene duplication and amplification. But although Cetus recruited university scientists, I do not know the degree to which it relied on licenses from universities or close links with university researchers to make its key discoveries (see Rabinow 1996).

21. McKelvey (1996: 103, 151). These licenses were from UCSF, the employer of Genentech's cofounder Herbert Boyer. UCSF agreed to license Boyer's inventions to his new company only after internal debate. Genentech also received nonexclusive licenses to other university inventions, such as the basic recombinant genetic engineering invention by Boyer and Stanley Cohen of Stanford.

22. According to Dibner (1988: 101) about 79 biotechnology companies (not only university startups) were formed in 1981 compared with about 17 in 1979. Dibner uses a broad definition of biotechnology company (companies working with new technologies of genetic engineering, monoclonal antibodies, large-scale cell culture, etc.) that includes some large pharmaceutical companies, subsidiaries of large companies and companies whose main business is in other fields. Among the notable biostartups founded between 1976 and 1980 were Biogen (1979 by researchers from Harvard, MIT, and the University of Zurich) and Molecular Genetics by researchers from the University of Minnesota. 1981 saw the founding of Chiron (Harvard), California Biotech (UCSF), and others.

23. Unless otherwise noted, support for the statements in this section are in Kneller (2003a) which describes the institutional and legal evolution of Japan's technology transfer system until 2003. This article is based on statistical analyses described in the article, documents cited in the article, and conversations

with a wide range of university researchers and administrators and government officials.

24. *National universities* account for Japan's leading centers of university R&D. Please refer to the Glossary at the end of this book for a further explanation.

25. i.e. Commissioned and Joint Research regardless of whether the sponsor is a private or government organization. See Glossary.

26. Known formerly as *kouhi* and more recently as *unei koufukin* funding, base amounts are usually less than US$ 10,000, and the majority is pre-allocated to utilities and other infrastructure expenditures.

27. In 1998, standard research allowances accounted for 46% of national university R&D expenses, but as noted in the previous note, most of these funds were earmarked for infrastructure costs. Donations accounted for 13% of the R&D budget, 25% of the budget net of the standard research allowances. Therefore, somewhere between 25% and 59% of R&D funds were attributable to non-project specific funding (Kneller 2003*a*).

28. Kneller (2003*a*) provides evidence for these estimates.

29. In the case of inventions arising under formal sponsored research agreements with companies, the companies could usually arrange to co-own the inventions with the government. Under article 73 of Japan's patent law, either co-owner can freely use such an invention. But it cannot license its rights, even nonexclusively. Usually this was not a disadvantage for large companies, (in fact co-ownership gave them a de facto royalty-free exclusive license) but it was for ventures, because ventures often need to transfer their IP in the course of business alliances.

30. Thirty-nine Japan Bioindustry Association respondents (almost all large or established companies) to a 1997 questionnaire, indicated that each had an average of 156 university relationships, the vast majority based on donations to individual professors. The average expenditure per relationship was less than US$ 10,000 (JBA 1998).

31. MEXT (2005).

32. See Hashimoto (1999) and Odagiri (1999) who describe the close prewar university industry linkages and how those linkages went underground (became informal) in the postwar decades. See Kneller (2003*a*) for descriptions of the ban on paid consultation, as well as restrictions related to sponsored research agreements, licensing, and using sponsored research funds to pay personnel expenses.

33. Many universities would accept at face value an inventor's assertion that an invention arose under donation funding and would not even require such inventions to be reported. A few universities did require all inventions to be reported, but these too did not question assertions that the inventions arose under donations or standard research allowances. All national universities had invention committees composed of faculty members (i.e. colleagues of the inventors) responsible for deciding the attribution of inventions, but these

usually met infrequently and to my knowledge never questioned the inventors' assertions.

34. See note 29.

35. *Project-specific* government support for university R&D is approximately three times greater than total industry support for university R&D. This does not take into account non-project-specific support, university salaries, infrastructure, etc. almost all of which are paid for by the government. See Kneller (2003a). In fact, official OECD statistics indicate that as a percentage of total university research support, industry accounts for only 2.5% in Japan compared with 6.8% in the US (National Science Board, Science and Engineering Indicators, 2004).

36. Baba, Yasunori, Shichijo and Nagahara (2004).

37. See Kneller (2003a). The terms under which the startup had to license back, from the pharmaceutical companies, the founder's own inventions while not excessive, were not trivial.

38. Monbusho (1998) and n. 2. One of the best documented cases of undeveloped university discoveries patented by private companies concerns a sample of 252 genetic engineering patent applications, each of which had at least one university inventor. Only 16% had issued as patents, and in the case of 62% examination by the Japan Patent Office had not even been requested. In a separate study, also by the Japan Bioindustry Association (JBA), 116 patent applications filed between 1992 and 1996 by 39 JBA member companies were identified as emerging from cooperation with universities. The companies felt that only 21% were for discoveries of practical use to the companies (JBA 1998 summarized in Kneller 1999).

39. Frequent reasons for companies not developing university inventions included their perceptions that the market was too small, and their intention to use the patents only as bargaining chips in case they were sued (or wanted access to another company's technology) or to prevent competitors from using the discoveries. Other reasons included inappropriate assessment by the inventors of the companies' needs, and lack of incentives for the university researchers to keep working with the companies on the inventions (because the benefits they would receive in terms of royalties, etc. would be minimal, even if the invention became a commercial success). Reference to the success of well-known US TLOs, such as those of MIT and Stanford, was frequent (Monbusho 1998). In other words, the advisory committees reasoned that universities could make better decisions than individual inventors about which companies should receive exclusive rights to university discoveries and could better insist on contractual provisions (royalties, due diligence clauses, etc.) that would increase incentives for the licensees to develop the inventions and for inventors to continue to cooperate with development.

40. However, there are some examples of small companies benefiting from consultations with professors in well-known universities (Chapter 4 and Kneller 2003*b*).
41. At least this is the impression from key advisory reports, such as Monbusho (1998), advocating the reforms.
42. [Daigaku nado gijutsu iten sokushin hou] (Law No. 52 of 1998).
43. *TLO* stands for technology licensing office or technology licensing organization. This is the general term used in Japan to refer to a university licensing, technology transfer, or technology management organization. It also has the same generic meaning in the US.
44. However, these cannot be used to pay salaries of permanent TLO staff nor the fees of outside patent attorneys. Many US TLOs rely on subsidies from their universities, yet over time more are becoming self-sufficient (based on conversations with US TLO officials). Also despite operating deficits, it seems that many US universities have decided that the long-term benefits (technology development, new company and job creation, and increased industry sponsorship of research) outweigh the shortfalls in license revenues. Whether the same reasons justify subsidies in the Japanese case remains to be seen. Another potential problem is that the METI/MEXT subsidies are distributed as equal size block grants, whereas in the US, decisions are made by the university administrations. Thus the US system probably facilitates better alignment of technology management with the goals of individual universities.
45. [Sangyou katsuryoku saisei tokubetsu sochi hou] (Law No. 31 of 1999).
46. Also the Japanese law authorizes, *but does not require*, Japanese S&T funding ministries to let grantees and contractees claim IP rights to the inventions they make under government funding. However, in the case of university inventions, METI has encouraged all agencies to apply the law and, with a few exceptions, all have complied (Kneller 2003*a*). The main exception is the ERATO Program administered by the Japan Science and Technology Corporation (JST), now part of MEXT. JST continued to retain ownership of ERATO inventions by university researchers following incorporation of national universities.
47. [Sangyou gijutsu ryoko kyouka hou] (Law No. 44 of 2000).
48. Exceptions are permitted to enable faculty with special expertise to serve as directors of TLOs and accounting firms, even though such activities are not directed toward commercializing the faculty member's discoveries. The names of nearly 3,000 faculty that had obtained permission to serve as managers or directors by the end of 2003 are available at http://www.mext.go.jp/a_menu/shinkou/sangaku/03121501.htm
49. Prior to this law, funds for commissioned and joint research could only be disbursed once a year on a fiscal year basis. Disbursements had to be approved by MEXT and the Ministry of Finance and thus funds were not available

between February and July. In other words, funds could only be used for equipment purchases and some travel, not for personnel.

50. [Kokuritsu daigaku houjin hou] (Law No. 112 of 2003).

51. MEXT (2002).

52. In 2003, 5 years after enactment of the TLO Law, the 35 approved Japanese TLOs applied for 1,679 Japanese patents (average 48 per TLO). In comparison, 109 US TLOs applied for 1,584 US patents (average 15 per TLO) in 1991, 11 years after enactment of the Bayh-Dole amendments (data from METI and AUTM).

53. In 2004, 38 Japanese TLOs issued 626 licenses, approximately 16 per TLO (METI). In 1991, 109 US universities issued 1,229 licenses, average 11 (AUTM).

54. Average royalties per royalty-earning license was on the order of US$ 17,000 in 2003, and this has probably not increased substantially. In comparison, in 1991 US TLOs received US$ 218.4 million in royalties on 2,602 royalty earning licenses—about US$ 84,000 per license. In 2004, this had increased to approximately US$ 121,000 per license. The difference may be due both to the US averages being inflated by a small number of 'blockbuster inventions (of which Japanese universities so far have none), to Japanese TLOs being hesitant to bargain hard with large companies for high royalties, and to some of the best inventions having been siphoned off under joint research agreements. See Kneller (2006).

55. Using the same definition for a *core* university venture (i.e., a new company based directly on university discoveries), the latest METI survey report says the overall total in Figure 3.1 (1,503 ventures) should be discounted by 44% to obtain the number of core ventures. In other words, according to METI there were 845 *core* university startups in mid–late 2005. METI kindly provided me lists of startups attributed to U Tokyo and Keio Universities in the 2003 METI survey, and I found out information about most of the companies on these lists (see Chapter 4). My independent analysis of these startups (See Chapter 4, Appendix 2) suggested a somewhat lower discount factor, leading me to conclude that the most appropriate discount factor is about 40%.

56. See Chapter 4.

57. Shane (2004) and others have argued that startups that are run by professional managers tend to do better than those run by academic founders. This sentiment is now common in Japan and many academic founders have yielded formal management authority to nonacademics. Nevertheless, it is still fairly common for the academic founders to retain de facto control, and to hear criticisms that companies are being directed more by academic curiosity than business goals.

58. Then within five years, 10 million yen must be deposited as paid in capital in the case of joint stock companies, 3 million yen in the case of limited liability companies.

59. The average amount of annual funding per joint research project in 2004 was around US$ 20,000, nearly identical to the average in 2000 (MEXT 2005).

60. *Large* companies are defined as having over 300 employees, *small* as having 21 to 300, and *very small* as no more than 20 (except in the case of retail and service businesses where *very small* is defined as no more than 5 employees). Most startups would fall in the very small category in their first years of business. The 1990–2002 data are from Nakayama, Hosono, Fukugawa and Kondo (2005). The 2003–4 data are from MEXT (2005) which does not give a breakdown by company size.

61. In contrast, a joint owner of a US patent can transfer rights his or her rights to a third party without the consent of the other joint owners, barring a contractual agreement to the contrary among the joint owners.

62. Based on conversations in Dec. 2004 with technology transfer officials at the National Institute of Advanced Science and Technology (AIST), one of Japan's major government research institutes, which, like most universities, also includes a clause to bypass article 73 in its standard joint research contracts.

63. On some occasions, companies that are coinventors on inventions insist that no patent application be filed, essentially converting the invention into a trade secret.

64. In cases of an invention that overlapped two of these categories, I assigned it one-half to each field—on rare cases, one-third to each of three fields covered by a single invention. The full analysis and results are described in Kneller (2006).

65. Activity-specific funding means funding other than the *operational and administrative subsidies*. These subsidies pay for full-time salaries and infrastructure, but leave little to support specific projects (i.e. equipment, stipends, travel, and so on). Activity-specific funding includes (in order of largest to smallest) MEXT grants-in-aid, Commissioned Research (mainly from government agencies), donations, and finally Joint Research.

66. Unlike many US universities, most Japanese universities permit joint and commissioned research between a startup and the founder's laboratory.

67. Applications are usually filed jointly by the company and university, with the company paying a majority of associated costs.

68. *International Herald Tribune-Asahi Shimbun* (2004). 'Seeking profit, firms leave basic R&D to universities', Jan. 15, 21.

69. Nationwide, the numbers of company researchers engaged in joint research in universities doubled from 1,398 in 1992 to 2,821 in 2002 (MEXT 2003). The rise actually predates the IP ownership reforms. Even under the donation system, the only way corporate researchers could engage in research in universities was under joint research agreements or nearly equivalent *commissioned researcher agreements*.

70. See Chapter 7.

71. Personal communications with US technology transfer officials in 2004 and 2005.

72. However, any comparison along these lines ought also to take into account consulting and startup formation.

73. Chabrow (2005).

74. These agreements usually pertain to open source software applications (Kauffman Foundation 2005; IBM 2006).

75. According to my observations, such professors will usually segment their research, collaborating with one company on a particular aspect and another company on another aspect.

76. Chapter 7 discusses possible reasons why preemption is less common in biomedicine.

77. More on this issue in Chapter 7.

78. As indicated in Figures 3.4 and 3.5, startups and other ventures also engage in joint research with universities, although on a smaller scale than large companies and disproportionately in biomedicine. However, startups sponsoring research in the founders' laboratories raise conflicts of interest issues that have not yet been resolved or even openly debated in Japan. For example, once a laboratory director has formed a company, is it appropriate that joint research agreements with the company enable most of the laboratory's discoveries to flow to that company, a process known as *pipelining* that is discouraged in the USA (Shane 2004)? What about the risk that the laboratory will be turned into the professor's company's laboratory, leveraging public research support and appropriating not only IP but also the energy and creativity of graduate students? To a degree, these risks exist in any collaborative research situation, but they are heightened when the collaborating company is also the professor's startup. US universities generally discourage such sponsored research, but many also deal with these issues with some degree of flexibility.

79. Other countries, notably Germany in 2002, have gone through similar transformations of their university IP ownership systems. It would be interesting to know whether the former system lives on through cooperative research in Germany, as it does in Japan.

80. The following summary of the situation at MIT is from Etzkowitz (2002).

81. These policies gave faculty the leeway they wanted in the case of consulting, but in the case of IP they gave MIT authority to own work related inventions, more than seventy years before the Japanese government (acting without strong backing from universities or their professors) would order often unprepared Japanese universities to claim such inventions.

82. Based on 2006 discussions with MIT researchers.

83. As of 2006, there seemed to be little effort by universities to assert control over their discoveries and over faculty relations with large companies in ways that might conflict with the interests of the companies. Their stance was accommodating rather than assertive. The main areas of disagreement concerned how much companies should pay in overhead (30% still being the ceiling, with

most payments retained at the department level) and whether companies that want exclusive rights to joint research inventions should pay, in addition to the exclusive license fee, for the universities to give up their right to practice jointly owned in inventions under article 73 of Japan's Patent Law. In case of disputes, inventors often side with companies.

84. In 1994, national universities received about 230×10^8 yen in commissioned research, in 2004, 772×10^8 yen, a more than threefold increase. The vast majority of commissioned research was from government-affiliated agencies, such as JST and NEDO (METI 2005 for 1994 data, MEXT 2005 for 2004 data) 10^8 yen $\cong$ \$1 million.

85. Some might question this initial premise, citing the large number of startups being created each year as shown in Figure 3.1. My continued doubts about the depth of entrepreneurship in Japanese universities relate to issues addressed later in this book and to the aforementioned weakness of startups (with some exceptions, mainly in biomedicine). It is also based on discussions with students and faculty, and personal knowledge of quite a few startups. Some faculty members are interested in founding companies, but many of these are also engaged in joint research projects with large companies. The interests of the large companies usually win out, although I know of one possible exception that I describe in Chapter 7. If a startup is formed, often its business scope is confined and/or it becomes a de facto subsidiary of one of the large joint research partners. As for students, most masters students want to work in large companies. While Ph.D. students may have more varied career goals, very few MS or Ph.D. candidates want to work in ventures—either ventures with high risk/return prospects or ventures with low risk/return prospects. I administer a survey to about half of the new graduate students in my research center each year, and consistently fewer than 10% say they would consider work in ventures to be desirable. Attitudes in less prestigious universities may be different.

86. For example, compromising core of academic values, scientific integrity, or the quality of graduate students' education, or harm to patients in clinical trials if precautions are sidestepped to enhance business prospects for a company whose therapy or device is being tested (see note 88).

87. See for example, the regulations for the University of Tokyo in Japanese at http://www.u-tokyo.ac.jp/per01/d04_10_j.html. These require reporting of consultancies, management positions, contractual relationships, and stock holdings in companies with which one has cooperative research, advisory or business relationships; as well as income from intellectual property and instances in which students have been sent to companies under cooperative relationships.

88. In 1999, a young man died in the course of gene therapy trials in the University of Pennsylvania. The subsequent investigation uncovered various shortcomings in trial procedures that contributed to his death. The principal investigator (PI) had founded a startup to commercialize the gene therapy technology

that was the subject of the trial. Although a link between the death and the PI's financial interest in the outcome of the trial was not clear, this focused attention on the need to prevent, or manage carefully, such conflicts of interest in the case of clinical trials (Weiss, Rick and Nelson, Deborah (2000). FDA halts experiments on genes at university; probe of teen's death uncovers deficiencies (*Washington Post* January 22, 1)).

89. Tokushima University (2006). The reporting requirements go a bit beyond those of the University of Tokyo regulations (note 87) in that they require reporting of donations and sponsored research. Perhaps more importantly, they recommend coordination with the institutional review boards (IRBs) that are responsible for reviewing research proposals involving human subjects to try to ensure the safety, privacy, and voluntary, informed participation of the human participants.

90. The effort was sparked in part by incidents abroad such as that mentioned in note 88 and also by 2003 revelations in a nationwide daily newspaper that researchers in a major Japanese university hospital involved in human testing of a new therapy owned by a startup of that university had received stock in the startup, which they sold just before the startup's IPO. As an illustration of the widely held critical stance toward faculty entrepreneurship, the newspaper revelations focused criticism on the fact that the researchers had made a profit. However, the university was aware of the stock holdings and the researchers sold their stock at the advice of the IPO underwriters in order to avoid the appearance of impropriety. (They could have sold their stock for more if they waited until after the IPO.) Absent from the initial media reports were concerns about the possibility that the researchers' financial interest in the outcome of the trials might have led them to take shortcuts in the planning or execution of the trial that might have compromised patient safety.

91. See, for example, the reporting requirements and decision process at the faculties of medicine of Tohoku University at www.med.tohoku.ac.jp/jimu/rinri/3.rinsyo.pdf and the University of Tokyo at http://www.crc.h.u-tokyo.ac.jp/doctors/documents/riekisouhanshinkokusho_000.doc. These both require reporting of any interest above 1 million yen (about $8500), and sharing of information with IRBs (see note 89).

92. Conflict of interest issues are far from resolved and management procedures are far from uniform in the USA. But there is more open debate at an institutional level, with deans of various medical schools actively involved in working out ways to manage a variety of often complex situations. See Kaiser (2002) discussing recommendations by the Association of American Medical Colleges. Also many US universities openly describe how specific conflict of interest issues will be managed.

93. Out of 40,804 students who graduated with masters degrees in natural science, engineering, agriculture, and pharmacology in 2004, immediately after graduation at least 78% of these (31,882) joined the labor force, and 13% (5,212) continued academic studies (mostly in doctoral programs). Information is

not available on the remaining 9% (MEXT Basic School Survey 2004). (To my understanding, most persons who study for doctoral degrees in Japan first obtain a masters degree).

US data show only that 61,639 students graduated from US universities with masters degrees in these same fields in 2002, while 19,640 graduate with doctoral degrees in 2003 (NSB 2006: A2-63, 75). Even if none of the Ph.D. graduates received masters degrees (an unlikely assumption, although it is more common for Ph.D. candidates to skip a masters degree in the USA than Japan), and thus the total number of first or second year US S&E graduate students is around 81,279 (61,639 + 19,640), the percentage of first year S&E graduate students pursuing doctoral degrees would be around 24%, higher than the estimate of 13% for Japan. If half of US doctoral recipients obtain a prior masters degree, then approximately 27% of first or second year S&E graduate students in US universities are planning to pursue doctoral level research—double the estimated percentage for Japan.

94. Between 1990 and 2004, the number of S&E doctoral graduates from Japanese universities (excluding social science) increased over threefold from 4,525 to 10,770 (MEXT Basic School Survey, various years). However, 4,077 (38%) of the latter figure were doctoral degrees in either medicine or dentistry, which are usually awarded to persons in their late 40s who are already junior faculty in medical or dental schools. These persons do pursue research to fulfill their degree requirements, and thus might be loosely equivalent to persons pursuing the Ph.D. component of a combined MD/Ph.D. program in the USA. But they are older, on average, and their career paths are already set. (Coleman's 1999 account of academic careers describes these extended doctoral programs as part of the indentured servitude future professors in medical and dental schools must endure.)

In comparison, the number of US non-social science S&E doctoral graduates (excluding medical doctors, MDs) in 2003, was 19,477, roughly two- or threefold the number of Japanese graduates in 2004, depending on how the Japanese doctorates of medicine and dentistry are counted. About 37% of these were awarded to foreign students as compared with about 13% of the Japanese doctoral degrees (NSB 2006: A2-122, 129).

Note, the Japanese figures in the above analysis do not include so-called *thesis doctorates* (ronbun hakasei) which are awarded on the basis of research done outside the university, typically in a corporate laboratory. These do not involve university graduate level course work. Supervision by the professor who approves the thesis is often minimal. In 2001, approximately 950 thesis doctorates were awarded in engineering compared to about 2,950 normal doctoral degrees, and in the sciences (including social science) about 200 thesis doctorates were awarded compared to about 1,350 normal doctorates (NISTEP 2004). MEXT is trying to reduce thesis doctorates and encourage all persons who want doctoral degrees to go through formal university programs.

95. There were about 5,250 Japanese postdoctoral researchers in S&E fields (excluding social science) in 2003 (MEXT Personnel Advisory Com. 2003). In contrast in 2003 there were 45,237 postdocs in S&E fields excluding social science and psychology in the US (NSB 2006: A2-103), more than an eightfold difference. Traditionally, postdoctoral positions have been considered undesirable in Japan, refuges for persons not capable enough to obtain permanent employment in universities or industry. However, as the number of Ph.D. graduates increase, the number of postdocs is likely to increase.

96. 6,305 of 19,640 doctoral degree recipients from US universities in engineering and the natural sciences in 2004 (32%) were female (NSB 2006: A2-75,77). Among 6,693 doctoral graduates in equivalent fields from Japanese national universities in 2004, 1,109 (16.6%) were female, roughly half the US percentage (MEXT 2004 Basic School Survey, p. 424–5).

    As for university faculty, among approximately 132,100 doctoral degree holding full-time faculty (instructor to full professor level) in engineering and natural science in US universities and colleges in 2003, approximately 30,700 (23%) were female (NSB 2006: A5-46–49). Among 36,772 full-time faculty (joshu to full professor) in equivalent fields in the central academic divisions of Japanese national universities in 2004 *plus all full time faculty* in special institutes affiliated with national universities and graduate-level-only national universities, 2,598 (7.1%) are female, roughly one-third the US percentage.

97. Some kouzas contained an instructor (*koushi*) intermediate in rank between a joshu and the assistant/associate professor. A koushi was expected to emphasize mainly teaching, and sometimes was not considered to be in line to fill a vacancy at the assistant/associate professor level. In 2007, titles are expected to change. Assistant professors (jo kyouju) will become associate professors (jin kyouju). Assistants (joshu) will become assistant professors (jo kyou).

98. MEXT's Priority Area Research projects (recently folded into the new Development of Innovative Seeds and the Promotion of Key Technologies Programs) and JST's CREST and ERATO projects tend to be of this type. See Appendix Table 3A.

99. The following MEXT programs tend to fund such projects: grants-in-aid for Specially Promoted Research, CREST, ERATO, Research for the Future, Centers of Excellence, Special Coordination Funds for Strategic Human Research Resources, Pioneering Research in new Fields, and Training for Emerging Fields. These accounted for 101 billion yen (36%) of MEXT's total competitive extramural research budget of 277 billion yen in 2002. As discussed in Chapter 7, university funding by METI and the Ministry of Public Management funding tends to involve multiple laboratories (consortium research) at least in cutting-edge areas of S&T. On the other hand, funding by the MHLW and MAFF is probably most often in the form of grants to individual laboratories. (See Appendix Table 3A and Kneller 2007.)

    Some of the programs mentioned above, such as Centers of Excellence and Special Coordination Funds, are non-project-specific funding to support research and education in a particular department or center. The committees that evaluate competing applications for such funding generally do not have in-depth expertise in individual areas of research and tend to make decisions based on a macro-level analysis of competing institutions (Kneller 2007). The disadvantage of this evaluation system is that it tends to perpetuate concentration of resources in a few prestigious institutions (discussed further below) and it leaves authority about which individual researchers and projects to fund in the hands of individual kouza and department heads. *It provides no alternative to the traditional system of recruitment and promotion for bright young energetic researchers who do not have patronage.*

100. In 2005, NIH accounted for about 66% of total US government support for academic R&D, and NSF accounted for 13% (NSB 2006: A5-11).

In 2002, NIH supported 34,613 *investigator-initiated basic research projects*, paying about US$ 365,000 per project, including overhead—about 75% of its extramural R&D budget. The same year NIH funded 1,261 *center projects* (e.g. comprehensive cancer centers to combine research and patient care) at a total average cost of US$ 1.74 million per center, about 13% of its R&D budget. In 2003, NSF funded 6,140 research grants, mostly for *individual* research projects in universities, at an average cost of US$ 135,000 per project. It also spent US$ 364 million to support about 300 research *centers* in US universities (e.g. collaborative engineering research centers) for about a 3 to 1 ratio of individual to center funding.

101. See the list of Competitive Research Funding Programs [Kyousou-teki kenkyuu shikkin seidou ichiran] issued by the Cabinet Office at www8.cao.jp/cstp/compefund/ichiran.html. Appendix Table 3A presents a modified version of this list. A considerable number of the non-MEXT grants-in-aid programs have an explicit applied emphasis and/or are open to applicants and coapplicants from industry. Even programs labeled as 'basic research programs' such as JST's CREST, PRESTO, and ERATO stress the need for research results to have practical applications and social contributions. (See the description of the main competitive funding programs open to university researchers in Kneller 2007.)

Perhaps the hypothesis that university research tends to be more applied in Japan than the USA should not be overstated. To put this matter in perspective, a colleague at the university of Tokyo recently remarked that when he applied for a large MEXT grant-in-aid, he had to decide whether to portray his proposal as application or basic research-oriented. He chose the latter and, somewhat to his surprise, got the grant. Also, I have reviewed lists of NSF awards in nanotechnology. Many of these seem to combine basic research and applications themes. The mission of NIH, the largest supporter of US university research, is research to improve health, so many of its projects have applications to health and medicine. NIH routinely issues requests for proposals (RFPs) that solicit applications in specific areas deemed to have high priority for science or health.

Nevertheless, the review in Chapter 7 of non-MEXT grants-in-aid projects indicates that in cutting-edge fields, such as nanotechnology, various fields of cellular biology, fuel cells, wireless networking, and communications, consortium research is very common. Projects in these areas frequently involve large companies and universities and usually aim for industrial applications.

102. Of course, creative applicants often can write applications that fall under the ambit of the predetermined research themes but nevertheless allow the applicants to pursue their own research ideas. However, according to Japanese colleagues who have applied to some of the programs that set forth specific research themes, sometimes the review process is dominated by a single research group that expects applicant to address lines of research that group considers to be important.

103. There are few systematic studies that look beyond metrics such as numbers of patent applications or of joint university–industry publications. As for anecdotal evidence, conversations with university and industry researchers prior to 2004 generally revealed negative perspectives, made more believable by the mention of exceptions that seem to prove the rule: e.g. amorphous silicon for solar cells and drug delivery systems. A professor in the field of IT, one of the few who in 2000 could claim significant commercial applications for his research, remarked that year that large government applied research projects are a distraction for Japan's most capable students and their professors— relatively easy money (at least for well-known professors) for projects that are not critically evaluated either before or after they take place. He said that, without such funding, researchers in IT would be forced to work more closely with companies and they would come to grips with problems that are of real importance to industry. Corporate researchers generally tended to agree, and said that they obtain greater benefit by sponsoring university research on their own.

On the other hand, perspectives of companies appear to be becoming more positive. Also, the ERATO program, in particular, has been carefully studied and the results have been praised in Japan and overseas (JTEC 1996; Hayashi 2003). Finally, I am impressed with the progress in various fields of engineering, IT, and materials science by some of the academic research teams funded by agencies such as JST and NEDO. These impressions are shared by exchange scientists (primarily from Europe) in these fields who have also attended presentations by the heads of these research teams. (These presentations are by better known researchers, and thus may not be representative of most recipients of such funds.) The research may indeed have a practical orientation, but in the process of developing practical applications, it is clear that fundamental scientific knowledge is expanding. What is less clear is the consistency with which industry is developing these discoveries. Almost all these researchers have industry collaborators (usually large companies) and in some cases it is clear that the companies have pushed forward rapidly with commercial development of the professor's research. In other cases, commercial interest (or merit) is less clear. One of the underlying questions of this book is whether a more supportive environment for startups would enable startups to develop some of these discoveries more rapidly, or whether the Japanese system of substantial funding for applied research coupled with close interactions with established companies provides a better environment for progress in basic science as well as the commercialization of university discoveries.

104. See Appendix Table 3A showing that these programs constituted 55.5% of total MEXT grants-in-aid in 2002. Kneller (2003) shows that in 1998, MEXT grants-in-aid (all types) constituted about 67% of total competitive funding for research in Japanese national universities.

105. NSF uses mail review by experts, follow up discussions by an assembled committee knowledgeable about the field, and written feedback to the applicants about the bases for decisions. The NIH peer review system goes even farther to ensure that projects are selected on the basis of merit and likelihood of scientific progress. This process is based on committees consisting of about twenty experts (attempting to achieve diversity in age, gender, scientific perspectives, etc.) who meet three times per year to review applications. During their committee service tenure, their universities accord them reduced teaching and administrative responsibilities. Deliberations incorporate a process of advocacy and open debate, and in the end a written rationale for the committee's decision is prepared for the applicant. According to my own experience at NIH and to observations of Japanese researchers who served on or observed NIH peer review committees, this process tends to bring out strong points and shortcomings that might not be initially apparent (Hayashi 1996; Suga 2004). In addition, each NSF and NIH committee is managed by a Program Officer (almost always a doctoral degree holder) with a strong scientific/technical background in the committee's field. This person ought to understand the frontiers of knowledge in the committee's field and where research priorities lie. She or he can give feedback to reviewers and discuss program goals with prospective applicants. Such in-house expertise and opportunities for dialogue are absent in the Japanese funding agencies with which I am familiar, at least the divisions that manage peer review and funding allocation. Nevertheless, the NIH system has been criticized for being unwieldy, time consuming, and still deficient in detecting novel research proposals. (See Kaplan 2005, although this article fails to substantiate the most serious criticisms and to show that alternative systems would likely be better.)

106. The decision process is more systematized and transparent, numbers of reviewers have been increased from 3 to 6 for many programs, names of review committee members are made public two years after their tenure ends, and applicants can receive their overall score in the event their application is denied.

107. See Chapter 7.

108. For an in-depth exposition, see Coleman (1999). See also Whitely (2003).

109. I am familiar with recruitment and promotion practices in only a few Japanese universities, but these include two of the leading national universities and one leading private university. Within each of these three universities, I know of one department that practices this form of open recruitment. But persons within these departments themselves say that they are pioneers within their universities. In other words, they are exceptions that prove the rule.

110. Such steps are under consideration in a few departments, but I know of no department that has implemented such procedures. However, such procedures are common in US universities.

111. See Appendix Table 3A for a list of all competitive funding programs and their sizes. Most of these are open to university applicants, and the MEXT programs mainly fund university research. Funding amounts are in units of 10^8 yen, which is slightly little less than US$ 1 million (the exchange rate having varied between 105 and 125 yen per US$ since 2000). A brief explanation of the Centers of Excellence Program follows in the text. A fuller description is in Kneller (2007) and also various reports issued by the Tokyo Regional Office of NSF.

112. Support for university research from JST's Basic Research Program (CREST, PRESTO, and ERATO), JSPS's Research for the Future and from METI/NEDO and all other ministries *other than* MEXT is generally classified as *commissioned research*. Contract research from private companies that does not involve company researchers working collaboratively in university laboratories is also classified as commissioned research. However, such funding probably accounts for less than 5% of commissioned research funds, at least in major universities. Most industry funding is either under Joint Research contracts or donations (Kneller 2003).

113. See Shinohara (2002).

114. According to discussion with university and government officials.

115. Kneller (2003, 2007).

116. See the classifications of the Carnegie Foundation for Advancement of Teaching at www.carnegiefoundation.org. These well regarded classifications are used in NSB (2006). As an indication that there are many other universities in the USA, the 199 research universities account for less than half (89,500 of 194,100) of full-time university faculty that hold S&E doctoral degrees (NSB 2006: A5-46).

In 2004, Japan had 87 national universities, 4 national academic research institutes under MEXT (such as the National Institutes of Natural Sciences in Okazaki listed in Table 3.2), 80 local government universities, and 542 private universities.

117. NSB (2006: A5-18, 19). It might be argued that, because the Kanto (Tokyo-Yokohama), Kansai (Osaka, Kyoto & Kobe) and Nagoya regions account for a high proportion of Japan's 127 million population, it is appropriate for leading universities in those regions to receive a disproportionate share of R&D

funding. However, a separate analysis shows that these three regions together account for 43–49% of population (depending upon whether metropolitan regions (43%) or entire prefectures (49%) are used as the basis for population counts) while universities in these regions receive 63% of MEXT grants-in-aid (Kneller 2007). This indicates, at least on a nationwide level, regional funding imbalances even in proportion to population.

118. Nakane, Chie. (1970). *Japanese Society*. Berkeley: University of California Press. This is one of the classic analyses of Japanese society. Nakane notes that the superior–subordinate relationship is not so much a requirement of personal loyalty, as it is a requirement to uphold the structure and stability of the group.

119. Doi (1971) asserts that *dependence* is a mutually recognized and accepted part of these relationships that increases their stability and palatability/appeal. If this is true, then freelancing might be seen as threatening the foundation of such relationships.

120. My impression is that group membership in Japan tends to be more time intensive than in the USA. Academic study groups (kenkyuu kai) organized by individual professors are quite common and meet regularly (once a month or more, usually at night), although usually not directed at a particular project or issue. University student study groups meet regularly for long hours, often in evenings or on weekends. Elementary school volleyball entails not only students but also parents devoting most of their weekends to team activities. Weekly university labor union meetings run late into the evening, oblivious to the fact that some of the representatives at those meetings have children and/or elderly relatives who need care at home.

121. See Chapter 7 and the works by Aoki and Chuma cited therein.

122. Although parents are still generally loath to request to leave evening meetings early in order to take care of family members.

123. Here are a few disparate examples: In Japanese dining halls for faculty and graduate students, members of one laboratory usually sit together and are rarely joined by outsiders. The doors to most laboratories and faculty offices are closed. Unless involved in common projects, communication among graduate students and junior faculty even in the same laboratory is not close. Groups involved in particular projects tend to stick to themselves. If a graduate student or even a junior faculty member has a research question, she or he will generally ask his or her supervisor for help. Approaching other members of the laboratory or going outside the laboratory seems relatively rare, according to overseas researchers who can compare laboratory environments in Japan with those in America, Europe, and China. Work related social functions rarely involve spouses or friends from outside the group. A few negative comments about a particular person by a senior professor will lead other academics, not only in his or her kouza, but also outside persons who in one way or another acknowledge his or her authority, also to cease

communication with that person. Managers and lead researchers in a venture company, who know a range of outside contacts who might help the venture, will not contact those persons unless they feel they have a clear go-ahead from the head of the venture. (I strongly suspect the same applies to large companies.)

Are these observations unique to my experience, or to foreigners in Japan, or to the University of Tokyo? Perhaps—but based on my observations, I think not. It would be helpful for anyone who doubts these conclusions to present evidence that shows the opposite.

124. In this regard, see Yamagishi, Cook, and Watabe (1998) and other writings by Yamagishi suggesting that general trust (the tendency to trust another person regardless of whether he or she is bound by the same stable social relations, i.e. is a member of the same family or work group) is lower in Japan than America, and this is due largely to the closed nature of key social groups in Japan, particularly work-related groups.

125. This issue is dealt with again in the following and final chapters.

REFERENCES

Baba, Yasunori, Yarime, Masaru, Shichijo, Naohiro, and Nagahara, Yuichi (2004). 'The role of university–industry collaboration in new materials innovation: evolving networks of joint patent applications'. (Paper presented at the International Schumpeter Society Conference 'Innovation, Industrial Dynamics and Structural Transformation: Schumpeterian Legacies', Milan, Italy, 9–12 June.)

Barnett, Jonathan M. (2003). 'Private Protection of Patentable Goods', Research Paper 28, Fordham University School of Law. Available at http://ssrn.com/abstract-434585

Bayh-Dole25, Inc. (2006). 'The Bayh-Dole Act at 25'. Available at www.bayhdole25. org/whitepaper

Branstetter, Lee and Ogura, Yoshiaki (August 2005). 'Is Academic Science Driving a Surge in Industrial Innovation? Evidence from Patent Citations', NBER Working Paper Series No. 11561. Cambridge, MA: National Bureau of Economic Research.

Chabrow, E. (2005). 'U.S. Universities Loosen Grip on Tech Rights to Keep Vendor Grants from Going Overseas', *InformationWeek*, Dec. 20.

Coleman, Samuel (1999). *Japanese Science: From the Inside*. London: Routledge.

Doi, Takeo (1971). *The Anatomy of Dependence*. Tokyo: Kodansha International.

Dibner, M. D. (1988). *Biotechnology Guide U.S.A.* 1st edn. New York: Stockton Press.

Eisenberg, Rebecca S. (1996). 'Public Research and Private Development: Patents and Technology Transfer in Government-Sponsored Research', *Virginia Law Review*, 82: 1663–727.

Etzkowitz, Henry (2002). *MIT and the Rise of Entrepreneurial Science*. London: Routledge.

Hashimoto, Takehiko (1999). 'The Hesitant Relationship Reconsidered: University–Industry Cooperation in Postwar Japan', in Lewis M. Branscomb, Fumio Kodama, and Richard Florida (eds.), *Industrializing Knowledge: University–Industry Linkages in Japan and the United States*. Cambridge: MIT Press, pp. 234–51.

Hayashi, Masao (pen name: Hakuraku Rokubiro) (1996). *NIH and American Research Funding [Amerika no kenkyunhi to NIH]* Tokyo: Kouritsu [in Japanese].

Hayashi, Takayuki (2003). 'Effect of R&D Programmes on the Formation of University–Industry–Government Networks: Comparative Analysis of Japanese R&D Programmes', *Research Policy*, 32: 1421–42.

IBM (2006). 'Understanding the Emerging Innovation Landscape: Enabling University and industry research' (Open collaboration principles overview).

Japan Bioindustry Association (JBA), (1998). *The Usefulness of University Research to Promote our Country's Bioindustry* [Daigaku nado no kenkyuu seika wagakuni no baioindasutori no shinkou ni yakudateru tameni]. Tokyo: JBA, unpublished report in Japanese.

Japan Technology Evaluation Center (JTEC) (1996). Japan's ERATO and PRESTO Basic Research Programs. Available at http://www.wtec.org/loyola/erato/toc.htm

Kaiser, Jocelyn (2002). 'Proposed rules aim to curb financial conflicts of interest', *Science*, 295: 246–7.

Kaplan, David (2005). 'How to Improve Peer Review at NIH', *The Scientist*, 19(17): 10.

Kauffman Foundation (2005). Twelve Leaders Adopt Principles to Accelerate Innovation (Press Release), Dec. 19.

Kneller, Robert W. (1999). 'University–Industry Cooperation in Biomedical Research in Japan and the U.S.', in Lewis M. Branscomb, Fumio Kodama and Richard Florida (eds.), *Industrializing Knowledge: University–Industry Linkages in Japan and the United States*. Cambridge: MIT Press, pp. 410–38.

Kneller, Robert W. (2003*a*). 'University-Industry Cooperation and Technology Transfer in Japan Compared with the US: Another Reason for Japan's Economic Malaise?' *University of Pennsylvania Journal of International Economic Law*, 24(2): 329–449.

—— (2003*b*) 'Autarkic Drug Discovery in Japanese Pharmaceutical Companies: Insights into National Differences in Industrial Innovation,' *Research Policy*, 32: 1805–27.

—— (Summer 2006). 'Japan's New Technology Transfer System and the Pre-Emption of University Research by Sponsored Research and Co-Inventorship', *Journal of the Association of University Technology Managers*, XVIII(1): 15–36.

Kneller, Robert W. (2007), 'Prospective and Retrospective Evaluation Systems in Context: Insights from Japan', in Richard Whitley, Jochen Gläser and Kate Barker (eds.), *The Changing Governance of the Sciences: The Advent of Research Evaluation Systems*. Dordrecht: Springer.

Latker, Norman J. (1977). Statement before the Subcommitte on Science, Research and Technology of the U.S. House of Representatives (May 26).

Matsuo, Yoshiyuki (1997). Chiteki sangyou e no toushi de aru kakenhi wa jinrui no atarashii bunka o umidasu [Grants-in-aid as an investment for producing intellectual property gives rise to a new human culture] in Kobayashi, Tetsuo (chief editor). 1997 Daigaku Rankingu [1997 University Rankings]. Tokyo: Asahi shimbunsha shuppan kikaku shitsu] Asahi Newspaper Corporation Publication Project Division.

McKelvey, Maureen (1996). *Evolutionary Innovations: The Business of Biotechnology.* Oxford: Oxford University Press.

METI (2005). Industry–University Cooperation: Issues and Themes [Sangaku renkei ni okeru genjoui to kadai] (January).

—— (2006). Basic Survey Report on University Ventures [Daigaku hatsu benchyaa inkansuru kiso chousa jisshi houkokusho] (May, in Japanese). Produced in cooperation with Kabushiki Kaisha Kachi Sougou Kenkyuusho (listed as official report author).

MEXT Basic School Survey (various years) [Gakkou kihon chousa houkoku sho]. Tokyo: Kokuritsu Insatsu Kyoku.

MEXT (2003). Actual Situation in 2002 Related to Joint Research with Industry in National Universities [Kokuritsu daigaku nado no (Kigyou nado tono kyoudou kenkyuu) no heisei 14 nendo no jisshi joukyou ni tsuite] (issued 31 July, in Japanese).

—— (2005). University–Industry Cooperation: The Actual Situation in Universities in 2004. [Heisei 16 Nendo: Daigaku nado ni okeru Sangaku Renkei Jisshi Joukyou ni tsuite] issued June (in Japanese, available at http://www.mext.go.jp/b_menu/houdou/17/06/05062201.htm).

MEXT Personnel Advisory Com. [Kagaku gijutsu gaku jutsu shingikai, Dai 13 kai jinzai i-inkai] (2003). The Way the Postdoctoral System Ought to Be [Postudoku seidou no arikata ni tsuite] (unpublished report).

MEXT Academic Advisory Committee, IP Working Group (2002). 'Intellectual Property Working Group Report' [in Japanese: Kagaku gijutsu gakujutsu shingikai, gijutsu kenkyuu kiban bukai, sangakukan renkei suishin i-inkai, chiteki zaisan wakingu gu-rupu, Chiteki zaisan wa-kingu gu-rupu houkokushou] (issued Nov. 1).

Monbusho (1996). Full citation: Science and International Affairs Bureau of the Ministry of Education, Science, Sports and Culture (Monbusho). 1996. *Research Cooperation Between Universities and Industry in Japan.* Tokyo: Monbusho (in English and Japanese).

—— (1998). To construct a new technology transfer system linked to patents: Report of the full study committee for the promotion of industry–university collaboration and cooperation [Sangaku no renkei.kyouryoku no suishin ni kansuru chousa kenkyuu kyouryokusha kaigi matome: tokkyo nado ni kakawaru atarashii gijutsu iten shisutemu no kouchiku wo mezashite] (issued 30 March, in Japanese).

Murray, Fiona (2004). 'The Role of Academic Inventors in Entrepreneurial Firms: Sharing the Laboratory Life', *Research Fukugaua [nc change] Policy*, 33: 643–59.

Nakayama, Yasuo, Hosono, Mitsuaki, Fukugawa, Nobuya, and Kondo, Masayuki, of the National Institute of Science and Technology Policy (NISTEP) 2nd Theory-Oriented Research Group (2005). University–Industry Cooperation: Joint Research and Contract Research (NISTEP Rept. No. 119 of 2005) (in Japanese, available http://www. nistep.go.jp/index-j.html).

National Science Board (NSB) (2006). *Science and Engineering Indicators 2006* (two volumes). Arlington, VA: National Science Foundation.

National Institute of Science and Technology Policy (NISTEP, under MEXT) (2004). *Science and Technology Indicators: 2004* [in Japanese: Kagaku gijutsu shihyou]. Tokyo: NISTEP.

Odagiri, Hiroyuki (1999). 'University–Industry Collaboration in Japan: Facts and Interpretations', in Lewis M. Branscomb, Fumio Kodama, and Richard Florida (eds.), *Industrializing Knowledge: University–Industry Linkages in Japan and the United States*. Cambridge: MIT Press, pp. 252–65.

Powell, Walter W., Korput, K.W., and Smith-Doerr, L. (1996). 'Interorganizational Collaboration and the Locus of Innovation: Networks of Learning in Biotechnology', *Administrative Science Quarterly*, 41: 116–45.

Rabinow, Paul (1996). *Making PCR: A Story of Biotechnology*. Chicago, IL: University of Chicago Press.

Shane, Scott (2004). *Academic Entrepreneurship: Academic Spinoffs and Wealth Creation*. Cheltenham, UK: Elgar.

Shinohara, Kazuko (2002). Toyama Plan—Center of Excellence Program for the 21st Century, NSF Tokyo Regional Office Report Memorandum #02–05 (June 21).

Suga, Hiroaki (2004). *Diligent American Scientists* [Sessatakuma suru Amerika no kagakusha tachi]. Tokyo: Kyouritsu (in Japanese).

Tokushima University (2006). Policy making guidelines concerning conflicts of interest in clinical research [Rinshou kenkyuu no rieki souhan por-ishii sakutei in kansuru gaidorain] (issued March in Japanese, available at http://wwwip.ccr.tokushima-u.ac.jp/servlet/default.asp?MNO=33).

Uekusa, Shigeki and Takaoka, Kano (2005). Financial indicators of national universities and the availability of these indicators [Kokuritsu daigaku houjin no zaimu shihyou to sono kanousei]. *Research on Academic Degrees and University Evaluation* 1 (March, a publication of the National Institute for Academic Degrees and University Evaluation, in Japanese).

Whitley, Richard (2003). 'Competition and Pluralism in the Public Sciences: The impact of institutional frameworks on the organisation of academic Science,' *Research Policy*, 32, 1015–29.

Working Group on Ethics and Conflicts of Interest in Clinical Research [Rinshou kenkyuu no ronri to rieki souhan ni kansuru kentou han] (2006). Guidelines Concerning Conflict of Interest Policies for Clinical Research [Rinshou kenkyuu no rieki souhan porishii sakutei nikansuru gaidorain].

Yamagishi, Toshio, Cook, Karen S., and Watabe, Motoki (1998). 'Uncertainty, Trust, and Commitment Formation in the United States and Japan', *American Journal of Sociology*, 104 (1): 165–94.

Zucker, Lynne G. and Darby, Michael R. (1996). Star Scientists and Institutional Transformation: Patterns of Invention and Innovation in the Formation of the Biotechnology Industry', *Proceedings of the National Academy of Sciences*, 93: 12709–16.

4

Up the Rocky Road: Venture Case Studies

INTRODUCTION

This chapter looks directly at Japanese high technology ventures to describe their present situation and future prospects. Over the past ten years, the environment for these companies has improved greatly. Formation of university startups rose sharply from 1998 to 2000 due largely to the legal changes described in Chapter 3. Now the rate of formation seems to have reached a respectable plateau of just over 100 new startups per year.

However, other policies have aimed to help all ventures, not just startups. These include increased access to government funding and private venture capital, advice on business and IP matters, and tax breaks for all ventures. Appendix 1 summarizes the most important of these up through 2004.

In addition, venture capital (VC) and public equity financing have matured. VC companies are now more likely than in the past to make equity as opposed to loan investments, to invest in newly formed companies, and to invest in technology as opposed to service-based startups. Special public equity markets have opened that have less stringent listing requirements and thus are more suited to the needs of startups and the need of VC investors to recoup their investments (for an exit mechanism).[1] Several private VC companies focus on early stage investments in high technology ventures, mainly in the life sciences. Government institutions contribute approximately half of the investment capital of several of these VC companies. In addition, the government has its own VC organizations that often invest in technology startups. Several traditional VC companies have created special funds to invest in technology ventures, and some of these have negotiated agreements with individual universities under which they screen inventions for possible startup funding opportunities.

As for the companies themselves, systematic data are not available except in the case of university startups and life science ventures. However, at least until 2002, the rate formation of new manufacturing SMEs had been declining and was considerably lower than the rate of exit.[2] For ventures that are neither startups nor focused on life science, the case studies later in this chapter provide some insights.

But as for startups, 38 percent are biomedical-related, a share that continues to increase. Software is next (30%) followed by machinery and devices (17%).[3] In general, these companies are quite small, although this might be expected because of their young age.[4] Biomedical startups as a group have received more investment (including a substantially larger proportion of venture capital investment) than startups in any other industry, and their total sales are approximately the same as for startups in other industries.[5] Seven life science startups have had initial public offerings (IPOs) of their stock.[6] A few of the biomedical startups have drugs in clinical trials.

Yet the average size of even the life science startups is small. Moreover, in this field there is evidence, for other ventures as well as startups, that both total employment and average numbers of employees are low compared to the US industry *at comparable periods of development*. Specifically, as shown in Table 4.1, total employment in Japanese ventures engaged in the development of biomedical therapeutics is about one-third of total historical employment in US therapeutic-oriented bioventures *of equivalent age*, and thus much less than total employment in the US therapeutic bioventure industry as a whole, even it its early days.[7] In addition, the average number of employees in Japanese therapeutic-oriented bioventures is also low compared to historical averages in US therapeutic bioventures of equivalent median age. Table 4.1 shows these comparisons for the approximately 113 Japanese therapeutic bioventures existing in 2004[8] and for US companies of the same median age existing in 1987 and 1998/99.

This suggests that Japanese bioventures are not growing as fast as their US counterparts, and that total sector employment is considerably less than in the USA. It suggests that access to human resources and the capital to employ these resources are problems for Japanese ventures in biomedicine, the field in which Japanese ventures are most active.[9] The survey data on startups tend to confirm that access to human resources is probably their main challenge.[10]

Table 4.1. Employment in therapeutic-oriented Japanese bioventures in 2005 and US therapeutic ventures of equivalent median age in 1987 and 1999[11]

	All Japanese therapeutic ventures in 2004, employment assessed 2005	All US therapeutic ventures formed 1981–6, employment assessed 1987	All US therapeutic ventures formed 1992/4–97/8, employment assessed 1998/99*
No cos.	113	33	98
Total employees	Slightly over 1,463	3,814	3,900
Mean employees	Slightly over 13	116	40
Median employees	9.5	50	25

Methods and explanations in Notes at n. 11.

A Japanese venture capital (VC) company that invests in both US and Japanese bioventures has two portfolio companies founded on Japanese university discoveries with Japanese CEOs but headquartered in California. When asked the reasons for moving those companies to the USA, the reply was that resources, particularly human resources (postdocs, technicians, and academic networks), are more abundant in the USA.

To obtain a clearer picture of the status of startups, I identified almost all the startups from the University of Tokyo, Keio University, and AIST formed prior to the end of 2003 and obtained information on their core technologies and various business indices. Appendix 2 presents these findings.

Three, all biomedical, have had IPOs. Four had revenue exceeding US$10 million in 2004.[12] But aside from a few such exceptions, the startups are small. Life science ventures predominate and are better capitalized than other companies. Next in prominence are software companies, with six having revenue over US$1 million in 2004. Approximately half the startups seem to be developing new technologies, while the remainder provide services or utilize technologies that already have been developed.

The most surprising finding is the small number of startups developing novel or broadly applicable engineering or materials/chemical technologies, especially considering that all three institutions are strong in these fields. Also, few startups in these fields have substantial sales.[13] These observations are consistent with the assertion in Chapter 3 that valuable discoveries outside of biomedicine are flowing directly to established companies via joint research and this is limiting opportunities for startup formation.

However, to really understand the challenges faced by ventures and their potential for growth it is necessary to understand the experience of actual companies. The remainder of this chapter presents ten case studies of ventures in fields such as engineering, materials, and software and ten case studies of biomedical ventures.

PART I: CASE STUDIES OF NON-LIFE SCIENCE VENTURES

The following ten case studies summarize information obtained primarily from interviews, all of which took place in 2004.[14] Because some of the companies requested anonymity or had concerns about confidentiality, all companies are identified by pseudonyms and the information presented reflects confidentiality requests. Unless otherwise indicated, information is current as of mid to late 2004. Readers who are not interested in the details of these companies can skip to the salient points from these case studies at the end of this part.

Interface Chip

Interface Chip's founder had worked on integrated circuit (IC) chip design in major Japanese and US electronics companies before founding Interface Chip in 1991. Interface Chip focuses on designing specialty ICs to control flat-panel displays or to provide the interface between digital devices (e.g. DRAMs and flash memory chips) and analog or mixed signal systems. Rather than pushing the limits of miniaturization, it designs circuits with micrometer rather than nanometer channel widths, but with high performance characteristics that it customizes to the needs of individual electronic manufacturers. It aims to remain at the forefront of technology. It does not want to design chips that will be mass produce with falling profit margins.

Initially negotiating with potential Japanese customers was difficult. They wanted a working prototype device in order to be assured of quality and performance capabilities. The first breakthrough came one year after founding when Interface Chip entered into a joint venture with a large Korean electronics company to develop ICs for its products. In the early 1990s, Korean companies were eager to partner with small Japanese companies to learn about Japanese technology and how to acquire technologies through foreign partnerships. Sales of its own-brand chips began in 1997, manufacturing being contracted out to various foundries in Asia. Its next breakthrough was a sales agreement with a major Japanese electronics manufacturer. Now major clients include most of Japan's major electronics manufacturers, its original Korean partner and at least one major European electronics company.

Its initial investors were mainly non-Japanese Asian electronics companies and subsequently Japanese electronics companies. It had an IPO on Jasdaq in 2001. Its 2003 sales were over US$100 million and trending up. Its annual profit was nearly US$10 million and trending up. It is one of the most successful Japanese high technology ventures.

Interface Chip employs about fifty engineers, most in their 20s to early 40s. They are organized into teams of five, and sales are attributed to specific teams. This is the basis for performance bonuses. Most have been lured away from large companies, bringing with them several years of experience. It has about 100 applications for every vacancy. 'In large companies, engineers work in just a narrow area. In our company, they participate in the big picture and develop their full range of talents.'

Interface Chip has commissioned and joint research projects with particular university professors whose work is relevant to its business. It does not send researchers to universities, nor does it use university advisers. It has not used government SBIR-type or collaborative R&D funds. 'Our engineering priorities change too often. The application process for government grants is too long and we cannot commit our engineers' time to government funded projects that are no longer of key importance for the company.'

Interface Chip patents its basic circuit designs. These patents form its core IP. Its technology would be easy to imitate if reverse engineering were permitted and its know-how were known. Therefore, both patents and trade secrets are necessary to prevent copying.

As to reasons why large companies are hesitant to collaborate with ventures, I received the following answer. 'Big companies are conservative and risk averse. Their chain of command and approval has many levels. Good ideas, including collaboration proposals, are often killed along this chain. In addition, big companies are not rigorous in cost accounting for in-house R&D. If 40 percent is the minimum acceptable rate of return for external investments, then 50 percent should be the minimum accepted return on internal investments in order to cover overhead costs. But instead, large Japanese companies tend to be lax about committing funds to internal R&D projects.'[15] In addition, large Japanese companies are surveying overseas technology resources, particularly in India, China, and the USA, but they discount Japan because they believe there is insufficient entrepreneurship in their home country.

As for the most significant challenges facing ventures in general, the respondent mentioned difficulty obtaining financing and good management, but, above all, *family bias against work in ventures*. 'Wives who expect their husbands to work in large companies are the biggest problem.'

Chip Connect

After graduating from secondary school, Chip Connect's founder joined a major Japanese electronics company and worked on printed circuit boards (PCBs). In 1987 he founded Chip Connect as a designer and manufacturer of PCBs. His clients were US, European, and non-Japanese Asian contract manufacturers of PCBs for the founder's former employer. He began to concentrate on developing new interfaces between miniaturized PCBs, printed wire boards (PWBs) and IC chips, for use in mobile phones and other devices requiring compact three-dimensional configurations of these devices. Some revenue came from licensing its PCB designs to these clients.

Chip Connect's breakthrough came in 1993, when it started a joint venture with the founder's former employer to develop a new semiconductor packaging and interconnection system. Patents were held jointly with the large electronics company, which was a source of concern for the company. The joint research ultimately failed to produce the desired breakthrough. Nevertheless, it gave the company credibility in the eyes of one of Japan's largest traditional VC companies, which invested in Chip Connect and helped it conduct an IPO.

Since then it has developed other chip packaging and connection technologies. It has in-licensed complementary technology from a US venture. Its first major customer was a large Korean electronics company, which licensed its designs. It currently has about ten licensees. Eighty to ninety percent of its roughly US$15 million annual revenue comes from licensing plus technical support. However, it has recently built its own manufacturing plant to supply its proprietary copper foil interface material to its licensees. Sales of this material make up the remainder of its revenues.

Since the IPO, it has obtained financing from several banks. It does not obtain support from the government via loans or research grants, believing that access depends on insider or political support. Nor does it engage in government sponsored university–industry collaborations. However, it has obtained help from two professors in national universities who have expertise important for the company.

Since the company's products are easy to reverse-engineer, patent protection is essential, and the company is always concerned about infringement. Also, reflecting on its previous experience with its joint venture partner and that of other venture companies, it does not want to hold patents jointly with a large company because that would give the large company too much control over its technology.[16] As of mid-2004, it held four Japanese and four overseas patents and had applications pending for about fifty Japanese and forty overseas patents.

The company knows that a large Japanese electronics manufacturer holds similar patents. Despite the possibility for overlap between the two company's portfolios, it feels its patent position is sufficiently strong to allow it to counter any attempt by this other company to encroach on its core technology and customers, and to this date the other company has not done so. In other words, Chip Connect uses its patents as a defensive shield. They also serve to attract investment and customers, and simultaneously put the latter on notice they cannot copy Chip Connect's designs without a legal fight.

The company out-licenses its know-how in addition to its patented technologies. For applications of its technology, the company's know-how is essential. Information in the patents simply would not be sufficient.

Somewhat surprisingly, most of its new researchers are recent social science (especially economics) and humanities graduates. Apparently for the sort of design work the company does, these persons do quite well. Much of the design work involves pattern and three-dimensional designs for thin metallic tape that provides the interface between PCBs, PWBs and IC chips. The company currently has about 100 employees. Recruitment of new employees has not been a problem.

Its main suggestion for reforms to help high technology ventures was 'lower corporate taxes and reduce patent costs'.

Chip Detect

Chip Detect was founded by four researchers at a major Japanese electronics company, including a laboratory division manager and a young engineer who had developed a new method to detect flaws in IC chips. The electronics company applied for patents on this method and related device components, but then canceled development work. The electronics company assigned the patent rights to the spin-off and also provided substantial initial funding, not in return for royalties or immediate equity, but rather for options to purchase the venture's stock.[17] From its inception, METI backed formation of this spin-off, hoping to emulate the success of some Silicon Valley companies with a mixture of public and private funding. A VC company with ties to the Development Bank of Japan and METI contributed startup funding. The VC company also seconded one of its officers to be Chip Detect's chief financial officer.[18] A CEO with experience in Japanese and overseas chip manufacturing equipment ventures was recruited. Two rounds of private VC funding followed within two years of founding, with a lead participant being a well-known American VC company. The company has received nearly 1 billion ¥ in SBIR-like funding from METI.

Although it has no sales yet, three companies (one Japanese and two overseas) are evaluating its prototype machine and helping to improve it. The biggest challenge the company faces is to improve the prototype to a working model. It has some income from consulting services related to IC chip quality. It is also developing another technology to protect the quality of IC chips. 'Big companies are interested in our products. This is a change from the past. Autarky is diminishing', in the words of the CEO. However, relations with the parent appear strained and the degree of collaboration and support the parent has offered has been minimal.

In addition to eight pending patents assigned from the parent, Chip Detect has applied for patents on twelve of its own discoveries. Because its products can be copied, patent protection is vital. It says it could not have obtained VC financing without them.

Chip Detect employs about twenty engineers, scientists, and managers. Like Interface Chip, it hires engineers in their 20s to early 40s from large companies. Their experience is valuable and training costs are kept low. Sometimes headhunters are used for recruitment. The CEO believes labor mobility is increasing. 'Unlike the past, people are now willing to leave big companies, mainly because they find our work more interesting and challenging. Stock options are also a draw.' However, funds are insufficient to hire all the R&D engineers the company feels it needs.

One university researcher is cooperating with Chip Detect, but it relies primarily on in-house R&D.

Nanofilm Applicator

Founded in 1999 by a Chinese engineering graduate from a small Japanese university, Nanofilm Applicator makes machines for depositing nanometer-thin light filtering films. Such films are used in optical fibers for signal transmission and data processing and also in the dense wavelength division multiplexing modules (DWDMs) incorporated in many optical-electronic products such as digital cameras and DVD players.

The company's first clients were Taiwanese fiber optic cable manufacturers. Many Japanese companies would not even consider looking at its machines, preferring instead to rely on their own in-house vacuum coating processes. From the perspective of the venture, there was great hesitancy on the part of middle and senior level managers in large companies to be the first to adopt a new technology. If it turned out badly, their careers would be damaged, but if it turned out well, there would not be any significant reward. This problem was compounded by the venture always having to deal with managers in the central research laboratories of the large companies. Even if it could convince the research laboratory managers of the merits of its products, it had to rely on these managers to convince persons in the large companies' central offices to adopt the new technology.

In 2002, after the technology had proved its merits in overseas sales, but also as the market for fiber optic cable collapsed, Nanofilm Applicator began to shift sales to companies making DWDMs and digital camera lenses. Increasingly these companies have been Japanese electronics, glass, and photo-optics companies. In 2003, sales to Japanese companies for the first time accounted for the majority of the company's revenue, US$24 million that year.

Many Japanese companies have their own thin film deposition technology. But Nanofilm Applicator believes its machines are more efficient and give better quality results, and it would not be cost effective for its customers to try to duplicate the company's technology. The company estimates it has about 10 percent of the world market for vacuum application of thin film optical coatings and related instruments. Its main competitors are specialized US and German companies, which each employ different application methods.

Although the company has nearly twenty pending or issued patents, the basic technology behind its machines is almost 50 years old. Thus its strength lies not in what is claimed in its patents but in its processing and quality control know-how. It believes this is also the case with its competitors. These are guarded as trade secrets and tacit knowledge and constitute greater barriers to would be competitors than its patents. Unlike Chip Connect, it does not license its know-how. Its business is based on selling its machines and related services.

The company's initial capital came largely from a Chinese angel living in the USA. In 2001 it attracted VC investment, and currently about half its equity is held by VC companies. Its largest investor is a large US financial services company, but aside from advice with respect to corporate governance issues, the VC investors have adopted a hands off policy. Business decisions and the task of finding customers have been left up to the company. Recently, the company has also obtained financing from major Japanese banks. It is not planning an IPO in the near future. It has applied unsuccessfully for government R&D funding. It attributes its lack of success to its positive balance sheet, but also has doubts about the process by which recipients are selected to receive government support.

In building its business, this company turned to a subsidiary of a large electrical machinery company to codevelop a new ion source for its machines. This collaboration was successful.

Nanofilm Applicator has about eighty employees, twenty to thirty of whom are involved in R&D. A significant proportion of these are Taiwanese or mainland Chinese. The company feels it is difficult to attract capable Japanese engineering graduates to venture companies and thus it has had to turn to Chinese engineers. Also major Japanese universities are not interested in the applied research problems that are important to the company. Collaborations occur with local private Japanese universities and major universities on Taiwan and the Chinese mainland. Personnel exchanges of about two weeks each occur both ways with the Chinese universities. Looking toward the future, the company knows it will have to innovate to anticipate changes in technology that will affect demand for its machines. Such innovation will require more human resources and outside collaborations.

Despite the problems this company has encountered breaking into the Japanese market and despite its reliance on Chinese researchers and US capital, the founder of this company remarked that Japan is probably the best place in the world for manufacturing. His company's machines incorporate products from hundreds of other companies and his company's machines are used in turn by companies that use components and equipment from hundreds of other companies. This utilization of components from various manufacturers requires great coordination among companies and attention to quality. Having familiarity with high technology manufacturing in Japan, the USA, Taiwan, and mainland China, he thinks the environment for this type of development and manufacturing is best in Japan.[19]

As for policies that would help venture companies such as his, the CEO immediately answered, 'lower corporate taxes'.

Molecule Visualizer

Molecule Visualizer was founded in 1999 by a senior GRI scientist with previous experience in a major photo-optics company. The founder was approaching the mandatory retirement age of 60. Formation of the company enabled him to continue his research on molecular structure and atomic level microscopy and to develop commercial applications. He was inspired to form the company by frequent inquiries about the industrial applicability of his research while he was still at the GRI—queries that came almost exclusively from foreign companies[20]—and the eagerness of METI to aid promising startups that might commercialize university or GRI technologies.

From the outset, this company has been blessed with access to the fruits of large government research projects and public as well as private sector funding to promote R&D in startups. Business plan development began in 1997, two years before founding.[21] By that time, new techniques for visualizing and measuring biomolecules were emerging from a large GRI project commissioned by NEDO in which the founder was one of the lead investigators. That same year, a major national university and a venture capital consortium led by Japan's largest VC company established a fund to aid high technology enterprises in the area around the university.[22] The fund decided on Molecule Visualizer as one of its first investments, and became the lead investor in a blue ribbon consortium of banks, insurance companies, and other VC funds.

Currently the consortium members hold 60–70 percent of Molecule Visualizer's stock while the founder holds about 20 percent. The VC investors are represented on Molecule Visualizer's board of directors. However, they do not oppose the founder's decisions, and the founder effectively has full decision-making authority.

Since establishment, Molecule Visualizer has received numerous subsidies, contracts, and grants from government-affiliated organizations.[23] In other words, the company first leveraged the founder's central role in a large GRI research project to obtain private investment capital to start the company. Then it leveraged its status as a high technology startup to obtain public funds aimed at promoting venture companies.

In 2004, those public funds provided Molecule Visualizer with just under 20 percent of its approximately US$7 million revenue that year. Its other sources of revenue were molecular level measurements done on a contract basis for government and industry laboratories, sales of atomic force microscopes manufactured by a new US bioinstrument company, and sales of its own instruments to visualize and measure biomolecules. In addition, it is receiving funding from a Japanese prefecture to develop instruments for genetic analysis of food quality. Despite these revenues, it is still in deficit. It considers such public funding crucial, asserting that if it had to rely on private capital investment, the technologies it is developing simply would have languished.

Molecule Visualizer has seventeen employees, all but two with technical backgrounds. Five of the seventeen have doctoral degrees. Some of the employees have financial backgrounds to prepare for an IPO within the next few years. Some of the employees were recruited from big electrical machinery companies, some from GRIs. Headhunters have been relied on. Neverthless, recruitment of skilled personnel is a major problem. None of the employees are simultaneously employees of the GRI, reflecting an effort to separate company affairs from those of its GRI parent.

There are currently no joint R&D activities with large companies. Molecule Visualizer is concerned that such cooperation would enable large companies to absorb Molecule Visualizer's know-how and technology. For this reason, and also in order to earn future license revenue, Molecule Visualizer places high priority on obtaining patents. It has fourteen pending Japanese applications and some foreign applications. It has renewable five-year exclusive licenses to key patents held by NEDO on inventions by the founder when he worked at the GRI.

Molecule Visualizer has many advantages and it hopes to exploit these to develop new instruments for atomic level measurements and manipulations of molecules and also to develop new uses for such instruments. This is a competitive field and Molecule Visualizer's success will depend on its ability to innovate rapidly.

Big Crystal

Big Crystal was founded in 2000 by a scientist who started his career in Hitachi Metals, which sent him to Stanford for advanced training where he absorbed entrepreneurial ideas. He returned to Japan to work for a materials science GRI that had been developing high purity crystals since the 1980s. In the 1990s he was a key developer of a new process to make large high quality single crystals for use in lasers and other optical devices. The GRI tried to interest several large Japanese companies in the technology hoping that one would undertake large-scale manufacturing. However, these companies perceived the technical risks too high and market size too small.

Among the small number of Japanese scientists who were experts in the production of large pure crystals, the founder was the only one who wanted to start a company, believing that the potential market justified establishment of a startup. A friend at Stanford offered to invest US$2 million. The company was established in 2000 with seven venture funds contributing about half of the paid in capital. Prominent among these was a venture fund under the Tokyo Municipal Government and the SMRJ. Only about 40 years old, the founder left the GRI to devote all his time to the company and contributed about 10 percent of the initial capital. Another small manufacturer of crystal producing machines contributed most of the remaining 40 percent of initial capital and also provided land and laboratory space, essentially becoming Big Crystal's sister company.

In 2001 Big Crystal sold three machines for manufacturing large crystals (at least two to foreign buyers) and most of its revenues came from overseas companies and universities. In 2002 it sold five machines, and the source of the majority of its revenue shifted to Japanese companies and universities. Major Japanese electronics and photographic film companies began to turn to the startup for their high quality optical crystals. Previously they had manufactured these in house, but they appreciated the high quality and economies of scale the startup offered. Total revenue in 2003 from the sale of seven crystal producing machines and associated services was about US$2 million. By 2004, the founder and CEO was saying 'we are flooded with inquiries about joint development from large Japanese enterprises.' However, overseas sales were important for initial revenue and establishing credibility among Japanese companies.

The company has ten employees, seven actively engaged in R&D. Most of these joined the company at its founding. Recently, a considerable number of applications have been received from older persons laid off or forced into early retirement by large companies.

In 2003 Big Crystal entered into a strategic partnership with a ceramics company, ElectroCera, affiliated with a major Japanese electronics manufacture. ElectroCera had been contemplating moving into optical materials but considered the chance of success quite low if it were to rely on in-house R&D. It purchased 10 percent of Big Crystal's stock, mainly from the sister company.[24] Big Crystal's CEO believes that ElectroCera is interested in accessing both Big Crystal's technology and the research network of its CEO. In return, ElectroCera provides Big Crystal production know-how, skilled personnel, and financial resources.[25] However, the two companies have agreed that the ceramics company will not compete with Big Crystal in the latter's current main line of business, high purity large crystals. Big Crystal credits venture funds and the patent attorneys they introduced with helping to negotiate favorable contract terms with ElectroCera.

According to Big Crystal's founder, there have many cases of large companies appropriating the technology of venture companies. It relies on manufacturing know-how held as trade secrets as its main protection against infringement. It is reluctant to apply for patents that might disclose this manufacturing know-how. Nevertheless, since its founding, Big Crystal has applied for six patents, many covering methods to improve the yield of the manufacturing process.

Most of Big Crystal's IP rights involve about thirty inventions nonexclusively licensed from the GRI and the Japanese government. The situation regarding these licenses is complex. JST, which managed IP on behalf of the GRI, only issued nonexclusive licenses. However, under the rules governing inventions by employees of this GRI, the inventors could have half ownership rights. Because the founder is a co-owner, he has substantial control over licenses to other companies[26] and this has probably helped Big Crystal's ability to attract VC investment. Nevertheless, many large Japanese companies have a policy not to adopt an outside company's technology unless there are alternative suppliers. In other words, they would not license Big Crystal's technology and buy its machines if no other company could be a potential source of this technology. Therefore, Big Crystal has accepted other companies obtaining licenses to the patents covering its core technology.

Big Crystal also has licensed inventions from Osaka University and other universities. It has cooperative research contacts with many university laboratories and GRIs. It is also engaged in consortium research projects involving universities and large companies, many of these organized by METI/NEDO. IP to emerge from these projects is jointly owned by Big Crystal and the other research partners. For Big Crystal, the main attraction of cooperation with universities is economic. It can obtain research results at low cost, and when government funding is involved, the cost is even cheaper. It also receives small business support grants from the national and provincial governments.

The founder has established another startup that is also working on machines to produce crystals. However, because the GRI researchers are not interested in entrepreneurial work, much less in resigning to work in the startup, the startup is turning to Big Crystal's founder for management help. The founder believes there are many valuable discoveries arising in GRIs, but these inventions are not being developed because of insufficient support for the handful of GRI researchers who are interested in becoming entrepreneurs.

Fine Molded Plastic

Fine Molded Plastic (FMP) grew out of a family business originally established in 1969 in eastern Osaka, a region that has been the home for many high technology SMEs. Established as an independent company in 1996, FMP's business has relied on manufacturing molded plastic components for camcorders and other consumer electronic products and also for optical signal pickup devices used in digital video cameras and DVD players. Thus its major customers are electronics and photoelectronics companies.

However, like many other manufacturing SMEs, it faced a crisis due to Japan's economic downturn and the major electronic companies outsourcing components manufacturing to companies in China and other Asian countries. Largely in response to government promotion of R&D and startups in nanotechnology and biotechnology, FMP is trying to reorient its business to research intensive, high value-added products in these fields, although current revenue is still from sales of conventional plastic parts. Its main R&D focus is the development of ultrafine nozzles for ink jet printers, and also for applying wiring patterns on printed circuit boards (PCBs) and minute quantities of biological samples or reagents on chips for genetic or protein analysis. FMP's new core technologies include techniques to make metal molds to cast these nozzles and new polymer (plastic) resins that have low viscosity when poured into the molds but are also durable after the mold sets.

FMP has about fifty employees, including about twenty in manufacturing and an equal number in R&D. Some of the researchers are recent Ph.D. graduates from Osaka area universities. Others are transferees from the laboratories of large companies. Interdisciplinary skills are important, the company believing that discoveries arising from mono-skilled researchers can be copied too easily by overseas competitors. Its products involve the fusion of materials, metals, ion beam, reliability testing, and other technologies.

R&D expenses amount to 40 percent of sales, which were US$5.5 million in 2003. However, traditional plastic components still accounted for almost all its sales. Its advanced jet heads still are not ready for general marketing.

FMP is engaged in many joint R&D projects with large companies as well as with universities. The former usually involve developing applications of FMP's core technologies. However, big companies are interested in cooperation only when an application is near at hand. Also, they are not interested in cooperation related to improving production processes, marketing, or customer feedback. A senior FMP executive described the attitude of big companies as, 'Let's get the good stuff!' In other words, from the perspective of this venture company, the philosophy of large companies is still autarkic, 'let's innovate through in-house R&D rather than ongoing partnerships with SMEs'.

Projects with universities are oriented toward developing new core technologies or future applications of existing technologies, for example, applications to gene and protein sequencing and to PCB wiring patterns. Some of these collaborations have received public funding.

FMP has also benefited from direct government R&D support. However, although this support has been valuable in the past, FMP has decided not to receive any more. Too much time is consumed in applying for these grants and in completing mandatory annual reports. Also these subsidies are for research and early stage development work, not for product commercialization which is where FMP feels it must devote most of its efforts. Its CEO remarked that too often receipt of public funds becomes an end in itself for venture companies, and this hurts the companies' efforts to develop and commercialize new technologies (a sentiment echoed by Interface Chip).

Investment by a major trading company accounts for two-thirds of FMP's paid in capital. A senior executive of the trading company is on FMP's board of directors and provides management advice. From 2004 a trading company employee will be seconded to FMP to assist in management. This commitment is an example of the prominent role that some trading companies are beginning to play in the promotion of Japanese ventures.

FMP has fifty issued Japanese patents or patent applications and about eight foreign patents or patent applications. The costs of patent applications and patent maintenance force the company to be selective concerning which discoveries it patents. FMP regards patented inventions as complementary to production know-how. In other words, patents and trade secrets are complementary mechanisms for protecting its technologies from being copied by rivals.

FMP's biggest problem with respect to IP is that large companies can afford to apply for many more patents. In particular, the numerous process patents[27] of large companies restrict FMP's ability to manufacture or use its own discoveries. One way to counter this problem is to engage in collaborative R&D with these larger companies. As a result, the larger companies sometimes adopt its technologies and purchase its products. Whether this in the long run will enable the larger companies to duplicate or leapfrog FMP's key technologies, or lead to FMP's technologies becoming widely used throughout the industry while FMP preserves its innovation edge and market position, is an open question.

In any case, FMP's patent applications have served to attract potential customers and collaborators. Companies that approach FMP regarding collaborative relationships are well versed in the content of FMP's patent applications. As indicated above, negotiations become serious when the other party realizes that FMP has unique patented technologies such as plastics with low thermal expansion coefficients or micromachining and composite materials technologies to make reflective microspheres.

As with other venture companies I interviewed, it is wary about large companies copying its technologies, sometimes under the pretense of negotiating a cooperative relationship. However, the only specific situations it mentioned were foreign engineers from overseas subsidiaries of Japanese companies who would be sent to FMP and other Japanese venture companies on educational or quality inspection trips.[28]

Fresh Air Catalyst

Formed in 2002, Fresh Air Catalyst (FAC) is another GRI startup, but its history shows that not all startups from major GRIs or universities receive generous support. Rather than R&D, it is focused primarily on sales of catalysts[29] developed in the GRI that can purify air by absorbing contaminants. Its main products include catalysts embedded in building construction materials or air purifiers to neutralize impurities such as aldehydes that may affect health or cause odors. The founder was a GRI scientist in his mid-40s who has expertise in manufacturing powders containing the catalyst. The company was formed largely in response to requests from outside companies for samples of the catalyst. Since the GRI researchers could not manufacture the catalyst in sufficient quantities, FAC was formed to do so.

In 2003 total sales were about US$400,000. It has sent samples of its products to over 500 companies. It is engaged in cooperative development with about thirty. As an example, one of these is a major construction company with which FAC is developing a coating material for interior finishings.

A prefectural industrial research center lets one of the FAC employees use its electron microscope. The GRI continues to support the company by loaning it laboratory space, reducing normal license fees by 75 percent, and granting the founder conditional permission to work as a director of FAC.[30] Otherwise there are no collaborations with academic institutions.

The company has eight employees, three of whom are responsible for manufacturing the catalyst powder. Most of these were recruited by personal contact with the founder. In addition, five persons are hired on a contract basis to help with management.

To date, all the capital has been contributed by the founder and the current CEO. It has not received funding from private venture capital or angels. According to the founder's frank assessment, private investors say that the employees are motivated by their own research interests but not by business concerns ('The employees just want to enjoy their work.'). Private investors might be willing to invest if they could have a role in determining corporate policy, but so far FAC has not been willing to accept this condition. Neither has FAC received subsidies from government agencies, although it would be willing to accept these.

FAC's IP consists entirely of nonexclusive licenses covering its core technology from the GRI. At least one other company has received a similar license. However, FAC believes that the patents do not disclose all the know-how necessary to manufacture the catalyst.

In the future it wants to pursue R&D jointly with the GRI to develop a new generation of air purifying catalysts. Currently it has neither the human nor financial resources to pursue such development. It hopes to obtain its own patents, but it is concerned about the application and maintenance costs.

Internal Search Engine

Internal Search Engine (ISE) was founded in 2001 by a 40-year-old software engineer, who had spent his previous career developing supercomputer control and information management systems in one of Japan's largest computer manufacturers. Early in his career, his employer sent him to Stanford University for graduate studies, where he became interested in forming his own business, similar to the case of Interface Chip's founder.

Perceiving the need for better information management and retrieval systems within businesses based on networked personal and minicomputers, the founder has focused his company on developing customized information search engines that will link various types of internal corporate databases and file systems, while accommodating their various security systems. Most of ISE's customers are large Japanese manufacturing and service companies. It has installed its information retrieval system on over 500 main corporate servers. Revenues have increased year by year and are expected to be about US$7 million in 2004.

When the founder decided to leave the large computer manufacturer, almost all Japanese electronics companies felt they were in a crisis state. His former company is known for encouraging its employees to form their own *independent* companies. This policy is intended to stimulate the development of technologies lying dormant in the company and to better manage its human resources—in the expectation that those who leave will create a friendly network that will benefit the parent in the long run. Company managers proactively approach employees with entrepreneurial inclinations to ask if they would like to form a company, a practice that the CEO thinks is rare in other large manufacturers.

In ISE's case, its parent nonexclusively licensed two key software patents to ISE[31] and contributed 49 percent of its approximately US$100,000 initial paid in capital. Since formation, the parent's share of ISE's now US$1.3 million paid in capital has fallen to about 35 percent. The parent holds one of the four directorships. However, the parent lets the founder determine company policy. It does not try to dissuade ISE from cooperating with other major electronics companies that compete with it. If ISE fails, the founder will not be able to return to his former job. From the parent's perspective, its main interest in ISE is the value of its stock holdings and the annual royalties that ISE pays for use of the two patented software inventions.

Nevertheless, informal ties to its parent continue to be important. ISE has access to the parent's affiliated companies and it can make use of its sales network. It has more collaborative R&D projects with the parent than any other company, because its employees are familiar with their counterparts in the parent.[32]

This laissez-faire approach by the parent toward its spin-offs is still unusual in Japan. More often, the parent will maintain a dominant equity position and control the board of directors. It will have de facto veto power over key business decisions. Many key personnel will be transferred from the parent and may be able to return to the parent should the spin-off fail. In other words, as discussed in Chapter 6, the typical spin-off receives more ongoing support from the parent but also has less independence.

Despite this entrepreneurship-friendly approach by the parent and ISE's success so far, few of the founder's colleagues are following his example and forming their own companies. He attributed this in part to the traditional pension system in large Japanese corporations. Under this system, a person who retires before age 50 receives only a small one-time payment from the company.[33] Between ages 50 and 60, pension rights accumulate rapidly reaching a plateau at 60, part of a policy to require all but senior managers to retire. However, this company changed its pension system in 1999, so that now retirement benefits are accumulated linearly for persons hired after 1999. Similar pension reforms have been adopted by other major electronics companies.[34] Thus for young employees, loss of unvested pension benefits should no longer be a major reason not to form a venture company.

However, social pressures against joining small companies also play a role. Echoing the perspective of Interface Chip and other companies, the founder said that parental pressure for young graduates to join large, established organizations is still intense and pressure from wives and other family members for husbands and sons to keep working in large companies is a major barrier against more persons joining venture companies. The founder credited his experience at Stanford with providing sufficient inspiration to swim against these social currents. But few others are following his example.

Indeed hiring capable employees is probably the main challenge facing ISE. Its staff has grown from seven to twenty-five since founding. About twelve of these are technical personnel, five from the parent, five from other companies, and two recent university MS graduates. ISE receives many applications from persons in their 40s, a few of whom are suitable for management positions. Nevertheless, it is difficult to find good managers who have both business and technical experience. It needs more young software engineers, but applicants are few. 'Not enough talented young people are willing to work in venture companies.'

When asked about the possibility of hiring Chinese, Korean, or other foreign young engineers, the founder noted that the parent was hiring such persons, but he was concerned about their willingness to remain with his company and the possibility of proprietary technology leaking to foreign rivals. He said that, so far, not many employees had left ISE.

ISE is engaged in collaborations with several major Japanese electronics companies, including its parent, mainly on ways to match its software with the manufacturers' hardware. It collaborates with other venture companies, particularly on ways to bundle and market software. Cooperation with universities so far is not important for its core R&D. However, it is engaged in a government-sponsored project with a major private university. For ISE, the main benefit of this collaboration is access to government funds, although contact with professors and graduate students is also helpful.

ISE has applied for two patents on its own inventions. These patents and the licenses it received from its parent are important to prevent infringement of its discoveries by competitors and also to obtain private venture capital.[35] Roughly one-third of its paid in capital is from five venture capital funds. The lead investor among these five funds conducted a thorough review of ISE's IP prior to its investment decision. It insisted on a clear division of ISE's and the parent's assets and argued strenuously (but ultimately unsuccessfully) that ISE have an exclusive license to the two patents it licensed from the parent.[36] Ultimately it decided to invest in ISE after it was persuaded that ISE's main competitors would not receive nonexclusive licenses to these patents—at least until ISE was able to apply for patents on its own related inventions.

Phoenix Wireless[37]

Phoenix Wireless was formed in 1990 by a young electrical engineering graduate interested in the applications of biological neural networks to signal processing and who had first worked in an image processing venture. The company began developing hardware and IC chips to handle wireless communication. It excelled in miniaturized systems that consumed low power, systems such as matched filters that could process and switch between analog and digital signals.

Because of this expertise and its applicability to mobile communications, the head of a large partially-state-owned mobile communications company (SoMoCo) invited Phoenix Wireless to become a key participant in a pioneering project organized by SoMoCo to develop third generation (3G) wireless communications technology. As part of the participation agreement concluded about 1995, SoMoCo was granted co-ownership over some of Phoenix Wireless's inventions arising within the scope of this project. Large electronics companies that had close relationships with SoMoCo's parent were also involved. Two of these companies had supplied SoMoCo and its parent with IC chips and other electronic components and one, in particular, stood to be major supplier of chips for the new 3G wireless system. According to Phoenix Wireless, its IC chips offered several advantages over those of other companies[38] and it was viewed as a rival upstart by the two large electronics companies.

About 1999, the leadership of SoMoCo changed. According to Phoenix Wireless, the previous head was somewhat of a maverick who had supported the participation of a small company with innovative technology. But he was replaced by a more traditionally minded person, who had a long association with the two large electronics companies and felt a bond of loyalty to them. These two companies convinced the new head of SoMoCo to diminish the profile of Phoenix Wireless in the project, and in effect require that Phoenix Wireless participate in the project only through their own business and engineering staff. At this point Phoenix Wireless began to consider collaborations with other companies on different projects. However, because some of its key patents were co-owned by SoMoCo and other companies, it could not engage in collaborations that might involve licenses of its patented technology to new collaborators.[39] SoMoCo, on its own or at the behest of its large partners, would probably have blocked such transfers. In addition, with large rivals free to practice some of its important discoveries, its competitive position was undermined. Thus Phoenix Wireless found its business options limited in the field it had pioneered.

Phoenix Wireless withdrew from 3G R&D about 2001 and began to reorient its business to communications networking and content delivery. It had a successful IPO, but by 2002 its stock had plunged to one-tenth its peak value. It acquired a PHS[40] company and in 2004 was trying to exploit opportunities to use PHS phone and Internet communication. After investing in this new line of business, it was still in the red but hoped to regain profitability.

Salient Findings from the Nonbiomedical Case Studies

Although my sample is small, some points recur so frequently that they likely apply to many other ventures.

1. Foreign companies have been the initial lifeline for many of these companies as large domestic companies hold back from engaging young companies.

Interface Chip, Nanofilm Applicator, Chip Connect, and Big Crystal all obtained their first big breaks through sales to foreign (mainly Korean, US, and Taiwanese) companies. These companies also happen to be among the most successful companies in my sample in terms of sales. Even Chip Detect and Molecule Visualizer, two companies working on pre-commercial development of advanced instruments, relied on foreign companies for their sales of prototypes or for the signal that commercial interest existed in what was previously academic work.[41]

This is consistent with Japanese companies' tendency to innovate autarkically and to be cautious when adapting new technologies from companies with which they do not have long-standing relationships. Only when the potential of outside technologies has been validated by foreign companies do large Japanese companies feel inclined to use them. Although there are indications that large companies are becoming more open to dealing with independent ventures, the degree of openness seems insufficient to significantly boost growth opportunities for high-tech ventures.[42] In contrast, large US companies are probably more open to collaborations with ventures in ways that promote the ventures' independent growth.[43]

From a policy perspective, if Japan wants to promote the growth of its independent ventures, it should maintain liberal policies regarding trade and international investment (including the transfer of IP rights), even with respect to companies it perceives as foreign rivals. For their part, overseas investors and high technology companies should realize that independent Japanese ventures with promising technologies are likely to be receptive to business partnerships.

2. Overseas experience was important in the decision of the founders of nearly half these companies to become entrepreneurs.

The founders and CEOs of both Big Crystal and Internal Search Engine had worked in big companies which sent them to Stanford where they caught the entrepreneurial spirit. The CEO of Interface Chip worked in Hewlett Packard in California before founding his company. The CEO of Nanofilm Applicator grew up in China and soon after coming to Japan realized the potential of combining engineering talent from his own country with high quality Japanese production processes.

3. Only a few persons leave large companies to work in ventures. Nevertheless, large companies are the main wellsprings of managerial and research talent and technology for these venture companies.

Despite the still pervasive system of lifetime employment, people are leaving the security of big companies to join ventures and bringing with them skills and innovative ideas. This limited flow of human resources and technology probably is driving the most successful ventures. For example, three of the four leading ventures in terms of sales, Interface Chip, Chip Connect, and Internal Search Engine (ISE) were formed by entrepreneurs who left established companies. Chip Detect, although it has virtually no sales, was formed in the same manner, and in the process quite a few researchers left the parent to join the venture.[44] Interface Chip, ISE, and Chip Detect also recruit most of their R&D personnel from large companies. Molecule Visualizer and Fine Molded Plastics recruit partly from established companies. In drawing managerial and research talent from large companies, ventures obtain employees with relevant experience and they do not have to expend resources on training.

However, the number of scientists and engineers in large corporations who would seriously consider leaving to join ventures is still small.[45] Most of the ventures are having trouble recruiting the talent they need.[46] In order for more ventures to grow quickly, the system of lifetime employment in high technology manufacturing companies will probably have to change. As discussed in Chapter 7, this seems unlikely in the short term. In ten years or so, this may change as the pension reforms mentioned above begin to apply to a cohort of experienced workers. Also as the number of Ph.D. S&E graduates increase, an increasing number will find employment in large companies. As they gain experience, they may be more willing to move to ventures and may become an important source of skilled researchers and managers for ventures.[47]

4. In contrast, the ventures born out of GRIs[48] (and by extension universities) tend to be small. For various reasons, universities and GRIs may be even less fertile environments for high technology entrepreneurship than established companies.

5. IP is vital for high technology ventures, and, despite shortcomings, the IP system can protect their core interests.

Seven of the ten companies said that patents or exclusive licenses are very important, because they prevent copying by rivals and help them to raise VC. Protecting against infringement also attracts customers and secures international markets.[49] More than half the companies want knowledge about their IP positions to be widely known.[50]

Protection of know-how as trade secrets is also important to prevent copying. But on balance, the interviews suggested that patents are at least as important as trade secrets, especially if the company is trying to obtain venture financing.[51]

I expected to hear more complaints about the patent system. There are fewer patent attorneys per capita in Japan compared with the USA, which might increase barriers to small companies prosecuting infringement suits.[52] Awards in infringement cases are probably low compared to the USA,[53] which might diminish the value of patent protection.[54] Other researchers have reported cases of large companies stealing technology from SMEs, apparently feeling immune from threat of legal prosecution.[55] There are documented reports of SMEs not taking legal action in the case of infringement or acceding to unfavorable licensing terms with large companies because they do not have funds to engage in litigation or to amass large numbers of patents to use as ammunition or bargaining chips in litigation.[56]

All the companies I interviewed were concerned about rivals copying their technology. But despite the problems noted above, all thought that patent protection was necessary to prevent such copying, although a few considered trade secrets to be a more important method.[57] Some of the companies are concerned that by disclosing their technology in patent applications, they provide competitors with clues how to design around their patents. Others complained about the cost of obtaining patent protection. However, these concerns are heard with respect to most other patent systems.

A sample of technology-focused IP cases brought before the Tokyo District Court did not suggest that small companies were involved in a disproportionate number of suits either as plaintiff or defendant. In particular, small companies were not alleging infringement by large companies with disproportionate frequency, nor were they being sued by large companies for infringement with disproportionate frequency.[58] Nevertheless, the case of Phoenix Wireless is a warning that large companies may take advantage of opportunities to use the patent system to squash small companies that they perceive to be rivals and that co-ownership of their patents by large companies places ventures in a vulnerable position.[59] Also prior use rights under Japanese Patent Law are broad in comparison to at least US law. Large companies and their affiliates can disregard a venture's patents if they can show that they were using or preparing to use the patented technologies at the time the venture filed its patent applications. This encourages large companies to protect their technologies by trade secrets rather than by patents and weakens the strength of venture's a patents in any dispute with large rivals. Large electronics companies are even urging that the requirement to show actual prior use be abolished.

To put the importance of patents in perspective, it was clear that these companies believed their competitive advantage rested mainly on their technical expertise and ability to satisfy customers. Also, when I asked, 'What prevents potential competitors from copying your technology?' with one exception, they did not respond, 'Our patents, which enable us to sue infringers.'[60] Instead most said they feel they have substantial expertise in a particular technical niche and it would not make business sense for large companies to replicate their technologies, even though may have the resources to do so. Nevertheless, almost all the ten companies view their patents as a means to stake out their fields of technical expertise and business focus and as a deterrent against encroachment by competitors.[61]

6. Private VC companies have played an important, but not always lead role, in the early stage financing of many of these companies.

At least six of the ten companies received VC investments. Although in some cases, the VC investors were public investment corporations and in others they were 'betting on a favored horse',[62] in the case of Chip Connect and Nanofilm Applicator, private VC companies (in one case an overseas company) appear to have taken the lead in supporting companies that previously had little outside investment. Even when the VC companies were 'betting on a lead horse' they sometimes provided valuable support to the target ventures.[63]

7. Trading companies[64] are playing an important role in supporting some ventures.

Their level of direct equity investment in startups may be exceeded only by that of independent and government-affiliated VC funds and manufacturing corporations.[65] Sometimes they second their own personnel to bolster management of their portfolio companies, as in the case of Fine Molded Plastics, and to assist universities in founding startups. Although they generally do not have in-depth technical or entrepreneurial expertise, they have extensive marketing networks as well as networks among manufacturing and financial companies.

8. The record of government support for high-tech ventures is mixed.

The most successful companies in terms of sales (Interface Chip, Nanofilm Applicator, and Chip Connect) expressed skepticism about the value and fairness of government support programs. Even Fine Molded Plastics, which has obtained such support in the past, has decided that it is not worthwhile to continue to receive it, citing distraction from its main business R&D focus and the time necessary to satisfy paperwork requirements. Nonetheless, two ventures that seem to be based on interesting pioneering technologies, Chip Detect and Molecule Visualizer, have benefited greatly from government assistance. These companies probably could not have been created or advanced as far as they have without such support. Whether these will ultimately prove to be wise investments of public funds remains to be seen.

9. Many ventures have a few focused collaborations with universities.

Most of the ten companies are collaborating with a few professors whose work is relevant to them. Sometimes the outcomes have been very helpful.[66] But sometimes the main attraction has been the ability to leverage government funding earmarked for university–industry R&D partnerships.

PART II: CASE STUDIES OF BIOMEDICAL VENTURES

I have been studying Japanese bioventures since 1999. In 2000, I interviewed nine of these companies, a significant proportion of the total at that time. I have continued to follow them and have had in-depth interviews with many more. This part presents studies of ten bioventures. With some exceptions, I have purposely selected companies that are well-known so that this book does not reveal inside information. Nevertheless, I hope these case studies illustrate some of the issues common to many bioventures. I have also selected companies that present a cross-section of activities such as instrument development, research services, drug discovery technologies, and actual drug discovery and development. However, the latter three categories are not distinct. A common strategy is to focus initially on providing drug discovery or other contract research services, in the course of which the bioventure will identify a few drug candidates that it will develop on its own. Unless otherwise indicated, information on the ten companies is current as of late 2004. However, the concluding 'salient points' are based on the situation in late 2006.

Bioventures Focused Primarily on Drug Development

Gene Angiogenesis

Gene Angiogenesis was founded in 1999 by a faculty member of one of Japan's leading national universities to develop hepatocyte growth factor (HGF) as a drug to increase new blood vessel formation near arteries that have been blocked by atherosclerosis. It is also developing a new method of delivering the gene that codes for HGF to cells near blocked arteries. The founder did some of his early related research at Stanford University, where he became imbued with entrepreneurial culture and began to think seriously about forming his own company. Soon after forming the venture, he recruited a senior manager from the pharmaceutical industry who was instrumental in helping to build the company.[67] There was little VC investment in this company prior to its IPO in September 2002. The main sources of funding were a small chemical company that is now producing a gene therapy kit containing the new delivery vector and a large Japanese pharmaceutical company.[68] The latter alliance involves substantial support for the bioventure while enabling it to maintain its independence. It is almost unique in Japan.[69] In addition, the bioventure benefited from skilled researchers in the national university hospital who worked on projects related to the venture.[70]

The IPO in September 2002 was very successful, and within six months the venture's market capitalization had risen to over US$700 million, quite high in comparison to US bioventures that had IPOs about the same time. This was probably only the third Japanese biomedical IPO since 1980 and only the second since the beginning of the reforms described in Chapter 3. Moreover, it was the first ever 'university venture' IPO in any technology field. So far it is the only IPO for a bioventure focused from birth on drug discovery and development. It triggered widespread interest in university-based bioventures. It led to the establishment of biomedical VC funds in a large number of financial institutions and even some manufacturing companies. However, the establishment of the first biomedical VC funds had occurred earlier and by the time of this IPO, a number of these funds already had other bioventures under their wings (see Chapter 5).

Currently, Gene Angiogenesis's HGF gene therapy drug is in the final phase of clinical testing needed for approval to treat peripheral artery disease. It has also begun clinical trials to treat ischemic heart disease. If the drug is shown to be safe and effective, and especially if it turns out be widely used to treat patients, this will be a triumph not only for medical science but also for Japan's fledgling bioventure industry. Even if the drug does not pass these final hurdles (and considering the pioneering nature of the therapy, there are many possibilities for things to go wrong), it ought to be an inspiration to other bioventures and the scientists and financiers who support them, because it shows that the ingredients for success can be brought together in Japan.

Looking toward the future, the company is trying to acquire new drug leads.[71] It reportedly has had difficulty obtaining additional new drug leads from its home university.

Cycling Against Cancer (CAC)

Cycling Against Cancer was formed in 2000 by MD/Ph.D. graduates from two relatively unknown universities, who felt that the control senior professors exert over scarce research funds prevented them from pursuing new methods of cancer therapy within their home universities. They set up their laboratory and headquarters remote from either university but close to the family business of one of the founders, which provided initial funding and also management advice. This venture's main focus is the development of drugs to disrupt the growth cycle in cells, drugs that will increase the effectiveness of mainline cancer drugs to trigger the self-destruction of cancer cells. Early stage clinical trials have begun.[72]

Lacking close ties with a major university laboratory, CAC has had to be more self-reliant than Gene Angiogenesis. In 2003 the company employed twelve persons, including six scientists (four with Ph.D.s). One is a foreign national. Three have pharmaceutical industry experience suggesting some movement of R&D staff from established companies to bioventures.[73] The company maintained that human resources have not been a major bottleneck, although it would like to employ more persons with postdoctoral level experience. The management team does not have pharmaceutical experience, but in 2005 it employed a CFO experienced in the management and financing of bioventures.

Cycling Against Cancer appears to be leveraging well the resources offered by Japan's fledgling VC funding industry. By 2005 it had gone through three rounds of VC financing and raised over US$40 million. The lead investor in its initial funding rounds was a Japanese VC company that is targeting early stage bioventures and is committed to helping build the companies in which it is the lead investor. That VC company holds one of the outside directorships of CAC. The other outside directorship was held by another investor, one of Japan's major trading companies. Although lacking technical expertise, such trading companies have business networks that have been helpful to ventures such as CAC. At the peak of Japan's biotech IPO bubble in 2003–4, these lead investors seemed to be counseling a deliberate long-term development plan, rather than pushing the company toward an early IPO.

CAC regards patent protection as absolutely essential to its business. The various VC investors did not invest until after the company had commissioned a search of patent literature to try to make sure there were no competing patents. It has vigorously pursued Japanese and international patents and has recently been issued US and European patents covering its main drug and also a screening method to find similar drugs.

Unlike Gene Angiogenesis, Cycling Against Cancer has not found partners among Japanese pharmaceutical companies. The CEO remarked that Japanese pharmaceutical companies 'are skeptical about new technologies, while US companies won't consider you unless you offer something different'. CAC is hoping for an alliance with a major pharmaceutical company that would market its drug, but it does not want to lose control over its main technology, except under favorable terms.

CAC has entered into a collaboration with a major optico-electronics company to develop a test to screen patients' cancer cells to determine the stage of cell growth at which various anticancer medicines are effective. It is collaborating with a medical university to screen various compounds for their ability to stop cell growth at particular points in the cell development cycle. It is also cooperating with the pharmacology department in a local university to prepare its drug for clinical trials.

Leave Development to Us (LDU)

Leave Development to Us (LDU) is one of the first of an increasing number of companies whose main drug development strategy is based on in-licensing therapeutic compounds not being actively developed and then outsourcing the work necessary to complete development and market the compounds. It was formed in 2001 by a businessman with many years experience in pharmaceuticals. The company has only a small full-time staff. It does not have laboratory facilities. However, it has two drugs in early-stage clinical trials in the USA. It has about ten other active drug development projects. One is for a new class of drugs it designed in-house to treat ulcers and prevent gastric cancer. Some of its candidate drugs originate in universities. However, the majority originate in corporate laboratories, but for a variety of reasons the companies stopped development and later transferred rights to LDU. Most of its drug leads are of Japanese origin.

LDU has a network of university and corporate researchers who can refine drug candidates. It has connections with companies in Japan and overseas that can manufacture drugs to GMP[74] standards and with other companies that can perform the preclinical animal trials necessary to satisfy the requirements of FDA and MHLW to begin human trials. Finally it has connections with CROs that will manage clinical trials either in Japan or overseas.

By mid-2004, after two rounds of VC financing, LDU had raised enough private funding to take its first two drugs into clinical trials.

Bioventures Primarily Focused on Instruments, Contract Research, or Drug Discovery Technologies

Mother of Bioventures (MBV)

Before there was Gene Angiogenesis, before there was Cancer Answers, there was Mother of Bioventures (MBV), one of the first Japanese bioventures in modern times and the first to have had an IPO. MBV was formed in 1969 by a graduate of the social welfare department of one of Japan's main national universities. After working as a technician for three years in a respected prefectural cancer research center, she organized seven antibody experts (mostly immunologists in their 30s, many with US training) to form a company specializing in making customized antibodies for diagnosis and research. In 1977 it began to focus on diagnostic antibodies for autoimmune diseases. By 2000 MBV had 90 percent of the Japanese market for clinical applications in this field. That same year, it had about US$28 million in annual sales and about 190 employees, 40 in R&D and most of the rest in manufacturing or sales. In 2004, its sales were about US$37 million and trending up, and its profits were approximately US$1.5 million.

MBV attributes its success to moving quickly into a relatively new market, having good quality control, offering its customers total product and service support, and reaching out to the larger biomedical community. It also has recently acquired a new method for making human antibodies.[75]

Until 1981 the company was entirely self-financed with capital contributed by employees and their families. In 1985, it received funding from an angel investor who owned a golf course. In the 1990s it began to receive equity investments from a small chemical and diagnostics company, an insurance company, and some other financial companies. Because the company usually ran deficits, banks would not lend it money. In 1996 the company had an IPO on Jasdaq, the first IPO for a Japanese bioventure. The company has over fifty patent applications, and it considers patent protection for its technology very important.

MBV has R&D collaborations with over ten other biotechnology companies and universities both in Japan and overseas (mainly in the USA). However, aside from the recent purchase of the genetic reagent and diagnostics arm of a large metals company[76] and collaboration with a spin-off from a midsize pharmaceutical company,[77] MBV appears to have no significant collaborators among pharmaceutical or other established Japanese companies. When asked why, the response, like that of so many other Japanese ventures, was 'Japanese pharmaceutical companies do not know how to evaluate our technology.'

MBV also has agreements covering exclusive distribution rights and cross licenses with overseas pharmaceutical and biotechnology companies, some of which are its main competitors. It is trying to build links to overseas researchers. It has established a US diagnostics subsidiary and a separate US research and sales subsidiary. It has sent some of its own researchers for training in US and Canadian universities. However, as of 2000 only 5 percent of its sales were overseas.

MBV has been actively reaching out to the academic research community and newer bioventures. It sponsors annual symposia on immunology and cell biology that are attended by leading scientists. These and similar meetings let the company know about targets against which it should develop antibodies.[78] It provides quality control testing for hospitals and research institutions, including nearly 150 overseas. Along with an international licensing consulting company spun off from a Japanese pharmaceutical company, it has established a VC company that is one of the most active in supporting early stage bioventures. Finally, MBV has itself entered into alliances with some of the newer Japanese bioventures.

DNA Extractor[79]

DNA Extractor was established in 1985 as a provider of maintenance services for clinical testing equipment. However, it transformed itself into an R&D company developing machines that can extract DNA from cells for subsequent analysis more quickly and using smaller sample sizes than traditional methods. The first machines were built in 1996, but the company's big break came in 1999 when a large European pharmaceutical company began to purchase its machines. Since then it has added more customers, almost all overseas.[80] It raised about US$15 million in an IPO in 2001, the first biomedical IPO since the beginning of the reforms. This cash infusion enabled the company to develop complementary technologies that are combined in a single machine that extracts DNA, purifies it, amplifies it, and then detects whether it contains particular genes of interest. By the end of 2003, over 1,500 of the DNA extraction instruments had been sold. Worldwide sales were approximately US$20 million and rising, and the company had turned a profit. Informal collaborations with university researchers have been important for development of the company's technologies.

Designer Mouse[81]

Designer Mouse, founded in 1998, arose from collaboration between a medical school professor[82] who had invented a new method to make knockout mice, and the heir to a company in western Japan whose main business of breeding laboratory animals had fallen on hard times. Knockout mice are grown from embryos that lack particular genes as a result of genetic manipulation. The functions of these genes usually are not known, but studying the knockout mice enables determination of their function. Thus, knockout mice are useful tools to discover genetic causes of various diseases, design drugs to treat these diseases (by studying the effect of the proteins the genes ought to produce), and predict how the presence or absence of particular genes might alter the effects of particular drugs. The technique invented by the professor is quicker than previous conventional methods for making knockout mice.[83] Several of the professor's university colleagues have also been collaborating with the company. The key patents on which they are inventors have been assigned to Designer Mouse.

Designer Mouse's survival depended on being able to sell to pharmaceutical companies. In its early years, many pharmaceutical companies expressed skepticism about its technology and business methods. They thought they could duplicate its technology in their in-house laboratories. Some thought of using the mice primarily to screen existing prototype drugs rather than for targeted drug discovery. They generally did not like Designer Mouse's stipulation that it co-own the genes it discovered in the course of commissioned research. They did not like the company's demand for milestone payments and royalties on the sale of drugs discovered using its mice. However, gradually the company began to make deals with pharmaceutical companies. As of 2004 it was selling mice to at least one Japanese pharmaceutical company. Three other companies designated particular genes they wanted knocked out, and Designer Mouse made mice, tested them, and then provided information on the functions of the genes to the companies.[84] Universities and GRIs are also among its customers.

Because its own laboratory cannot produce enough mice to meet demand, it has licensed its technique of producing knockout mice to two pharmaceutical companies and also a GRI. Revenue for 2004 was over US$5 million, nearly double its revenue for the previous year. It has over seventy employees. However, expenditures have also been increasing and as of 2004 the company had not turned a profit.

The initial equity investors were the cofounder's family's company and several insurance and traditional VC companies. It had an IPO in December 2002. Most of the funds it raised have been plowed back into its R&D.

Gene Chip Research (GCR)[85]

Gene Chip Research (GCR) was founded in 1999 by one of Japan's foremost genetic researchers, who had recently reached the mandatory retirement age in one of Japan's leading national universities. He remains keenly interested in the genetic basis of cancer. By clarifying what genes are turned on and off during the progression of cancer and other diseases, he hoped to find out the functions of many genes and their relationships to disease. He also intended to put this information in a database as a resource for Japanese biomedical researchers, who previously could only obtain such information from expensive proprietary US databases.[86] His strategy was to form a company that would develop chips containing gene sequences that would be useful for genetic research and possibly also clinical diagnosis, chips that would be more accurate and less expensive than chips available from other companies.[87]

The founder (and CEO as of 2004) obtained initial funding from an angel near his former university and also from a subsidiary of a major electronics company specializing in biomedical instruments and related software development. This subsidiary designed and manufactured the machines to make and to read the chips, and it also markets the chips. This company owned nearly 70 percent of GCR's stock prior to an IPO in March 2004 and about one-third thereafter.[88]

VC companies did not contribute to initial funding of the company. The subsidiary of the large electronics company has remained the most important source of capital. The founder commented, 'Generally, Japanese VC companies want to recover their investments too soon.' Also, pharmaceutical companies were little interested in the founder's research until the Japanese government began in 1999 to provide large-scale funding for academic genomics research via the Millennium Project.[89] Even since then, collaborations with pharmaceutical companies have been few. However, GCR has been awarded quite a few government commissioned research projects. The founder has a wide academic network. Persons in this network have supported the company both in terms of R&D (e.g. contribution of samples and related medical data) and as investors. The company has numerous collaborative research relationships with academic institutions.

Sales of the chips have steadily climbed from about US$3 million in 2000 to about US$11 million in 2003, but these levels remain far below its large overseas rivals.[90] Revenue from contract research also increased from about US$3 million to US$6 million over the same period. GCR's main customers include the subsidiary of the electronics company (which resells to the wider research community), a small pharmaceutical company, a private biomedical research institute, a large academic medical center, and one of Japan's main GRIs. The value of its stock increased over threefold at the IPO, however, the post-IPO market capitalization is low in comparison with other bioventures that have had IPOs.[91]

In 2004, the company had about twenty-six employees including four Ph.D.s. Employment has grown little since 2000, when the founder remarked that it was very difficult for bioventures to recruit capable researchers. Reliance on the founder's personal connections was essential to recruit researchers.

Unique among all the bioventures in these case studies, GCR does not apply for patents.[92] The company obtains its revenue through contract research and sales of chips.

In summary, Gene Chip Research has been a vehicle for the founder to continue his lifetime research into the genetic basis of disease. It also provided an entrée for the subsidiary of a large electronics company to leverage public funding to enter the bioinformatics and bioinstrument business. Gene Chip

Research seems to have established a growing presence in a niche market for gene chips used in research. But other laboratories are making similar chips for research and GCR's chips are usually not used for clinical screening of patient samples.[93] Also, the company does not have significant sales outside Japan.[94] Finally, the subsidiary of the electronics company has not built on the progress of GCR to broaden its product base.[95]

Cancer Answers

Cancer Answers was founded in 2001 by one of Japan's leading geneticists and cancer researchers who is still active in a major academic medical research center. Already leader of a large-scale government project to identify and analyze variations within human genes,[96] the founder established Cancer Answers to identify genes that are linked to the progression of various cancers as well as other diseases such as diabetes. The company's business would initially focus on contract research to identify drug targets and molecular markers that might be used for diagnostics to predict individual responses to particular drugs.[97] As a second stage goal, the company would develop drugs aimed at some of the targets its own research had identified.

Like Designer Mouse, Cancer Answers had to find customers among pharmaceutical companies. For the first year it struggled to raise funds and find customers. However, beginning in 2002 it began to sign agreements with pharmaceutical companies, and it has now become one of the most successful bioventures in this respect. By mid-2004 it had secured contracts with at least five Japanese pharmaceutical companies to help them discover drugs[98] and with at least four Japanese companies to help them develop diagnostic tests.

Some of Cancer Answers' success is owed to its second CEO, the same person with high level pharmaceutical industry management experience who guided Gene Angiogenesis through its early years and put it on the road to a successful IPO. He joined Cancer Answers as a board member in 2002, shortly after handing over leadership of Gene Angiogenesis to another person. One year later, after securing the company's three initial contracts, he became CEO.

However, within the biomedical and venture communities it is widely believed that Cancer Answers' customers are attracted not just by the specific research the company will perform, but also by the prospect of increased access to one of the leading academic laboratories in the country.[99]

At its founding in 2001, Cancer Answers' lead investor was a VC fund that was also lead investor in some of Japan's other early biomedical university startups. Twenty months later, it had the most successful IPO in the history of Japanese bioventures and also of Japanese university startups. Indeed, it was one of the most successful recent IPOs for any bioventure worldwide. Within a few days of the IPO, its market capitalization was about US$1.3 billion.[100]

In 2004, Cancer Answers had between forty and fifty employees, almost all of whom are technicians or attending to business matters. Its scientific strength rested largely on university researchers who are not employees. The company has many patent applications related to genes. If these genes are discovered in the course of contract research, Cancer Answers usually will license rights to use the genes to the research sponsor. This provides it with a revenue stream and also enables it to retain some control over its core discoveries.

Computer Drug Discovery (CDD)

Computer Drug Discovery (CDD) was formed in 1999 by a biochemist with over twenty years' laboratory experience in a major Japanese pharmaceutical company followed by several years in a government-founded genome research corporation. CDD focuses on 'rational drug discovery and design', in other words, designing drugs based on information about the structure and function of genes and proteins. It relies on the expertise of its scientists who typically have backgrounds in genetics, protein chemistry, structural biology, and bioinformatics. It uses public genomics and proteomics databases and also a proprietary database and software package licensed from a UK company. CDD is developing its own data bases along the same lines. It has powerful computers but no 'wet' laboratory. Most of its current revenue comes from contract research to identify drug targets or potential drugs. It also licenses the databases already mentioned. Like many other research tool companies, it hopes to set aside a few promising drug targets it discovers to eventually develop into drugs on its own.

In 2000, the company estimated that only about three Japanese pharmaceutical companies had sufficient expertise to do such computer-based rational drug design. It considered the rest to be potential clients. However, they tended to pursue drug discovery by traditional methods such as screening and synthetic chemistry. The only collaborations with pharmaceutical companies were small scale, usually directed at identifying genes that may be important in various diseases. CDD also had collaborations with university medical departments and GRIs and has received government funding for at least one project. Its total revenue for 2000 was just over 80 million yen (about US$750,000).

By 2004, CDD had established more substantial collaborations with two of Japan's eight largest pharmaceutical companies. Its twelve-month revenue climbed to about US$2.5 million by early 2004, and it began to show its first quarterly profits. It had twenty-two employees, compared with nine, four years earlier.[101] It also has had what it describes as successful collaborations with US and European pharmaceutical companies. It has expanded its range of GRI collaborators, cooperating with one of Japan's major cancer research centers on diagnostic markers. On its own, it has identified candidate drugs and it has collaborations with two pharmaceutical companies to validate these compounds and to develop similar compounds.

CDD has had to work hard to obtain clients and funding. From its perspective, Japanese pharmaceutical companies are gradually beginning to outsource some of their data mining and drug discovery work, but they still tend to look down on Japanese bioventures and to assume they should turn to foreign bioventures for drug discovery partners. Lacking close ties to an elite university or a large company, a specific physical product for a niche market, and a manager with high-level pharmaceutical industry experience, CDD's situation is more typical of that faced by other research tool bioventures than that of the companies discussed previously.

CDD's initial investors were its founders, two of Japan's largest mainstream VC companies, and a major insurance company. Typical of early stage VC funding, these initial VC investments were small.[102] By 2004, CDD had not attracted substantial equity financing from other sources. The company continues to obtain project grants from government agencies.

Reverse Targeting

Reverse Targeting was formed in 2002 to develop a new method for using existing drugs to identify the molecular targets (receptors) of those drugs in the body, and thereby to develop other drugs that are more effective or have fewer side effects. The core technology was developed jointly by a physician-geneticist and a materials scientist in one of Japan's leading national universities and a polymer chemist in one of Japan's leading private universities. The core technology involves nanometer dimension beads to which molecules of a known drug are attached. The drug-covered beads are exposed to fluid containing extracts from cells.[103] In this way, the proteins from the cells that bind to the drugs[104] can be isolated and identified, and then the genes coding for the proteins can often be identified. In this way, the mechanism of action of a drug can be elucidated, and this may allow development of drugs that are more effective or safe than the original drug.

The management team is relatively young. The CEO has scientific training and worked for many years as a biotechnology expert and consultant in one of Japan's major securities firms. He played a key role in that firm's efforts to promote biotechnology investment. In this capacity, he became familiar with both the management and technology of many bioventures, as well as with the

VC companies investing in biotechnology. The chief technical officer (CTO) is a Ph.D. scientist with twelve years' experience in pharmaceutical R&D. He works closely with the company's scientists and tries to keep them focused on commercially directed research, often a challenge when the founders are respected academic scientists.[105]

The company has its own laboratory facilities rented from a suburban municipal government on the periphery of one of Japan's large urban centers.[106] In 2004, Reverse Targeting employed about fifteen persons including, in addition to its three-person management team, five MS or Ph.D. level scientists who shouldered the main burden of the company's scientific work.[107] Three of these were hired from established chemical and pharmaceutical companies, indicating, as in the case of Cycling Against Cancer, that relatively young scientists are sometimes willing to move from large companies to startups.[108] Most of the other employees are technicians.

Reverse Targeting has been fairly successful at obtaining VC funding. Initial investors included the CEO's former employer, two VC companies focusing on biomedical investments,[109] and the Tokyo Metropolitan Government's VC fund. Other independent and bank-affiliated VC funds followed suit and by mid-2004, the company had raised about US$10 million in paid in capital. It has also received government grants for research projects.

The company has begun to attract contract research from established companies. It licensed its beads to one of Japan's largest pharmaceutical companies. Then it signed a contract to screen the compounds of a chemical company. A private research institute has begun to evaluate its technology. Most recently it has signed an agreement with a US biotechnology company to screen its compounds for antiviral or antibacterial properties. However, most of its contracts are less than US$100,000 annually. Contracts with annual income streams above US$100,000 are still rare for Japanese bioventures.

Reverse Targeting is also pursuing its own drug discovery research by trying to clarify the body's natural receptors for several drugs that have been on the market for many years but whose mechanism of action is still not known.

Salient Points from Bioventure Cases

The above case studies, as well as information about other companies not presented, suggest the following points.

1. The unwillingness of pharmaceutical companies to engage bioventures is hurting them.

Most bioventures have struggled to find customers and development partners among pharmaceutical and other large companies. The exceptions are closely linked to elite university laboratories, are affiliates of large companies, have CEOs with high level pharmaceutical management experience, or are providing clinical medical services.[110] Aside from the company that invested in Gene Angiogenesis, I know of only two other major pharmaceutical companies that have significant partnerships with Japanese bioventures aimed at developing a bioventure's drugs.[111] For bioventures focusing on drug discovery technologies, it is not difficult to find customers who will pay a few US$10,000 to $100,000 for their services for one or two years. But it is rare for an established company to engage in a substantial collaboration involving over US$100,000 annually. When they arise, long-term partnerships are likely to be with non-pharmaceutical companies, perhaps representing efforts to move into a new field.[112] Most of the bioventures with which I am familiar say that this tendency of large companies to ignore them or to try to extract from them the lowest costs or most favorable licensing terms for a one-off deal is a major problem.

However, pharmaceutical companies typically maintain that the technologies and human resources of Japanese bioventures do not merit their support.[113] Therefore it makes sense for them to partner with overseas biotechnology companies and domestic and overseas universities, but not domestic bioventures. I cannot conclusively assess the accuracy of this perspective. A frequently-voiced shortcoming is that many bioventures have just a few therapies under development and their scientific or technical focus is narrow, often derived from the research of one key scientist. However, it is not clear that recently formed US bioventures are much different.[114]

Have US or European pharmaceutical companies behaved differently from their Japanese counterparts? There are indications that they have, even leaving aside well-publicized acquisitions in the hundreds of millions of dollars. In 1987, most young therapeutic-focused US bioventures had managed to establish at least one partnership with a pharmaceutical company. About half of these 'partnered' bioventures had licensed at least one of their therapies to a pharmaceutical company. Licensing and marketing agreements aside, the vast majority of 'partnered' US bioventures had at least one ongoing relationship with a larger company such as joint research, a joint venture, or an equity investment.[115] These latter types of partnerships suggest an ongoing supportive relationship with the large company—relationships that are very rare for Japanese bioventures.[116]

In addition, US and European pharmaceutical companies that are equivalent in size to large Japanese pharmaceutical companies each had more than twenty licensing partnerships with bioventures and an equivalent number of research or equity partnerships.[117] In contrast, partnerships between Japanese pharmaceutical companies and bioventures were lower by about a factor of five. The point is not that US and European pharmaceutical companies are five times more likely than Japanese counterparts to partner with bioventures.[118] Rather it is that US and European pharmaceutical companies do license discoveries from bioventures across a wide spectrum of development stages.[119] Also they enter into multiple year research partnerships with bioventures, sometimes on early stage technologies. On the other hand, early stage partnering that exceeds US$100,000 a year is rare between Japanese pharmaceutical companies and Japanese bioventures.

This still begs the question whether the Japanese bioventures' dearth of substantial partnerships is their fault or the fault of autarkic Japanese pharmaceutical companies. As of 2006, overseas pharmaceutical companies had entered into few partnerships with Japanese bioventures,[120] reinforcing the plausibility of the claim that Japanese bioventures have little of value to partner for. However, there are real barriers in terms of communication and different expectations regarding management, that make transoceanic partnerships difficult.

I am familiar with several Japanese bioventures that have not secured partnerships but that appear to have promising therapies under development. Their therapies are still in early stage development,[121] and some have only one or two drugs in their pipelines. Under some conventional guidelines it would still be too early to license their drugs. But US and European pharmaceutical companies have been flexible concerning in-licensing and R&D alliances with bioventures. From my perspective, it would be reasonable for overseas pharmaceutical companies to invest in some of these Japanese bioventures, provided good lines of communication were in place and there was mutual understanding on key management issues.

Regardless of which side is most to blame, the negative perspective of Japanese pharmaceutical companies on Japanese bioventures is a self-fulfilling prophesy. If large companies shun ventures believing that their capabilities are weak, promising ventures will have difficulty obtaining funding from other sources to improve their capabilities. Moreover, the bioventure sector as a whole will be a less attractive place to work and a less attractive target for investment financing. The comparison between Japanese and US bioventure sector employment at the beginning of this chapter is consistent with this vicious cycle already occurring.

The CEO of one of the bioventures featured in the case studies offered the following insights into the reasons for large companies' autarkic stance: Unlike many of their overseas counterparts, the home offices and laboratories of Japanese pharmaceutical companies have few employees responsible for finding and evaluating outside technologies and for managing collaborations with outside organizations. Moreover, their organizational structures do not facilitate building alliances with small companies.[122] No one with significant influence in these companies has a career that depends on establishing and maintaining alliances with Japanese companies, particularly small independent companies. In contrast, some overseas pharmaceutical companies have close to 100 employees responsible for establishing and maintaining alliances, primarily with bioventure companies. The policies of such companies are to actively seek out and maintain mutually beneficial alliances that will allow the pharmaceutical companies to increase their pipelines.[123]

2. Most bioventures arise from universities and many of these continue to depend on them for key discoveries. But universities are becoming stricter about separating company from academic research.

The dependence on universities is particularly great in the case of bioventures focused on drug development. With a few exceptions, such as Cycling Against Cancer, biostartups rely on the founders' university laboratories for most new drug discovery research. However, biostartups usually are required to obtain their own laboratory space for R&D specific to the needs of the startup. Many universities now have incubator facilities but rent is only slightly below commercial rates.[124] Startups and their founders are now cautious about graduate students working on projects intended to benefit the startup, but nevertheless, the academic research of graduate students often overlaps with the interests of the startups.

Finally, although the distribution of biostartups is skewed in favor of universities that receive a disproportionate share of funding, promising biostartups are arising from lesser known universities.

3. Behind the scenes, founding professors still control many startups.

Based on my knowledge of about forty companies, it is quite common for the founding professor to exert more management control than in the USA. One likely reason is the dearth of qualified managers with relevant scientific, technical or business experience. This means the venture has to rely on the founder to make technology-related decisions. Another reason is that venture capital investors lack expertise, influence, and inclination to insist on managers who will follow their directions rather than the founder's. However, the trend is toward professional managers to control the companies. The founders of Gene Angiogenesis, Designer Mouse, Cancer Answers, and Reverse Targeting are examples of university founders or cofounder having stepped aside and ceded real control.

4. Most bioventures are short of skilled personnel, but beginning around 2004 they began to hire persons with industry experience.

At the beginning of 2004, very few bioventure managers had high level pharmaceutical industry experience, and few of their research scientists had industry experience. The inability to recruit from pharmaceutical companies was a handicap in many ways. In addition to skilled managers and researchers, most bioventures needed persons experienced in developing partnerships with pharmaceutical and biotechnology companies, and, in the case of those developing drugs, in planning the preclinical and clinical trials needed to obtain marketing approval. However around 2005, in part as a result of mergers among pharmaceutical companies, bioventures began to recruit persons with pharmaceutical experience. Most of the bioventures with which I am familiar that were formed after 2004 do have CEOs with managerial-level pharmaceutical experience. Many of the larger bioventures have also been able to recruit mid-career research scientists from pharmaceutical companies. This is one of the most favorable recent developments for bioventures. The flow of skilled people from large companies is probably even less than in the case of the non-life science ventures, but it is a start and the trend seems to be increasing slowly.

5. Government support is pervasive.

Most of the companies with which I am familiar have received government support, typically in the form of cofunding for private venture capital investment and government commissioned research (often involving funds earmarked for the support of ventures). For some bioventures, government funding probably constitutes the majority of revenue. This raises concern about overdependence on such funding. However, government funding is insufficient to support a sustained, competitive development program. Growing companies must struggle hard to obtain investments, contracts, and sales from private sources.

6. Venture capital is vital for the growth of bioventures.

Venture capital (i.e. equity financing) has been the main source of early and mid stage financing for most bioventures, particularly those aiming to develop drugs. In 2004, few bioventures, other than those that had had IPOs, had revenue from sales or contracts over US$1 million. Of the pre-IPO bioventures with annual revenue over US$1 million, two at most were focused on development of therapeutics.[125] In the face of low sales and contract revenues, equity investment has been the main source of financing sustaining drug discovery and development. As IPO requirements have become more restrictive the reliance on VC funding has become even greater, as described in the next chapter.

7. Patents are important.

Patent protection has been important for almost all the bioventures, especially to obtain VC funding for companies without much revenue. However, even companies focusing on analytical or drug discovery technologies believe patent protection is important to obtain outside funding and protect their markets.

APPENDIX 1: GOVERNMENT MEASURES TO AID VENTURE COMPANIES[126]

Establishing Public Stock Markets with Listing Requirements Suitable for Venture Companies

1998: Relaxation of listing requirements for the over-the-counter securities market (Jasdaq) making it a viable vehicle for public issuance of stock in venture companies.

1999: Mothers opened as the third section of the Tokyo Stock Exchange with reduced listing requirements.

2000: Nasdaq Japan launched (but closed in 2002).

Loans, Research Subsidies and Other Grants for Venture Companies

1963: Small Business Investment Corporations (SBICs) established as public VC firms. They can provide up to 10 million yen of initial capital to small businesses without collateral or guarantees. Various local governments are also providing financial support.

1975: Establishment of the Venture Enterprise Center, which guarantees loans to SMEs and also collects data on venture investment.

1989: Enactment of the *Newly Incorporated Businesses Law*, under which companies recognized by MITI/METI are given preferential funding to promote their growth and foster new industries. There are about 150 companies receiving such assistance.

1995: Revision of *Law on Temporary Measures to Facilitate Specific New Businesses*, which provides direct equity investment and debt financing and also loan guarantees by government-affiliated corporations.

1995: Enactment of the *Small-and Medium-Size Business Creation Activity Promotion Law* under which companies recognized by local governments are given preferential access to funding to promote their growth and foster new industries. There are almost 5,000 such companies. Also under this law, venture business assistance foundations have been established in prefectures and certain cities. National and local governments deposit funds to these foundations, which in turn invest through venture capital companies in promising businesses in their respective areas. In addition to these indirect investments, the foundations occasionally directly invest in or lend to venture businesses.[127]

1998: The *Law for Facilitating the Creation of New Businesses* authorized government loan guarantees to startups without requiring them to provide collateral or guarantors when borrowing through the Credit Guarantee Association. It also established the SME Technical Innovation Plan, Japan's version of the US SBIR Program. Under this system, government ministries and agencies commission R&D projects to small- and medium-sized companies from among competitive project proposals. One of the main selection criteria is the potential for commercializing R&D results. This law was amended in 2000 to enable additional support to venture businesses planning IPOs.

2004: Consolidation of the Japan Small and Medium Enterprise Corporation (JASMEC), Japan Regional Development Corporation, and Industrial Structure Improvement Fund, into the Organization for Small and Medium Enterprises and Regional Innovation, Japan (SMRJ), to rationalize many of the METI-affiliated programs to promote SMEs and regional business development. Government VC companies such as the Tokyo Metropolitan Small and Medium Enterprise Investment Corporation are now under the SMRJ umbrella.

Incubators

Publicly financed or cofinanced business incubators have been established in many locations, some with national, some with local government support.

Measures to Aid VC Funds

1997–1999: Prohibition on private pension funds investing in venture companies relaxed and then abolished. Restrictions on public pension funds making such investments are, however, still in place (see Chapter 5, notes 17 and 19).

1998: The *Limited Partnership Act for Venture Capital Investment* further reduced restrictions on pension funds investing in VC firms. However, public pensions still cannot make high risk investments, such as investments in most VC funds. This act also legally guarantees the limited liability of nongeneral partners in VC funds, and also the right of limited partners to obtain management information on the funds.

Government investment corporations under the SMRJ can invest in VC funds as a limited partner.

2004: VC companies can apply to list their shares on public stock exchanges. This in theory allows them to raise more funds for investment. However, as a condition for public listing, they will have to disclose certificates of value, that is, investments, which many would be unwilling to do.

In addition, the Development Bank of Japan, a large government-affiliated investment bank not under the SMRJ umbrella, invests directly into some VC companies. Also the Small Business Investment Corporation and other bodies under the SMRJ provide office space to some VC companies.

Tax Incentives to Encourage Private Investment

1997: *Angel Tax Incentive* implemented. Individuals investing in startups can deduct losses over a three-year period. Also in years where there are net gains from venture investments, individuals are taxed on only one-forth of the gains. Since 2002, these benefits also apply to investments in venture companies made through special METI-approved limited partnerships. However, all of these benefits apply only to investments in ventures *officially recognized* by METI. Also extensive paper work is required. As of 2002 only 226 persons had invested in a total of only 15 ventures under these provisions.[128]

Facilitating Incorporation

2003: Under the *SME Challenge Law*, joint stock companies (kabushiki kaisha) and limited liability companies (yuugen kaisha) can be established with only 1 yen paid in capital. Within five years of incorporation, paid in capital should be 10 million yen for joint stock companies and 3 million yen for limited liability companies, the levels required before passage of this law.

Facilitating Corporate Restructuring

These measures are intended to promote risk taking by entrepreneurs and investors and to direct resources to the most viable businesses.

1997: *Commercial Code* revised to simplify the process of mergers and acquisitions.

2000: *Commercial Code* revised creating a legal framework for corporate divestitures, easing the formation of spin-offs.

2000: *Civil Rehabilitation Law* enacted. Modeled on Chapter 11 of the US Bankruptcy Reform Act, this allows companies in trouble a chance to recover or exit in a way that is likely to protect the interests of creditors and the resources of the business.

Placing IP in Trust and Securitizing IP

These steps should allow venture companies to raise capital more easily and also create a secondary market for IP. Creation of such a market should encourage investment in venture companies, because if a secondary market exists and the venture goes bankrupt, the IP can be auctioned to a company that can use it.

2004: *Law on Trust Companies* amended to permit IP to be held in trust and to permit companies other than financial institutions to hold IP in trust. The amendments specifically permit TLOs to hold IP arising in universities in trust. Also it specifically permits venture companies to avail themselves of trust services for their IP.

Measures to Increase Labor Mobility

1999: *Employment Security Law and Worker Dispatching Law* revised to permit greater use of temporary staff, and to permit private employment agencies to play a larger role in recruiting workers.

2002: Defined contribution (401k-type) pension plans approved. By 2004, 845 companies, including many large prominent high technology companies, had implemented such plans. See Chapter 7.

Incentives for Employees of Venture Companies

1998: *Commercial Code* revised to permit employees to be paid, in part, by stock options.

2002: Additional amendments to the *Commercial Code* expanded the types of employees that can receive stock options and the percentage of their salaries that can be paid in options.

Improving University–Industry Technology Transfer

1998: *TLO Law* (see Chapter 3)

1999: *Japan Bayh-Dole Law* (see Chapter 3)

2000: *Law to Strengthen Industrial Technology* (see Chapter 3)

2004: *National University Incorporation Law* (see Chapter 3)

2004: TLOs allowed to hold IP in trust

In addition, national universities may have affiliated but legally independent TLOs, VC funds, and incubators, all of which are eligible for government support.

Professional Support and Training

Various national and local government professional support and training initiatives exist. Some are built into the measures mentioned above. Some are incorporated in the Special Coordination Funds initiative discussed in Chapter 3. Suzuki (1999) noted that government efforts to construct a support system to foster venture business firms modeled on the support systems available in Silicon Valley have not achieved positive results. Since then training programs in fields such as IP and technology management, as well as professional assistance programs to aid TLOs, venture companies, etc., have increased substantially.

APPENDIX 2: UNIVERSITY OF TOKYO, KEIO UNIVERSITY, AND AIST STARTUPS[129]

In order to better understand the situation facing university and GRI ventures, I compiled lists and data tables of most of the companies established between 1995 and 2003 that are affiliated with the University of Tokyo (UT), Keio University (Keio), and AIST, one of Japan's leading national universities, private universities, and GRIs, respectively.[130] For each institution I distinguished between ventures that are (*a*) based on discoveries originating in the institution, (*b*) founded by a scientist/professor or student of the institution, or (*c*) whose CEO or board of directors chair is a professor in the institution; and (*d*) those with more distant connections to the institution.[131] Those classified under (a), (b) or (c) are listed in Table 4A2.1.

Formation Year of UT, Keio, and AIST Ventures

Startup formation surged in 2000, while rates of formation appeared to be decreasing slightly by 2003. But viewed separately, the trends differ, with startup formation at Keio falling off after 2000, while AIST formation rates continue to rise (Table 4A2.2).

Technology Field of UT, Keio, and AIST Ventures

Table 4A2.3 shows the distribution of startups across technology fields.

Life Science

Twenty-seven (35%) of the seventy-seven startups based directly on discoveries from these institutions are focusing on the life sciences. Such companies account for the largest number of startups from UT and AIST, but not Keio, where software startups predominate. The breakdown for the three Japanese institutions combined appears

Table 4A2.1. University of Tokyo, Keio University and AIST startups, formed 1995–2003

University of Tokyo ventures		Keio University ventures		AIST ventures	
Name	Yr. Inc.	Name	Yr. Inc.	Name	Yr. Inc.
Software					
Elysium	1999	Lattice Technology	1997	Best Systems	1998
I-Transport Lab	2000	V-cube	1998	Evolvable Systems Research Inst	2000
COMS Engineering	2000	A Priori Microsystems	1998	Grid Research	2002
		CMD Research (Computational Market Dynamics) [Symplex]	2000	Sensor Information Lab	2002
		Newrong	2000	Smartec	2003
		KBMJ	2000	Materials Design Technology	2003
		Wide Research	2000		
		InternetNode	2000		
		Accelia	2000		
IT hardware					
Alnair Labs	2001			Biomolecular Institute of Metrology	1999
CellCross	2002			Environmental Semiconductors	2003
				Cyber Assist One	2003
				Moving Eye	2001
Other engineering					
LLC Measurement Science Laboratory	1995			LLC ElectroLab	2001
NanoControl	2000			General Robotix	2002
UTCE	2001				
Lazoc	2002				
Fluidware Technologies [Pentax]	2002				

Materials and chemicals (including nanotechnology)

IIS Materials	1998	SNT (Shiratori Nano Technology)	2002	Photo-Catalytic Materials	1998
ASTI (Advanced Science and Technology Incubation)	2000			Eamex	2001
				Nonami Science	2002
				Top Techno	2003
				Mitsui Nanotech Research Institute	2001
Life science					
Inst. of Medicinal Molecular Design	1995	GBS (Gene & Brain Science)	2001	Human Sensing	2000
MediNet	1995	Beacle	2002	InfoGenes	2001
NRL Pharma	1998	Oxygenix	2002	GenoFunction [Hisamitsu]	2001
Effector Cell Inst.	1999	Human Metabolome Technologies	2003	LLC NMRDB Tech.	2001
Post Genome Inst.	2000			Advangen	2002
LLC Genome Based Drug Discovery Laboratory	2000			*Tokai Global Greening*	2002
				iGene	2003
Perseus Proteomics	2001			Bioimmulance	2003
OncoTherapy	2001			LLC SurgTrainer	2003
Inst. of Microchemical Tech.	2001			LLC Tainetsu	2003
iGene	2002			*Gene TechnoScience*	2001
BioMaster	2002			*Redox Bioscience*	2001
ReproCell	2003			*Medical Image Lab.*	2002
Y's Therapeutics	2003			*TransAnimex*	2002
Summit Glycoresearch [Sumitomo Pharma]	2003			*MacroTech*	2002
				ROM (Research on Microbes)	2003
				Assay and Systems	2003

(cont.)

Table 4A2.1. (Continued)

University of Tokyo ventures		Keio University ventures		AIST ventures	
Name	Yr. Inc.	Name	Yr. Inc.	Name	Yr. Inc.
Energy and environment					
Starlabo [Sumitomo Elec.]	1999	Eco Power [Ebara]	1997	Ikari Inst. of Environmental Science	1999
Geosphere Env. Tech.	2000	LLC Ecos Corp	2003		
Bio-energy Corp.	2001				
Other					
CASTI (Center for Advanced Science and Technology Incubation)	1998	AIP (Asia Internet Plaza)	1997	Patents Technology Development	2002
		LLC Coin	1997		
ASTEC (Advanced Science and Technology Enterprise Corporation)	2001	LLC Nurse Care	1998		
		School of Internet	2000		

Notes:

1. LLC = limited liability company, usually smaller than a joint stock company. All other companies are joint stock companies.

2. *Italics* = companies whose core technologies did not arise from AIST, but that are being incubated in AIST or have cooperative R&D with AIST related to core business.

3. **Bold** = company has had an IPO.

4. [ ] = company that has a controlling interest.

* iGene was formed in 2003 by a scientist who holds positions in both UT and AIST. I have counted this as one-half a UT startup and one-half an AIST startup.

Sources: see n. 130.

Table 4A2.2. Founding years of U Tokyo, Keio and AIST startups

	UT	Keio	AIST	Total
1995	3	0	0	3
1996	0	0	0	0
1997	0	4	0	4
1998	3	3	2	8
1999	3	0	2	5
2000	7	7	2	16
2001	7	1	5	13
2002	5	3	7	15
2003	2.5	2	8.5	13
Total	30.5	20	26.5	77

similar to breakdowns in US and even European universities where a large proportion of venture companies have a life science focus.[132] But the significance of this high proportion of life science startups is unclear. Does it suggest that venture companies may have a comparative advantage with respect to large companies in only a few fields such as life science, or are there more prosaic reasons (such as allocation of R&D resources) behind this predominance? This question arises again in Chapter 7. For now, it is sufficient to note that the proportion of life science ventures is roughly the same as the proportion of research funding, except in the case of AIST.[133]

The majority of the biostartups are not currently focused on the development of drugs or other therapeutic technologies. Instead they are pursuing what might be called 'platform', 'research tool', or 'medical aid' technologies.[134] In some cases they are primarily doing contract research, for example, to identify the role various genes play in cancer, or to clarify the interaction of various components of the immune system. Many of these 'service-oriented' companies plan to develop diagnostic devices using their core technologies and eventually to identify particular drug targets, for which they themselves will pursue drug development. In other words, their business model

Table 4A2.3. Distribution by field of U Tokyo, Keio, AIST and MIT startups

	UT	Keio	AIST	Total 3 Jpn Insts	MIT* (%)
Bio (excluding energy/environment)	13.5	4	9.5	27 (35%)	41
Software	3	9	6	18 (23%)	23
IT hardware	2	0	3	5 (6%)	10
Materials including nanomaterials	2	1	4	7 (9%)	11
Engineering, instruments	5	0	2	7 (9%)	14
Environment, energy	3	2	1	6 (8%)	
Other	2	4	1	7 (9%)	
Total	30.5	20	26.5	77 (100%)	100

* Data for MIT startups formed 1980–1996 from Shane (2004: 140).

Table 4A2.4. Financial status of U Tokyo, Keio and AIST startups

	UT (N = 30.5)	Keio (N = 20)	AIST (N = 26.5)	Total (N = 77)
Paid in equity capital ≥ 100 million yen	12 (9 in bio)	8	5	25 (32%)
Private independent VC company 1 of 2 lead investors*	9 (2 by univ. VC fund)	4	4	17 (22%)
Business corp 1 of 2 lead investors	3	6	3	12 (16%)
Bank 1 of 2 lead investors	1–3	0	1	2–4 (3–5%)
Insurance co. 1 of 2 lead investors	0–2	0	0	0–2 (0–3%)
Trading co. 1 of 2 lead investors	1	2	0	3 (4%)
Indep VC co., business corp., bank *or* insurance co. is 1 of 2 lead investors	14	10	7	31 (40%)
Reported revenue ≥ 100 million yen	9	6	3	18 (23%)
Reported revenue ≥ 10 million yen < 100 million yen	6	4	5	15 (19%)
Reported profits ≥ 10 million yen	4	4	0	8 (10%)
Employees ≥ 10	11	7	3	21 (27%)

* Includes JAFCO, which is affiliated with Nomura Securities, but not Chuo-Mitsui Trust Capital, which I classify as a bank investment. Does not include government investment corporations, such as Tokyo SME Investment Corporation, under the SMRJ (see Appendix 1).

is first to provide services, then to develop diagnostics and then to develop drugs. Based on a best estimate of where these various companies lie on the service to drugs spectrum,[135] I estimate that four to ten (14–36%) of the twenty-seven biostartups[136] are directing most of their resources to developing diagnostics or therapeutics. The main reasons many companies are not moving more directly to develop diagnostics or therapeutics are lack of capital, lack of appropriate management, and lack of clear leads for drugs or diagnostics.

Software

Eighteen (23%) of the seventy-seven core startups are focused on software development or sales. The largest number of Keio's ventures are software related. Five of these nine companies are developing Internet applications and are linked to the laboratory of one professor in the Keio's Shonan Fujisawa Campus. Two of these five are reasonably successful in terms of sales. In other words, over a limited period, one laboratory generated a critical mass of entrepreneurial talent and spawned a number of startups developing broadly applicable Internet applications, and some of these have been successful. In contrast, with the exception of two companies,[137] the UT and AIST software startups are focused on more narrow technical applications and their sales are low.

Engineering, Including IT Hardware

The number of startups from these three institutions combined is only twelve (16%), much less than the combined number of biostartups. The absence of startups in these two categories from Keio was unexpected, considering that its engineering department is well regarded. About half the engineering and IT hardware startups from UT and AIST are focused on either consulting or developing a particular type of instrument. It seems unlikely that any of these companies can grow substantially with their limited product and service repertoire. The other half seem to be developing technologies with various potential uses.

As suggested in Chapter 3, the small number of engineering startups may be due in part to most engineering inventions having been transferred to established companies. Roughly half of the inventions reported to UT's TLO are engineering related.

Materials

Seven (9%) of the seventy-seven startups are focused on materials technologies. With just a few exceptions, the commercial applications being pursued by these companies appear limited, despite nanotechnology and other materials research in Japanese universities being vigorous and well supported.[138] Probably most nano-materials related discoveries in Japanese universities and GRIs are transferred directly to established companies.[139] In contrast, a considerable number of materials related companies have arisen from US universities.[140]

Environment and Energy

Only three of the six startups I classified as focusing on this area appear to be developing technologies that might have widespread demand, for example, wind power and biofuels. The other three seem to be developing or selling software with rather specific applications or to be engaged primarily in consulting.[141]

Finances and Employment of UT, Keio, and AIST Ventures

Three of the startups had had IPOs by the end of 2005. All are life science companies. One is closely linked to one of UT's main genetics research laboratories and is currently the main vehicle for commercializing a significant proportion of that laboratory's research. The IPO was extremely successful and for several months following the IPO, the market capitalization of this startup was over US$1 billion, making it the most highly capitalized Japanese startup and one of the most highly capitalized recent startups worldwide. The second also had its IPO in 2003 and its market capitalization is approximately US$34 million. It provides immune cells and related services to hospitals and clinics and conducts related R&D. Although classified as a UT venture, its links are more distant. One of its current board members cofounded the company around the time of his retirement from the UT Institute of Medical Science. The third has close links to UT and was the only biomedical startup to have an IPO in 2005.

Two other startups have not had IPOs but have paid in capital in excess of US$10 million. However, both of these are majority owned by larger companies. One is in biomedical R&D and its founding scientist has positions at both UT and AIST. The other is building windmills in various locations in Japan and selling the electricity.

However, these companies are exceptions. The vast majority of the startups are small whether measured by capital, revenue, or number of employees. Most are capitalized at less than US$1 million, and less than half have any private company as a lead or significant follow-on investor. When a private company is the lead or main follow-on investor, it is more likely to be an independent VC company rather than a bank, insurance company, or nonfinancial business, indicating that independent VC companies are playing a lead role in the financing of startups.[142] Life science startups from UT and Keio tend to have more paid in capital than other startups, some indication that life sciences may be a favored field for startup formation and funding.

Barely 40 percent of the companies have annual revenue over US$100,000 and barely 20 percent over US$1 million. Also revenue figures sometimes consist substantially of government commissioned research. However, the two companies that have had IPOs have annual revenue over US$10 million, as do two software companies. Startups with revenue over US$100,000 seem to be evenly distributed over technology field and institution. So far, there is little indication that having a VC company as a lead investor (or even an established company as a lead investor) is strongly associated with high revenue. In other words, so far private risk capital (equity) investment does not seem to be pushing these companies to generate revenue. AIST's startups generally have less invested capital, revenue, and employees than those of UT and Keio. However, this may be due to AIST's ventures being younger and also to AIST's goal of serving as an incubator to fledgling companies.

Startups may have more resources than suggested by their low levels of invested capital, sales, and employment, because the R&D of the 'virtual' startups often takes place in the university or GRI laboratories of the founders. Generally, startup and academic research topics are distinguished from each other and persons working on the latter have to be paid by the startup. Nevertheless, there is often considerable overlap between a founder's academic research and research of value to the startup. Through joint inventorship and less formal mechanisms, there seems to be a tendency for many of the discoveries in the founder's academic laboratory to flow to the startup. When this happens automatically without the TLO or any other independent entity considering whether the startup is best suited to develop these discoveries, the process is known as pipelining. It is generally discouraged in US universities.[143]

Management and Alliances

Management and control of the seventy-seven startups is varied. Eleven were substantially controlled by an outside company. At least three of the nine UT and Keio software startups are headed by former students who are also founders. Four of the life science startups are headed by officers of small VC companies that were lead investors in them. The remaining fifty-nine (approximately 75%) were headed either by the founder or by managers hired from the outside. In general, the larger the company, the more likely it is to have a manager from the outside. Nevertheless, the founding scientists often retain great influence even though they have no official management position. VC investors defer more to the wishes of the founder than they do in the USA, and consequently often choose CEOs in consultation with the founders who will not challenge the founders' behind the scenes authority.

At least four startups were founded by university or AIST researchers who had just reached the mandatory retirement age of 60.[144] At least thirteen (17%) of the startups are closely linked with at least one other academic laboratory in a different institution. In other words, they serve as commercialization vehicles for networks of scientists spanning several institutions.

How Technically Innovative?

I estimate that slightly over half of the seventy-seven startups are pursuing the development of new technologies.[145] Some of the remaining startups may be innovative in utilizing currently known technologies or organizing their businesses.[146] Indeed, some of the startups that are not developing cutting edge technologies are among the most successful in terms of sales. The point of this rough estimate is that a significant proportion of the startups are pushing forward the boundaries of knowledge in ways that may have practical utility.

This said, in materials science, IT hardware, and other fields of engineering (areas in which all three of these institutions continue to be strong and to produce world class research results) few of the startups are pursuing groundbreaking technology development. In other words, in contrast to quite a few of the life science startups, few of the startups in these areas are the vehicle for commercializing pioneering technologies or technologies with widespread practical applications. Similarly, except for the startups that emerged from Keio's Shonan Fujisawa campus and pioneered mobile Internet systems, few of the software startups seem to be pursuing groundbreaking technologies or applications.[147]

Universities/GRIs as Incubators: The Special Case of AIST

Not only is AIST incubating a number of its own very young startups, it also serves as an incubator for at least nine other startups based on discoveries originating primarily in other institutions. Three of these are based on discoveries made in the biomedical departments of several Hokkaido universities that are relying on AIST's Hokkaido laboratory for R&D facilities. Four others are engaged in cooperative research with AIST laboratories that appears vital to their core business interests.[148] An eighth company is a nanotechnology R&D subsidiary of a large trading company, conducting joint research with AIST on nanoscale porous membranes on behalf of manufacturing companies in the same keiretsu.[149]

NOTES

1. The most important of these markets (that perform much the same function as the US Nasdaq) are Jasdaq (the over-the-counter market whose listing rules were relaxed in 1998), Mothers (a special section of the Tokyo Stock Exchange formed in 1999), and Hercules, the concessionary section of the Osaka Stock Exchange.

2. Between 1999 and 2001, the latest period for which data are available, 1.6 manufacturing companies were incorporated per year for every 100 manufacturing companies existing at the beginning of 1999. In contrast, 4.1 manufacturing companies ceased operations per year for every 100 manufacturing companies existing at the beginning of 1999 (JSBRI 2003).

3. METI (2006). Some startups belong to more than one field, so percentages would sum to more than 100. Also these percentages are for all 1,503 companies classified as university ventures in the 2005 METI survey, not the 845 companies actually based on university discoveries. In other words, these percentages are derived from a set of companies, many of which have only distant links to universities. See the discussion related to Figure 3.1 in Chapter 3.

4. The median respondent to one of the principal annual surveys of startups was only two years old at the time of the survey, the median number of employees was less than 5, median annual sales were about US$200,000, and median annual profits were approximately US$5,000 (Tsukuba, 2006).

5. The other industry categories being *IT* and '*other*' (METI University 2005: 28, 51).

6. In 2002: Anges MG (Osaka U) and TransGenic (based in part on Kumamoto U discoveries). In 2003: MediNet (based in part on U Tokyo discoveries), OncoTherapy (U Tokyo), and Soiken (Osaka Foreign Language U and Osaka City U). In 2004: Gene Chip Research (Osaka U). In 2005: Effector Cell (U Tokyo).

7. Note that the row 'total employees' in Table 4.1 compares total employment in Japanese therapeutic-oriented bioventures in 2005 with employment in just a subset of US therapeutic-oriented bioventures in 1987 and 1998/99. In 1987, there were sixty-five US biotechnology companies whose primary focus was therapeutics (including subsidiaries of larger companies and LLCs, but not pharmaceutical companies). They employed 9,731 persons, thus mean employment was 150. The largest of these therapeutic-focused bioventures was Genentech with 1,500 employees (Dibner 1988: 214–20). In mid-1999, there were approximately 420 therapeutic-oriented US biotechnology companies. Half were established in 1992 or later. Mean number of employees remained about 150 (Dibner 1999: 6). Thus total therapeutic biotech employment was nearly 63,000.

8. Which happened to have a median age of 4 years.

9. Biomedicine is the most active field for startups according METI (2006) and Tsukuba (2006). See also Chapter 3. Data presented in Chapter 5 on IPOs suggests it is also the most active field for all ventures.

10. At the time of incorporation, financing is generally considered to be the major problem, but after incorporation, hiring and retaining staff (both managerial and R&D staff) is the most severe problem for startups in all technology sectors, ahead of sales and financing, and substantially ahead of issues such as incubation facilities, lack of advisory services, and IP. Needs are highest for researchers and managers, and unmet needs are particularly high for managers (METI 2006: 45–7; Tsukuba University 2006: 52).

11. Methods and explanations for Table 4.1: JBA (2005) contains a list of all Japanese bioventures existing in 2004. I selected the 268 companies that had therapeutics listed as their first or second business focus and then randomly selected 30% of these to confirm the business focus and determine the number of employees using various public data sources. Fifty-eight percent either did not have therapeutics as a significant business focus, were subsidiaries of larger companies, were limited liability companies (LLCs, whose operations and business focus are small), or were established earlier than 1975, and I excluded these from further analysis. Among the remainder, 4 was the median age since formation, and 70–80% were based on university discoveries. (I could not find employment data on seven companies in my final sample. I assumed these companies each had fewer than 9.5 employees.)

   Dibner (1988, 1999) contains lists of US therapeutic bioventures, including numbers of employees, based on surveys conducted in 1987 and 1998/99, respectively. Selecting those companies established closest to Dibner's survey years, excluding LLCs and subsidiaries, and working backwards year by year until I had a set of companies that also had a median age of 4 years at time of each survey, gave me two sets of US companies that I used as comparitors to the Japanese ventures.

   * It is not clear from Dibner (1999) whether employment was assessed in 1998 or 1999. Thus I calculated separate employment averages assuming employment was assessed in 1999 (for which I included companies formed 1994–8 to obtain a median formation year of 1995) and in 1998 (for which I included companies formed 1992–7 to obtain a median formation year of 1994). The figures in this column represent an average of these two sets of calculations.

12. Two biomedical and two software.

13. Of twenty-five independent AIST, Keio, and U Tokyo startups not in software or life science, only six had annual revenues of at least US$1 million in 2004, and over half had no publicly listed revenue data.

14. The purpose of these interviews was to understand how these companies were formed, the source of their key technology, how they obtain capital, human resources, customers and research partners, and the importance of intellectual property. Interviews were free form but structured around these issues. Respondents were informed in advance about the issues I wished to discuss.

   One of these companies I identified and interviewed entirely on my own. In the case of the other nine, I turned to Fujitsu Research Institute (FRI) to identify companies that would be willing to meet with me and would provide a representative picture of the situation facing nonbiomedical ventures—or as close to a representative picture as possible in view of the limited number of interviews I could conduct. I chose FRI because I knew researchers there working on innovation and S&T policy issues. FRI first helped me identify suitable interview candidates on the basis of information from its databases, accounting firms it deals with, and newspaper reports. Together we selected companies in various fields of nanotechnology or IT. The list of appropriate interview targets was short. Because some companies declined interviews, we ended up interviewing almost every company on the FRI lists that would grant us an interview.

Initially, I intended to limit my interviews to companies that had not had an IPO in order not to select companies that already passed a significant milestone on the road to success. However, after a few interviews with pre-IPO companies (many of which had no significant sources of outside revenue and few collaborative relationships), I realized that I should also include ventures that had had IPOs in order to have an idea of the factors that can lead to success. So the case studies include some companies that are widely regarded as successful but others that are not, in order to understand the situation facing venture companies as a whole.

Of the nine companies identified by FRI, six were interviewed jointly by me and the FRI researcher working with me. Three were conducted by the FRI researcher alone, but only after we had done several interviews together so he understood the type of information I wanted. Except in the case of one respondent who spoke fluent English and who suggested that the interview be in English, all these interviews were in Japanese. Most lasted about ninety minutes. All of the companies were cooperative in answering questions during the interviews. Usually the companies responded to follow-up inquiries.

'Quoted sections' in the case studies are approximate translations of the comments of the respondent, usually the CEO or another executive officer.

Prior to publication of this book, I have not been able to update information on these companies. The case studies are not meant to predict the companies' futures, but rather to be snapshots in time showing the various challenges confronting venture companies and how they have tried to deal with them.

15. This explanation recalls another explanation I heard from a former employee of a major Japanese electronics company who now works as an investment consultant: Japanese managers in large companies may understand the technical issues of the projects they are working on well, but they are often not well informed about broader S&T or business issues. They often cannot perceive the potential benefits of new technologies proposed by venture companies or universities. They tend to underestimate the possible rates of return on outside investment and to support in-house R&D which they understand better, although the discounted rate of return is probably less.

16. Reference was made to the case of Phoenix Wireless presented in the text below.

17. A media report described this arrangement as 'an elaborate severance package'. As of 2005 the parent had not exercised its option to purchase stock. The venture has no other obligations to the electronics company and can sell its products and license its technology freely.

18. Unlike many VC funds that scatter investments of a few US$100,000s the VC company has made a few large investments.

19. This recalls the discussion in Chapter 2 about the importance of direct interaction between software developers and users and the resulting predominance of customized as opposed to standardized software in Japan. This stands in contrast to a more modular approach to software development, and innovation in general, in the USA.

20. In the founder's words, 'New ideas do not fly well with Japanese corporations.'

21. This makes it one of the earliest startups from an academic institution since the beginning of the reforms discussed in Chapter 3.

22. The fund was about US$10 million.

23. These include awards from the Small and Medium Enterprise Agency (similar to US SBIR awards), the METI-affiliated Foundation for the Promotion of Small Business Ventures (zaidan houjin chuushou kigyou benchaa shinkou kikin), NEDO and other organizations.

24. This brought the sister's share of Big Crystal's stock to below 20%. Japanese stock market listing requirements do not permit any single person to own more than 20% of a publicly traded company's voting stock. Big Crystal is contemplating an IPO, so this redistribution of ownership from Crystal Ceramics to the ceramics company is a preparatory step.

25. Other large companies were also interested in a similar alliance with Big Crystal, but the ElectroCera with its top-down management style was able to negotiate an agreement more quickly than the others.

26. Recall the discussion in Chapter 3 about article 73 of Japan's Patent Law that requires the consent of all co-owners before an invention is licensed.

27. *Process patents* refers to patents covering methods of manufacture, methods of use, etc., as opposed to a particular machine, instrument, or type of material.

28. It regarded technology dispersed in this matter as hurting not only the Japanese ventures but also the large Japanese parent companies because ultimately this enables foreign rivals to compete with the Japanese companies.

29. Primarily photoactivated titanium dioxide catalysts (similar to the Hashimoto-Furukawa surface photocatalysts mentioned in Chapter 3) embedded or encased in apatite or ceramic.

30. Nevertheless, the founder could not describe his company's products or mention the company by name at outside lectures. For this reason, he resigned from the GRI and took a professorship in a local university. However, he also had to resign directorship of the company because of regulations prohibiting recently retired government employees from taking jobs that involve close dealings with their former government offices. The position of CEO is now held by the president of a local tile manufacturing company.

31. The parent's policy is not to license its technologies exclusively—an issue that was of some concern to the venture companies investing in ISE.

32. Contrast these ongoing, mutually beneficial ties with the cold shoulder given Chip Detect by its parent.

33. Although if she or he had worked for more than twenty-five years, she or he would be eligible to receive payments under the mandatory basic national pension system (equivalent to the US Social Security System) that might pay in the order of 60,000 to 250,000 yen per month depending on previous earnings and whether one has a dependent spouse.

34. See Chapter 7, note 234 and accompanying text.

35. ISE currently has a domestic competitor in its core business field.

36. Apparently, even though the parent was not using these technologies, it was reluctant to license them exclusively to another company. This recalls the insistence of Big Crystal's customers that they would not use its technologies if Big Crystal's parent GRI gave Big Crystal exclusive licenses. Are such demands also common by American and European companies, or is this another manifestation of autarkic behavior unique to large Japanese companies?

37. I had heard about the IP aspects of this case from several sources, and I requested an interview mainly to confirm the IP aspect. The interview did not cover the range of topics I normally cover in interviews (sources of technology, researchers, managers, capital, R&D partners, and customers). Thus the scope of this case study is narrower than that of the others. Nevertheless, I offer this as an example of the types of problems high technology ventures can encounter in dealings with large companies.

38. One of these claimed advantages relates to power control and the use of matched filter technology to enable mixed digital-analogy signal transmission which requires less power than pure digital systems. Also some of Phoenix Wireless's discoveries were directed to alternatives to Qualcomm's controlling patents on code division multiple access (CDMA) technology. A third advantage asserted by Phoenix Wireless was that its technology could deal with various types of wireless technologies, thus increasing the ease with which handsets and content designed for one type of wireless system could be used in a different environment. This incompatibility between wireless systems continues to bedevil wireless communication today. Handsets and content streaming systems for use in Japan cannot be used in Hong Kong, Europe, or America.

39. At the end of 2004, Phoenix Wireless had over 120 issued Japanese patents and nearly 350 unissued Japanese patent applications. Even more significantly, it was the applicant or coapplicant on at least 165 issued US patents, a significant investment in terms of patent application costs. Many of these Japanese and US patents were applied for during the period of collaboration with SoMoCo, roughly 1995–2000. Most of the patents applied for during this period were assigned initially solely to Phoenix Wireless. However, approximately 15–20% were joint applications by Phoenix Wireless and SoMoCo.

Reviews by persons familiar with wireless technology suggest that some of Phoenix Wireless's patents probably represent real technical achievements and that the company's patent portfolio had commercial value. Its discoveries are almost certainly not an alternative to Qualcomm's technology, and any commercial system incorporating Phoenix Wireless's technologies would have had to incorporate technologies licensed from many other companies, mostly in the USA and Europe. Nevertheless, Phoenix Wireless's discoveries were not incremental inventions submitted by employees to improve their promotion chances (a common practice in Japanese companies) but inventions that the company (or its collaborators) had reason to believe had significant value.

An interesting sidelight revealed by the patent documents is that Phoenix Wireless's key inventors were Chinese. This is one of several examples of Japanese venture companies providing a base for Chinese and Korean scientists to contribute to the advancement of human technology and the Japanese economy.

40. Personal handy phone system (PHS) technology was largely developed in Japan in the late 1980s. Although shorter range than 2nd and 3rd generation mobile phone systems, it consumes less power, and often has higher rates of signal transmission. Popular in Japan in the late 1990s before the boom in 2.5 and 3G phones, it is now experiencing a revival among handset users in other Asian countries and also in Japan as part of its wireless infrastructure.

41. Two of the remaining three companies are limited by the nature of their business to primarily Japanese customers, Fresh Air Catalyst sells mainly to construction and interior finishing companies and Internal Search Engine provides customized data management services to local businesses. Only Fine Molded Plastics (FMP) produces products with worldwide applicability and yet relied mainly on Japanese companies for its early clients, but this company was formed in 1969 before most foreign companies had access to small Japanese manufacturers. Moreover it continues to experience difficulty getting big Japanese companies to collaborate in ways that help increase FMP's core capabilities.

42. A VC investor in two IT-related Japanese companies remarked that, although large manufacturing companies still rely on their in-house laboratories, they are more willing to consider outside technologies and to engage small independent companies. 'If they are interested, they will say, "Can you do such and such" and ask for a demonstration run or sample.'

The companies in my case studies that asserted most strongly that large Japanese companies are now less autarkic were Big Crystal, Chip Detect, and Internal Search Engine. But although Big Crystal says that Japanese companies are avidly seeking collaborations, it acknowledges that its first sales were mainly to overseas customers. Internal Search Engine has no choice but to market domestically and it has the backing and pedigree of its well-known parent. Chip Detect supplied two of its three prototype machines to overseas companies, it has strong government backing, plus it also has a pedigree related to a well-known parent. Phoenix Wireless was an exception, but its large partners ultimately excluded it from collaboration and blocked its growth.

There are media reports of large companies making noncontrolling investments in small companies and adopting their technologies. In 1998 Toyota made a 100 million yen (US$715,000) investment in Lattice Technology and adopted its 3D image processing technology for use in the design and inspection of auto parts. At the time, Lattice (which METI and I have classified as a Keio startup because its board chairperson is a Keio University professor) was 1-year old and in severe financial difficulty (*Nikkei Weekly* 2002). It has since obtained customers worldwide and its revenue has grown to nearly US$9 million in the fiscal year ending March 2005. However the fund that Toyota established to invest in ventures appears inactive, at least in Japan.

Also among all the seventy-seven U Tokyo, Keio, and AIST startups I identified in 2004, Lattice Technology stands out as being the largest independent startup that has received significant investment from a large nonfinancial company. The only other was IIS Materials, based on U Tokyo discoveries, which received investment from Sumitomo Electric, but for which no sales data are available.

43. See Thayer (2003) and Paull, Wolfe, Herbert and Sinkula (2003).
44. Moreover these are truly independent companies. Their parents exert no or minimal control and there is no return option for the employees should the ventures fail.
45. Recall that, despite the growth of ISE, none of the CEO's former colleagues in the parent company was interested in starting a venture. Also, personnel officials in ISE's parent report that the number of engineers, scientists, and managers who take early retirement is less than 40 per year out of an R&D, systems engineering, and business planning workforce of nearly 25,000.
46. Only Chip Connect indicated it did not have recruitment problems, relying mainly on recent social science graduates.
47. I do not have data on whether Ph.D. researchers in large companies are more likely to move to ventures than researchers with only masters or bachelors degrees. However, colleagues at the University of Tokyo, who have been teaching graduate students, also think that Ph.D. graduates who have several years' experience in large companies often become disillusioned in large companies and are among the most valuable recruits for ventures.
48. The employees of Molecular Visualizer, Big Crystal, and Fresh Air Catalyst in 2004 numbered seventeen, ten and eight, respectively. These companies recruited most of their initial employees from the parent GRI or through connections with the GRI.
49. The main potential competitors of these companies are often their customers or overseas companies.

50. Chip Connect specifically said it advertises its patent portfolio to attract customers. Interface Chip mentions the technical fields of its patents prominently on its web page. Chip Detect, Molecular Visualizer, Fine Molded Plastics, and Internal Search Engine also mention on their web pages their patented technologies.

51. I have in mind a narrow definition of 'trade secret protection', in other words, the sort of protection against theft of technical know-how that would be covered under trade secret protection legislation. I am not referring to trade secrets broadly to include the full range of a company's S&T expertise.

52. In 1999 there were about 4,100 patent attorneys (benrishi) in Japan, compared with about 5,800 registered to practice before the European Patent Office and 19,400 patent attorneys and agents in the USA (www.jpo.go.jp/shiryou/toushin/shingikai/shi-04.htm). On the other hand, because the Japanese system does not permit extensive discovery proceedings, *costs of litigating patent infringement in Japan are lower* than in the USA, perhaps by a factor of five according to an attorney familiar with infringement litigation in both Japan and the USA.

53. In 1998–2001, the average damage award by Japanese courts in infringement suits was 181 million yen (about US$1.6 million), a substantial increase over the 1990–4 average award of 46 million yen (about US$0.4 million). The number of infringement suits brought in the first instance in Japanese courts in 2001 was approximately 130, although it is not clear how many of these resulted in damage awards for the plaintiffs (JPO: 2002 Patent Administration [Tokkyo Gyousei]; and 2002 IP Activity Survey Results [Chiteki zaisan katsudou chousa gekka]). In comparison in 2001, 2,520 patent suits were initiated in US Federal District Courts. Only 325 of these proceeded all the way to a final judgment (most were settled out of court, as were many Japanese cases), and of these only 49 (15%) involved monetary damages awarded to the alleged victim of infringement. Of these forty-nine, only twenty-eight involved awards over US$1 million (Kerr and Prakash-Canjels 2003). Thus, on the basis of these statistics, it is not clear that median litigation damage awards are higher in the USA than in Japan. However, some courts awarded damages and some out of court settlements in the USA result in payments to the plaintiffs that are in the hundreds of millions of dollars. (See lists of such damage awards and settlements in Kerr and Prakash-Canjels (2003) and as compiled by Gregory Aharonian at www.patenting-art.com/economic/awards.htm) Such awards are much higher than was traditional in Japan, although recently a court awarded about US$80 million in infringement damages to the owner of patents on Pachinko machines.

54. Because a large company might decide it makes sense to infringe an SME's patents, knowing that the costs of prosecuting an infringement suit might bankrupt the SME, and even if the large company ultimately lost in court, the penalty would be light.

55. Ibata-Arens (2000) interviewed forty-three technology-focused SMEs (some of which were shita-uke companies) in the Tokyo, Kyoto, and eastern Osaka regions. At least three of these companies had experienced theft of trade secrets and know-how by representatives of large companies who gained access under false pretenses. According to Ibata-Arens 'nearly half had either personally experienced or knew another firm that had its technology appropriated by a large firm'.

56. See Cabinet Office (2005) and IIP (2005).

57. No company indicated it had been involved in patent infringement litigation or that its patents had been opposed by another company, so none had first hand experience with litigation costs and damage awards.

58. I obtained a list of sixty-five cases dealing with IP infringement brought before the Tokyo District Court between 1967 and 2005 and on which the District Court issued a final ruling. Although my analysis is not complete, at least the latest rulings do not suggest a pattern of large companies using infringement suits to harass small companies or the Court tending to rule in favor of large companies in large vs. small company disputes. Moreover, in the case of recent victories for small/new plaintiffs, my preliminary analysis suggests that awards are not excessively low, i.e., they may be large enough to have deterrent value.

    However, cases that go to final judgment provide only a partial picture of the IP litigation landscape because they do not include the larger number of suits that were settled or dismissed prior to a final judgment. To get a handle on these cases, I reviewed a list of all patent and unfair competition cases brought before the Tokyo District Court from 1994 through 2004 that identifies the parties, the nature of the case (including the nature of the technology, if any, at issue), and the disposition of the case. Examining a random 30% sample of the 481 cases brought in 2004 and finding out information about the parties in these cases did not suggest clear asymmetries in terms of whether small/new companies were more likely to sue or be sued, or to suffer adverse outcomes, in disputes with large/established companies. In fact, there probably are not many technology-focused IP litigations involving small/new vs. large/established companies. Of course, it is still possible that simply the threat of litigation will usually make a small company acquiesce to terms favorable to a large company.

59. It is another illustration of how the requirement of article 73 of Japan's Patent Law tends to benefit large companies at the expense of ventures or universities. The nontransferability of IP rights is less a problem for large companies, but ventures need to be able to license their discoveries to new collaborators. Also, because of large companies' advantage in research manpower, in any cooperative R&D situation it is easier for them to position their researchers as coinventors on discoveries that are primarily a venture's or a university's than vice versa (see Chapter 3).

60. The exception is a company focusing on software development, whose main competitor is another software company. The patents that are its main defense against this competitor are essentially software patents.

61. This contrasts with the assertion of Tang, Adams, and Pare (2001) that patent protection is not important for small business development and innovation in the European software industry.

    The same VC investor who, in the note at the end of salient point 1, noted that large companies are beginning to be more open to considering purchases of the products of small companies, agreed that IP was crucial for new small companies. 'It is crucial in negotiations with other companies. As more is disclosed, the more important are patent protection and written confidentiality agreements. The most promising companies have core technologies with wide applications. Clear IP protection allows the small company to protect its core business while licensing out specific applications. Trade secrets are also important. Smart companies always leave a lot undisclosed. Also small companies are building networks of colleague small companies, usually all Japanese. When they receive a manufacturing order from a large company, they preferentially subcontract to colleague companies so as not to risk leakage of undisclosed know-how.'

    The Japanese ventures' perspective on Japan's patent system stands in contrast to the results of a survey of SMEs in the UK, which concluded that 'The patent system is at best an irrelevancy for most small firms, and it can be a major indirect cost.' The respondents to this survey said the patent system is too expensive and complicated, does not provide protection against infringement (largely because of delays and red tape associated with obtaining patents and patent litigation) and reveals too much information to competitors (Macdonald 1998). Perhaps the difference is due to the companies in my sample having a clear focus on new technologies. Also, for a European company, a domestic patent secures a smaller market than in either Japan or the USA. Greater expense is required for European wide coverage.

    On the other hand, the findings of Blackburn (1998) and Cohen, Goto, Nagata et al. (2003) are more consistent with the stance of the Japanese companies in my sample.

62. In the sense that the target ventures were backed by considerable government funding or had supportive ties with a large parent.

63. As in the case of Chip Detect where the VC company helped find the CEO and provided advice on business development, and the case of Big Crystal where the VC company introduced patent specialists who, together with the VC company's staff, provided advice on negotiating contracts with outside companies. In both these cases, the key VC investors included public investment corporations or private companies with substantial government backing.

64. e.g. Itochu, Marubeni, Mitsui Busan, Mitsubishi Shouji, and Sumitomo Shouji.

65. See Table 4A2.4 in the Appendix to this chapter, which analyzes the top two sources of financing for each of the University of Tokyo, Keio, and AIST ventures.

66. During the interviews, it was usually difficult to determine the importance of these collaborations, since to do so would require technical discussions

about a company's product development strategy as well as the expertise of the university researcher. However, the following example shows that collaborations can benefit the ventures greatly: The CEO of Chip Connect mentioned that the faculty member who had helped Chip Connect was in my center at the University of Tokyo. I happened to be familiar with his research and I later confirmed that the collaboration was directed to overcome a particular problem regarding bonding and signal transmission/leakage between different conducting/insulating materials. This collaboration occurred in the late 1990s before the IPO, when Chip Connect was a small company. Chip Connect had sought out the professor. There was no prior association. The fact that a University of Tokyo professor and a little known small company would collaborate to solve an important problem for the company is a favorable sign. As a contrasting example, Nanofilm Applicator reported difficulty in interesting researchers in prestigious universities in its technical challenges, although it has managed to obtain help from Chinese universities. Other colleagues at RCAST have also collaborated with small companies and one, in particular, reports that small companies generally are more serious about developing the results of the collaborations than large companies.

67. One of this manager's achievements, in addition to launching clinical trials, securing initial funding, and beginning preparations for an IPO, was to retrieve and consolidate the founder's key IP that had been scattered among various Japanese pharmaceutical companies. Patent rights to some of the founder's key inventions had been passed to various Japanese pharmaceutical companies, in the informal way that so much Japanese university technology was passed to large companies prior to 1998. In this case, as in most other cases of such informal transfers of university discoveries to large companies, the large companies made little effort to develop the discoveries.

68. The venture also obtained an SBIR-like grant from MHLW just after its formation. The chemical company now markets a transfection kit, so that other laboratories can use the same vector (the envelope of a virus found in Japan) for other gene therapy experiments.

69. The pharmaceutical company has provided funding that enabled completion of preclinical studies and it is also funding the ongoing human trials in both the USA and Japan. In return, the pharmaceutical company holds a small amount of the venture's equity, and it has exclusive rights to market the drug worldwide to treat peripheral and coronary artery disease.

70. Some of these were physicians working toward their doctoral degrees in clinical medicine (see Chapter 3).

71. It is trying to acquire drugs by in-licensing from smaller Japanese bioventures and by using new screening systems it is developing in cooperation with established Japanese companies.

72. This strategy has attracted interest from other research groups. However, CAC is among the leaders in bringing drugs of this type to clinical trials.

73. Three of the Ph.D.s are founders. The other is a national of an Asian country who previously worked in a Japanese pharmaceutical company. The other two persons with pharmaceutical industry experience include a technician and a temporary employee with an MS degree.

74. Good Manufacturing Practice. This refers to quality pharmaceutical manufacturing standards established by MHLW, FDA, and other drug approval agencies.

75. These antibodies are made using a bacterial phage library containing genes for human antibodies. The library and the technology for generating new antibodies via hybridization were initially developed by a professor in an obscure public health college near the company's headquarters, who now works closely with MBV.

76. Establishment of this genetics laboratory by the metals company seems to have been another example of unsuccessful diversification by a large company into a completely different area, of the type discussed in Chapter 6.

77. MBV is also an equity investor in this spin-off, which specializes in research on glycoproteins and other biologically relevant sugar chains.

78. In the past, MBV would make custom antibodies for academic researchers free on request, which it would then advertise in its catalogs.

79. Information for this case study comes primarily from published information, although I have spoken briefly with the CEO about links with universities.

80. It does sell domestically to organizations such as the Japanese Red Cross that use the machines to extract DNA which is then tested for HIV, HCV, and other infectious substances. A major Japanese fiber manufacturer has also purchased some of the machines.

81. Information for this case study also comes from published information and contacts familiar with the company, not interviews.

82. A member of the medical faculty of a national university located in western Japan near the collaborating company.

83. *Nikkei Weekly*, 'Knockout mice propel gene research', July 16, 2001, p. 10.

84. For example, the information might relate to using the gene or its protein product as a diagnostic marker, or as a predictor of the safety and efficacy of a drug.

85. This case study is based on an interview in 2000. I have not conducted a follow-up interview, but I have gleaned relevant recent information from public sources.

86. At the time of the company's founding, concern was widespread in Japan and other countries that US companies such as Incyte, Celera, and Human Genome Sciences had a monopoly on such information. Not only had they applied for patents on many genes and gene fragments (whose functions were often unknown), but they also were charging substantial sums to each laboratory that wanted to access their databases. By 2004, demand for access to these databases had fallen, in part because information that could be gleaned easily from their use has already been exploited, and in part because

publicly funded research (with funding conditioned on the researchers plac-
ing their findings in public databases) has provided alternative sources of
information.

87. Data presented by the company on its web site claims that its chips are less
    prone to cross-hybridation (mismatched binding) and are cheaper than rival
    chips.

88. In the terms of the company's own words, the relationship changed from one
    of 'subsidiary' to 'affiliated' company.

89. The Millennium Project involves combined funding by various S&T min-
    istries. Biomedical fields that the project emphasizes are genomics, regener-
    ative medicine, diseases of aging, and newly emerging infectious diseases.

90. In 2003, sales by Affymetrix were US$223 million, up from US$128 million in
    2000. Affymetrix is the company that pioneered commercial gene chips and it
    remains the best known gene chip company in the world today. It has broad
    patents which it has tried to use to block potential competitors, with mixed
    success. Other US companies are marketing competing chips.

91. Perhaps the IPO was primarily a means for the subsidiary of the electronics
    company to recoup its investment (and make a tidy profit) while still retaining
    substantial control.

92. The company was founded in part to counter what was perceived as aggressive
    gene patenting by US companies. Partly as a matter of principle, it has not
    pursued patents on genes or gene sequences.

93. Queries to genetic researchers where I work suggest that they use a competi-
    tor's chips and are not familiar with GCR's chips.

94. This may be due to the company's hesitancy to challenge Affymetrix's broad
    patents.

95. A 2004 review of the products and services advertised under the subsidiary's
    web site indicates that most are related to GCR's chips.

96. These variations are known as single nucleotide polymorphisms (SNPs),
    which are variations between individuals in the sequence of nucleic acids that
    make up their individual genes. These variations often result in variations
    in the proteins encoded by the genes, and these variations sometimes affect
    the function of the proteins and thus health. The SNP project is a major
    component of the Millennium Project mentioned above at note 89.

97. A major component of 'personalized medicine'.

98. Mainly small molecule drugs and antibodies against various types of cancer.

99. One also hears that, because the venture has high *brand name* value, even if
    the alliances do not produce results for the pharmaceutical clients, the phar-
    maceutical managers who recommended the alliances would not be criticized.
    (They acted reasonably relying on the brand image of the laboratory and the
    university that stand behind the venture.) However, they would be criticized
    within their companies if they made an ultimately unsuccessful alliance with a
    startup from a lesser known university.

100. At the end of 2005, it had fallen to about one-quarter that value. However, the
     market capitalizations of most post-IPO bioventures were also significantly

below their immediate post-IPO peaks. At the end of 2005, Cancer Answers was still in the top half of post-IPO bioventures in terms of market capitalization.

101. In 2000 the company's biggest bottleneck was funding for personnel. It could almost match the salaries of pharmaceutical companies, but not their large retirement bonuses or the 'extras', such as the inexpensive relatively high quality housing that many large companies provide their employees. Stock options were the only monetary incentive CDD could offer prospective employees.

102. Fifty million yen or about US$450,000 for the two VC funds together. Some of this was not pure equity investment, but rather investments for convertible stock that had to be backed by tangible collateral (e.g. the founder's house).

103. i.e. cell lysates.

104. i.e. the receptors for these drugs.

105. It is sometimes a challenge to keep eminent founders from treating the venture's laboratories as their second academic laboratory. This applies to companies founded by retired as well as currently active academics.

106. Thanks to science parks and incubators established by national and local governments and some universities, the shortage of affordable laboratory space for bioventures has eased from nearly nonexistent levels around 1998. Also the government funded incubators have begun to clarify their IP policies so that startups can keep control over their IP. However, obtaining laboratory space is still a challenge. Reverse Targeting says the rent it pays is slightly below standard commercial rates.

107. The founders advise, but their main interest is academic research and thus they are not relied on to carry forward commercial R&D. This is one of probably a minority of university biostartups where the academic founders do not play a dominant behind the scenes management role.

108. The CEO attributed this mainly to consolidation in the pharmaceutical industry, which, although slow by US and European standards, is resulting in mergers and reductions in R&D workforces.

109. Including the VC company that was lead investor in Cancer Answers.

110. In the latter category are a number of CROs that manage clinical trials on behalf of pharmaceutical companies, and companies such as MediNet, that provide clinical services related to immune cell therapy.

111. In early 2006, Mitsubishi Pharma completed clinical trials of a drug delivery system licensed from LTT Bio-Pharma (a startup based on discoveries in Jikei Medical University and St. Marianna University School of Medicine) for prostaglandin E1 to treat peripheral artery disease. If the results are positive, FDA is expected to approve the drug for marketing in the USA in 2007. Mitsubishi Pharma also completed clinical trials and is now marketing a drug for hemorrhoids licensed from a ten person Okinawa biotech, Lequio Pharma.

Another of Japan's ten largest pharmaceutical companies has begun preclinical collaboration with a university biostartup on developing arthritis drugs, the main focus of this bioventure's efforts. The pharmaceutical company obtained an option for an exclusive license and in return provided an undisclosed amount of support. The bioventure described the negotiations as difficult.

I am excluding some collaborations where the bioventure is a spin-off or licensee of a pharmaceutical company. There undoubtedly are other collaborations of which I am not aware, but probably not many involving annual funding over US$100,000.

112. e.g. CAC's partner is a photo-optics company. One of MBV's earliest investors was a chemical company, as is Gene Angiogenesis's second partner. Gene Chip Research's partner is a subsidiary of an electronics company.

113. Not only are these sentiments heard at middle levels of pharmaceutical companies, they have been made by the directors of some of Japan's largest pharmaceutical companies. However, some pharmaceutical company scientists have said that some ventures have technologies that would merit further investigation by their companies, but it is usually difficult to convince management to engage these ventures. These responses are documented in Kneller (2003).

114. The circumstances under which Genentech, e.g., was formed, when genetic engineering was a wide open new field, do not apply today. Are there equivalent new fields that Japanese or US biotechs are just beginning to explore? I cannot speak for US companies. In Japan, I know of two broad new fields that biotechs are developing. Time will determine the merits of these technologies. Meanwhile, the biotechs are working hard to raise initial funding.

115. Specifically, in 1987 there were thirty-three therapeutically focused US biotechs that were less than seven years old (i.e. formed no earlier than 1981). Nineteen of these were listed in Dibner (1988) as having active partnerships with larger companies in 1986 or later. Of these nineteen companies, fifteen had joint venture, research, or equity partnerships, while ten had licensing, marketing, or acquisition (one case) partnerships.

In the 1991 and 1995 editions of *Biotechnology Guide USA* (the second and third editions of Dibner's survey of the US biotech industry), Dibner also lists partnerships for US biotech companies. Although I have not done the same analysis for the companies on these later lists, the basic pattern of large numbers of biotech firms involved in research and licensing partnerships with pharmaceuticals companies seems similar.

116. Based on my familiarity with over forty Japanese bioventures, approximately half of which are therapeutic oriented, as well as discussions with other persons familiar with Japanese bioventures. As in the typical Japanese case, some of the partnerships involving US bioventures mentioned in the previous note may involve only small levels of funding. However, evidence presented in Chapter 5 indicates that, in general, large company funding of pre-IPO bioventures is substantially higher in the USA than Japan.

117. These counts were obtained from one-time access to the rdna.com database in 2002 that allowed me to obtain lists and background information on most of the alliances of Schering-Plough, Bayer, and Abbott, three companies that

are somewhat larger than Takeda, the largest Japanese company. I counted partnerships active during any period between 1996 and 2000. *Research partnerships* include joint ventures and acquisition of equity in a bioventure. Some of the license partnerships also include a research component and vice versa. More information about the rdna.com database and this analysis is in Kneller (2003).

118. The facts that there were few Japanese bioventures and most of the Japanese partnerships were with US bioventures suggests that geography and barriers to communication may have been factors behind the lower Japanese numbers.

119. Some of their alliances are for drugs that are near the end of clinical trials, but others to access early stage drug candidates and drug discovery technologies that are still under development.

120. DNA Extractor being a notable exception. Korean pharmaceutical companies have invested in some biostartups.

121. None later than phase 2 human trials.

122. Even large companies only have around ten professionals in their licensing departments. They are usually preoccupied with large downstream deals and have neither time nor a mandate to build relationships with small Japanese companies. Also, the corporate budgeting process is geared to support alliances that will yield results (usually new drugs) quickly and that require millions of dollars in annual financing, more than most early stage bioventures can absorb.

In any case, proposals for outside collaborations must be vetted at several administrative levels. Penalties for failure usually outweigh the rewards of success for the persons who approve collaborations. This is partly because each person in the chain of command who approved an unsuccessful collaboration can be held accountable, whereas it is harder for most of those persons (especially those with most authority) to claim credit for a successful collaboration with a small company. Another reason is that problems with an alliance usually become apparent soon, while success is often not evident until after the persons responsible for initiating the alliance have been transferred to other jobs within the company.

This system of frequent internal rotations also provides fewer opportunities and incentives to build working relationships with bioventure partners, which are vital to the success of most alliances (see de Rond 2003).

Research managers often have discretion to authorize up to US$500,000 annually for outside projects, but they rarely exercise this authority, for reasons that are not completely clear.

123. One of the best known examples is Roche, which has about eighty employees in a special division dedicated to partnerships. Roche's two largest alliances are with Genentech and Chugai, Japan's eighth largest pharmaceutical company in 2002.

The relationship with Genentech began in 1980 when Roche licensed patents and know-how from Genentech that allowed Roche to develop alpha interferon. Ten years later when Genentech was in financial trouble, Roche invested nearly US$500 million in Genentech in return for 50% of Genentech's stock (Roche already held about 10% of the stock), an option to buy most of the remaining shares, and subsequently a ten year option (until 2005) to license non-US rights to any Genentech drugs that completed phase 2 human trials. Both companies agreed on the importance of Genentech retaining its independence, and Genentech has continued to operate independently. Although Roche is majority owner, Genentech can and does enter into alliances with other companies. Today Genentech drugs account for about 43% of Roche's prescription drug sales.

Since acquiring a majority of Chugai's stock in 2002, Roche's policy is also to preserve Chugai's independence. Roche has alliances with about 17 smaller biotechnology companies. Most of these were initiated since 2001, when Roche decided to put even greater emphasis on partnering and create a new division within the company dedicated to partnering. Currently half of its 20 top selling drugs result from outside partnerships, a proportion that Roche hopes will increase as the recent partnerships bear fruit. (Interview with Roche Sept. 2004; van Brunt 2004.)

124. According to managers or founders of several startups.

125. In 2004 I collected revenue and investment data on twenty-eight bioventures, including seven that had had IPOs. These twenty-eight are among the best-known and fastest growing ventures. The companies with the highest annual revenue (including government research contracts) during the most recently completed fiscal year were Mother of Bioventures, Gene Angiogenesis, DNA Extractor, Gene Chip Research, Cancer Answers, and Designer Mouse with revenues of approximately US$37, US$22, US$21, US$15, US$14, and US$5 million respectively. All these companies had had IPOs. Among the bioventures that had not had IPOs, the Institute for Medicinal Molecular Design (not featured in the case studies but with a business focus similar to that of Computer Aided Drug Discovery) probably had the highest revenue, about US$3 million. Among the other companies, no company focused on drug discovery and development had revenue over US$1 million, with the possible exception of one company focused on drug delivery technology which did not make public its revenue data. Of course these were early years for most of these companies and since 2004, revenue has increased for some—as have drug discovery and development expenditures and the need for additional equity financing.

126. Information for this list comes from various sources, but three useful references are Nakagawa (1999), Suzuki (1999), and Rowen and Toyoda (2002).

127. According to Suzuki (2001) many of these foundations and the companies they invested in are in trouble, the fundamental problem being lack of skilled people in the foundations and VC companies who can identify and foster promising companies. Yet Suzuki praises the foundations in some prefectures for recruiting persons with industry experience who are trying to help existing SMEs. These activities have since been consolidated under the SMRJ and some

SMRJ-affiliated investment institutions, such as the Tokyo Metropolitan SME Investment Corporation, are active investors.

128. *NIF News* (Aug. 2002).

129. Except as indicated, information is current as of the beginning of 2005.

130. These lists and some of the associated data were based mainly on lists compiled by METI for UT and Keio and from the web site of AIST. In the case of UT and Keio, I added a few companies that I know from other sources are based on discoveries from these universities. In other words, the METI lists probably cover most but not all the ventures closely affiliated with these institutions. The lists from METI and AIST provide the year of formation, address, CEO's name, and technology or business focus. I tried to find additional information on all the listed companies: the nature of their ties to UT, Keio, or AIST, sales (including names of major customers); paid in capital (including the names of major investors); staff and business focus. In most cases, I obtained such information from Internet sources. For sales and shareholder data, my main sources of information were commercial databases, such as the Japan Chamber of Commerce's (Nihon shoukou). In some cases, very little information was available, and in some of these cases, misclassification with respect to degree of affiliation with the university may be possible and information on sales or paid in capital might be old.

131. Thus, I included some ventures not initially founded on university discoveries but (*a*) whose main line of business evolved to one based on a professor's research or (*b*) whose main line of business depends greatly on ongoing collaborative research with a university laboratory. On the other hand, I excluded companies whose only ties with universities was collaborative research that supplemented their own in-house research, or one of whose key officers had recently been a university researcher (although I included ventures whose CEO or board of directors chair assumed those positions directly from a university position). In other words, for the purpose of this analysis, my definition of a startup was slightly more inclusive than a venture founded on university discoveries. In the case of the AIST ventures, I created a third category of companies that are being incubated in AIST but are not directly based on AIST discoveries.

132. See Table 4A2.3. Shane (2004) provides data for MIT and also cites data from several European countries showing the same phenomenon. MIT does not have a medical school. If it did, the proportion of life science ventures there might be even higher.

133. The proportion of life science startups from UT (33%) is comparable to the proportion of R&D resources available to UT's life science departments. For example, the total number of UT masters and doctoral students in medicine, pharmacology, agriculture, and other life sciences in 2004 was 2,769, compared with 4,598 in engineering, nonbiological science, and information S&T, that is, a 62% to 38% split. Also using breakdowns of commissioned research funding in 2002 by department, I estimate that of a total 7.53 billion yen attributable to either a life science or a non-life science, non-social science department,

2.35 billion yen (31%) went to a life science department. This calculation assumed that all 2.45 billion yen to IIS, which received more by far than any other department or center (including the university hospital), was for non-life science R&D, even though IIS indeed does substantial life science R&D. So 31% probably underestimates the share of R&D support in UT for life science R&D.

The high proportion of life science startups in the USA may be due in part to generous funding for academic life science R&D. Fifty-eight percent of all R&D expenditures in US academic institutions in 2001 were in the life sciences (S&E Indicators 2004: Appendix Table 5.5) indicating an even greater predominance of public funding for life science R&D in the USA compared with Japan.

Nevertheless, in the case of AIST, the proportion of life science ventures does indeed seem more than expected—at least on the basis of AIST's traditional focus on non-life science fields and the relatively low number of life science-focused institutions that make up AIST. Only four of AIST's current twenty-one institutes have a life science focus (see the list of AIST institutes at www.aist.go.jp/aist_e/organization/index.html). Prior to its reorganization in 2001, AIST had just one bio-focused institute, the Institute for Human Science and Biomedical Engineering. Nine of AIST's ten biostartups come from life science research centers or institutes. In other words, most of the AIST bioventures did not arise from researchers coming up with biomedical applications for research that was fundamentally nonbiomedical.

134. e.g. improved methods of replicating proteins, new types and applications of DNA chips, methods for transmitting diagnostic images (CT, MRI, etc.) to remote locations, or processing and supplying immune cells to hospitals and clinics.

135. Based on information available from the Internet and other public sources, plus my own knowledge of some of these companies.

136. Two to five of the fourteen UT biostartups, two to three of the four Keio biostartups, and zero to two of the ten AIST biostartups.

137. These are Elysium, whose second core technology, software for transmitting 3D image data, was developed by a UT professor; and Evolvable Systems Research, whose founder is from AIST and which is developing artificial intelligence and systems optimization software, as well as data compression systems using genetic algorithms. Elysium has nearly fifty employees and 2003 revenue of approximately US$12 million. Evolvable Systems Research had nearly US$2 million in paid in capital at the end of 2003 (quite high for any startup) but sales and employment data are not available.

138. At my research center in UT, materials science and its applications to ICs is probably the single most active area of research. Various reports on Japanese R&D in this field are available at www.atip.org. One survey estimates that the amount of Japanese government spending to support nanotechnology R&D is slightly higher than that of the US government, US$800 vs. US$780 million in 2003 (www.nanoinvestornews.com under nanotechnology facts and figures).

139. Recall the example of how pioneering research in titanium dioxide catalysts and surface coatings has been transferred informally to a number of established companies. Two of the seven materials startups are also developing applications of titanium dioxide catalysts, but the scale of their R&D seems small and their sales niches limited compared to the larger companies working on the same technology.

140. By one estimate the USA has over ten times the number of nanotechnology companies as Japan, over 500 vs. just under 50 (www.nanoinvestornews.com under nanotechnology facts and figures). Lists of the US companies available on the same web site indicate that many are based on university discoveries.

141. However, if one includes companies in other fields whose activities are related to energy or environment issues, then the total number in this category becomes thirteen (17%). Such companies include companies I have grouped under 'bio' but that are developing DNA chips to detect environmental estrogen disruptors or plant hormones to promote rooting and reforestation, an 'IT hardware' company that is developing environmentally friendly semiconductor materials, and 'materials' companies that are developing materials for solar cells and photocatalysts to neutralize air pollutants. On the other hand, some of the 'energy/environment' companies are working mainly with software or biological systems.

142. See Table 4A2.4. Only 22% appear to be backed to a significant extent by private venture capital. Only 40% appear to have significant investment from any private corporation (VC company, nonfinancial business, bank, or insurance company).

   Most of the data on investors and sales are from the Japan Chamber of Commerce's online database, which relies on data reported from the companies. Since companies usually want to let other potential investors know about sales and current investors, I assume that if no investors are named or no sales data listed, then these are either low or nonexistent.

143. Shane (2004: 289).

144. This is probably an underestimate, because in three of these four cases, I found out about a retiree's founders role from personal knowledge.

145. Based on information available over the Internet and my own scientific knowledge (plus familiarity with some of these companies), I estimate that thirty-seven to forty-four (or 48–57%) of the seventy-seven startups are in this category.

146. e.g. those developing titanium dioxide catalysts as air purifiers, building windmills, providing mobile or broadband Internet services, developing eye or ear sets so that wearers can receive local wireless visual or auditory signals, modeling water flow, simulating stock market prices, providing immune cells to hospitals, developing instruments that can position an object to nanometer precision, providing and analyzing earth sensing images, consulting on residential environmental issues, and involved in technology transfer and incubation.

147. However, I am not a specialist in materials, software, or engineering technology, and, aside from some familiarity with the research at Shonan Fujisawa and with some of the UT startups in these fields, outside of the life sciences I do not have inside information about these companies or the laboratories from which they emerged.

148. One of these is another Hokkaido University startup. Another is a startup involving collaborators from Kyoto University and research laboratories of the Ministry of Agriculture.

149. This particular research appears to have potentially diverse and important commercial applications. The ninth company, Assay and Systems, had links to the Research Center for Advanced Bionics within AIST, but these links are unclear.

REFERENCES

Blackburn, Robert (1998). 'How SMEs See Their Intellectual Property Rights', Summary Report to the ESRC (Economic and Social Research Council, UK).

Cabinet Office (of the Prime Minister), Intellectual Property Strategy Headquarters, Expert Study Committee on Strengthening the Basis for Rights Protection [Chiteki zaisan senryaku honbu, Kenri hogo kiban no kyouryoku ni kansuru senmon shousakai]. 2005. Plan to Promote the Intellectual Property Strategy of SMEs and Ventures [Chuushou , benchaa kigyou no chiteki zaisan senryaku no suishin housaku].

Cohen, Wesley M., Goto, Akira, Nagata, Akiya, Nelson, Richard R., and Walsh, John P. (2002). 'R&D Spillovers, Patents and the Incentives to Innovate in Japan and the United States', *Research Policy* 31: 1349–67.

de Rond, Mark (2003). *Strategic Alliances as Social Facts*. Cambridge: Cambridge University Press.

Dibner, Mark D. (1988). *Biotechnology Guide U.S.A.*, 1st edn. New York: Stockton Press.

——(ed.) (1999). *Biotechnology Guide U.S.A.*, 5th edn. London: Macmillan Reference.

Ibata-Arens, Kathyrn C. (2000). 'The Business of Survival: Small and Medium-Sized High-Tech Enterprises in Japan', *Asian Perspective*, 24(4): 217–42.

Institute of Intellectual Property (IIP) [Chiteki zaisan kenkyuujo] (2005). 'Report of the Study Committee on how to make Practical Use of Intellectual Property Arising in SMEs and Venture Companies' [Chuushou kigyou, benchaa kigyou ni okeru chiteki zaisan no katsuyou housaku ni kansuru kenkyuu kai houkoku sho] (issued in March).

Japan Bioindustry Association (JBA) (2005). 'Statistical Survey Report on Bioventures 2004' [2004 nen baiobenchaa toukei chousa houkokusho] Tokyo: JBA (in Japanese).

Japan Small Business Research Institute (JSBRI) (2003). White Paper on Small and Medium Enterprises in Japan (English version), pp. 87, 270.

Kerr, William O. and Prakash-Canjels, Guari (2003). 'Patent Damages and Royalty Awards: The Convergence of Economics and Law', *Les Nouvelles*, 38(no. 2, June): 83–93.

Kneller, Robert (2003). 'Autarkic Drug Discovery in Japanese Pharmaceutical Companies: Insights into National Differences in Industrial Innovation', *Research Policy*, 32: 1805–27.

Macdonald, Stuart (1998). 'What the Patent System Offers to the Small Firm, Summary Report', Prepared for the ESRC (Economic and Social Research Council, UK).

METI (2005). 'Basic Survey Report on University Ventures' [Daigaku hastu benchyaa inkansuru kiso chousa houkokusho] (June, in Japanese).

—— (2006). 'Basic Survey Report on University Ventures' [Daigaku hastu benchyaa inkansuru kiso chousa jisshi houkokusho] (May, in Japanese). Produced in cooperation with Kabushiki Kaisha Kachi Sougou Kenkyuusho [The Valuation Institute, Inc.] (listed as official report author).

Nakagawa, Katsuhiro (Dec. 1999). 'Japanese Entrepreneurship: Can the Silicon Valley Model be Applied to Japan?' (adaptation of presentation given on May 19, 1999 at the Asia Pacific Research Center, Stanford University).

NIF News (Aug. 2002 edn.) Toushi jiggyou kumiai shusshi nimo tekiyou: shin enjeru zeisei sutaato [The angel tax incentive now also applies to investments in investment cooperatives—the start of new angel tax measures]. Available at www.nif.co.jp/public/img/pdf/news02_08.pdf

Nikkei Weekly (2002). 'Big Boys Buying into Start-Ups', October 28, 1.

Paull, Robert, Wolfe, Josh, Herbert, Peter, and Sinkula, Michael (2003). 'Investing in Nanotechnology', *Nature Biotechnology*, 21 (no. 10, October): 1144–47.

Rowen, Henry S. and Toyoda, A. Maria (2002). 'From Keiretsu to Start-ups: Japan's Push for High Tech Entrepreneurship' (Asia Pacific Research Center, Stanford Unversity).

Shane, Scott (2004). *Academic Entrepreneurship: Academic Spinoffs and Wealth Creation*. Cheltenham, UK: Elgar.

Suzuki, Hiroto (1999). *Venture Business in Japan*. Background Report from Foreign Press Center Japan. Copy available from author.

Thayer, Ann M. (2003). 'Nanomaterials: Large Chemical and Materials Companies Target Small Nanotechnology Firms for Venture Investing, Collaborations and Product Innovations', *Chemical & Engineering News*, 81(no. 35, Sept.): 15–22.

Tang, Puay, Adams, John, and Pare, Daniel (2001). Patent Protection of Computer Programmes: Final Report (Submitted to the European Commission, Directorate of General Enterprises).

Tsukuba University, Industry–University Liaison Joint Research Center (2006). Survey of University Ventures and Promotion Strategies [Daigaku nado hatsu benchaa no kadai to suishin housaku ni kansuru chousa kenkyuu] issued March, in Japanese.

Van Brunt, Jennifer (2004). 'The Ties that (Don't) Bind', *Signals Magazine*, Sept. 14. Available at www.signalsmag.com

5

IPO or Bust: Venture Financing

Low levels of angel financing and the unwillingness of large companies to engage ventures have compelled greater reliance on venture capital (VC) and public equity markets to finance the growth of high technology ventures in Japan. This is especially true in the case of bioventures. As for nonbiomedical ventures, the successful ones have managed to obtain substantial revenue through sales. Yet even these usually depended to some degree on VC financing in their early years, as shown in Chapter 4. Thus the small number of nonbiomedical ventures, particularly startups, may be in part the result of the inadequacy of VC financing to support promising new companies. Fortunately, the system of VC financing has matured greatly since the mid-1990s, although problems remain. However, even if all problems were to be solved, VC financing probably cannot substitute for the lack of financing from sales, angels, and investment and joint research by larger companies.

Sales and Alliances

Sales and other forms of self-financing probably are the most important overall source of funding for high technology ventures, larger than equity investments by angels and definitely larger than VC and public (post IPO) equity investments. This has been shown in the case of recently formed US life science ventures.[1] Even in the case of therapeutic-oriented life science ventures, which tend to have high R&D costs and low sales revenue, alliances with other companies (primarily pharmaceutical companies) provide pre-IPO financing equivalent to, and in some years exceeding, VC equity investments.[2] Some drug-development-focused US bioventures will wait until they have entered into a multi-million dollar license contract or alliance with a major pharmaceutical company before having an IPO. While new high technology Japanese companies also try to finance growth through revenues, Chapter 4 indicates that the amount of financing they raise from sales and investments by other companies is small.[3]

Angels

After revenue from sales and contracts, equity investment by individual investors (angel investment) is the second most important form of financing for new high technology companies in the USA.[4] In 2005, approximately 230,000 business angels invested US$23 billion in US companies in their early stages of growth. Most of these funds were to bridge the gap between seed investments from founders, friends and family (typically US$25,000–US$100,000), and early stage VC funding (beginning around US$2–5 million). Individual angel investments typically range from US$100,000 to US$500,000. Investments by groups of angels in the US$500,000 to US$5 million range are becoming more common to meet financing needs of new companies beyond the startup phase.[5] These angel investments are slightly greater than total VC investment (US$22 billion in 2005). However, because approximately three-fourths of VC investment is second or higher round funding,[6] angel investment is about fourfold greater than VC investments as a source of funding for companies in their early growth stages.[7] Angels usually do not expect as high or as fast a return as VC companies. They often are persons who know the industry and technology well, and they expect to be actively involved with the companies they invest in. Although financial return is a motive, often too are mentoring, promoting entrepreneurship, advancing a technology, and benefit to community.[8]

New Japanese companies also benefit from investment by individuals not related to the founders. Among the companies featured in the case studies in Chapter 4, Mother of Bioventures (MBV) and Gene Chip Research (GCR) received important investments from individuals in their early years.[9] Amounts invested by Japanese angels per venture company are, however, less than investments by US angels.[10] But the most striking difference is that the number of business angels, persons who invest to foster the growth of entrepreneurial companies, is very small in Japan. One of the main surveys of angel activity identified only 360 likely angels nationwide and only 1,667 likely target companies.[11]

A 1997 law that reduced taxes on gains from angel investments and allowed losses from such investments to be deducted over three years (instead of just the year they were incurred) has apparently done little to promote angel investment. Reasons include the requirement that the target companies be

officially recognized by METI as venture businesses and burdensome paperwork necessary to qualify for these tax benefits.[12]

VENTURE CAPITAL: RISING TO THE CHALLENGE

With limited angel or corporate investment, Japanese ventures must depend
on government funding or private venture capital, unless they have sufficient
revenue from sales and contracts or can secure bank loan financing. Government funding accounts for about 20 percent of average R&D expenses, but a
considerably higher percentage in startups that actually receive government
grants or contracts.[13]

Both NEDO and JST have competitive funding programs for university or
company researchers who are planning to establish startups or spin-outs. Until
recently, however, university researchers tended to regard these as similar to
MEXT grants-in-aid and few viable startups resulted. In order to improve this
situation, the composition of the NEDO review committees has been changed
from nearly all academics to a mixture of business persons and academics.
Whether this will have the desired effect remains to be seen.

In view of the danger of overdependence on government assistance, the
burden is on VC to finance the healthy growth of ventures. Moreover, venture
capital must shoulder that burden beginning much earlier in the development
of venture companies than in the USA.

In the late 1990s, the environment for VC financing of high technology
ventures began to improve, largely as a result of public equity markets adopting flexible listing requirements. 'VC investments' shifted from loans, and
purchases of bonds and publicly traded stock (much overseas or in established
companies in the retail and service industries) toward pre-IPO equity investments in high technology companies less than 5 years old.[14]

Prior to 1998, all of Japan's principal stock exchanges required substantial sales and capitalization before a company could be listed—requirements
that most independent technology based ventures could not meet. In 1998
Jasdaq,[15] Japan's over-the-counter market, relaxed its listing requirements to
allow companies in operation less than ten years to list, even if they have
no sales, so long as they have positive net profits. The following year, the
Tokyo Stock Exchange opened its third section, Mothers,[16] whose only listing
requirement with respect to sales is that they be greater than zero. In 2000,
Nasdaq Japan became the third major market for venture company stocks
with listing requirements similar to those of Nasdaq USA. Nasdaq Japan
ceased operations in 2002, and its operations were taken over by Hercules,
a concessionary section of the Osaka Stock Exchange, modeled on Mothers.

Since 1999, private pension funds have been free to invest in ventures.[17] Nevertheless, fund managers have invested conservatively and the percentage of pension assets invested in private equity (which includes pre-IPO ventures) is only a small fraction of that in the USA.[18] Also in the case of pension funds for civil servants, individual ministries generally still must approve each fund's plans to invest in private equities and such investments are rare.[19] The amount of private pension money being invested in ventures is gradually increasing but most of these investments are unlikely to go to companies developing new technologies.[20] It is unlikely that in the near future, pension assets will be the boon to venture financing that it was in the USA following legal changes that reduced restrictions of pension funds investing in ventures.[21]

Overall levels of VC investment are still low relative to the USA and Europe. In 2004, VC investments in the USA, Europe, and Japan were approximately US$21, US$48, and US$1.8 billion, respectively.[22] On the positive side, VC raised for eventual investment in startups had increased to somewhat under one billion US$ by the end of 2004.[23]

CHALLENGES CONFRONTING VC FUNDING

In the USA, most VC companies are independent, but in Japan only around 12 percent are. Most are subsidiaries of banks, securities, and insurance companies.[24]

Many lack scientific and business expertise, and tend to base investment decisions mainly on financial criteria appropriate to traditional lending.[25] Their general partners or managers have few career incentives to invest energy in building a young company.[26] There is a tendency to follow the lead of a few VC companies that have the reputation of having technical and business expertise, and also to invest small amounts in each target company.[27] As a result, the influence of VC investors over venture founders and managers tends to be weak and they are also rarely in a position to offer constructive guidance to their portfolio companies.

VENTURE CAPITAL ANGELS

Notable exceptions to this pattern of investment are a few small VC companies focusing on early stage investments in high technology ventures and often taking an active role in assisting their portfolio companies to grow. I am familiar with five such companies focusing on biomedical ventures, two that have a diversified focus, and I know of at least one or two focusing mainly on IT. They are playing roles similar to angels in the USA. Their professional staffs usually number fewer than five but usually include persons with technical expertise[28] experience in larger VC companies or corporate management. Most of these small VC companies have track records of starting companies, securing cofinancing, forming management teams, developing business plans, and advising on key business decisions.[29] At the height of the bioventure IPO bubble in 2004, they avoided the temptation to push for early IPOs.

The largest of these hands-on VC companies has raised a total of about US$100 million deployed in three funds.[30] But more typically, these companies have raised one or two funds, of about US$20 million each. When they are lead first round investor, they typically invest up to US$1 million. They raise money from government investment corporations, medium- or small-size pharmaceutical companies, insurance companies, trading companies, traditional VC companies, banks, and food, chemical, and a few other manufacturing companies (listed in approximate order of importance).[31] Except for the largest of these hands-on VC companies, raising capital has been difficult.

They use various methods to find investment opportunities. The best known often rely on entrepreneurial scientists to come to them. Some require new employees to bring with them at least one promising investment opportunity. A few of the VC companies actively seek out entrepreneurial researchers with promising discoveries.

As an indication both of the seriousness of these companies' commitment to their main portfolio companies, and also of the shortage of skilled venture managers, several of these hands-on VC companies have assigned some of their top managers to head some of their main portfolio companies. Usually these assignments are temporary until they can find a qualified outside manager. But as of 2006, at least two of the largest pre-IPO bioventures are headed by persons who were former officers in hands-on VC companies.

But despite these efforts by the hands-on VC companies, so far no particular type of investor seems to have a better track record than others in identifying investment opportunities and nurturing them to the point of a successful IPO or sustained growth. The main early investors in the bioventures that had IPOs before 2005, represent a cross section of investors.[32] Some of the portfolio companies of the hands-on VC companies are on a promising growth path, but their futures are still uncertain.

The small hands-on VC companies are facing competition for access to entrepreneurial university researchers and their discoveries. Some large traditional VC companies, insurance companies, securities firms, and trading companies have established special relationships with individual universities that allow them to screen invention reports and interview entrepreneurial inventors to identify startup opportunities. In effect, these companies have an option (subject to the inventors' and university's veto) to be the lead investor in startups to emerge from those universities.[33] One consortium of traditional VC investors established a fund of approximately US$80 million, primarily to invest in startups from the University of Tokyo. Doubts have been expressed about the ability of such funds to evaluate technical and business opportunities and to provide practical guidance to their portfolio companies. In any case, privileged access for a few VC companies to the entrepreneurs and their discoveries in key universities risk foreclosing opportunities for investors that may have greater competence, and thus reducing competition for high quality VC investment services.

FEBRUARY BLOSSOMS: THE FIRST IPOs

Between 2002 and 2004 nine technology-focused biomedical ventures had IPOs.[34] Except for three earlier life science IPOs[35] these represent the first wave of IPOs for biomedical ventures, and the first IPOs for university startups in any field. Indeed, they were among the few IPOs for new, independent high technology-oriented companies in any field. Thus the question, 'Do these IPOs represent a breakthrough heralding sustained growth for new high technology companies?' was on many persons' minds. These IPOs were remarkable because most of the companies were quite young and they had raised, on average, less than US$10 million in pre-IPO funding. Nevertheless, the IPOs in 2002 and 2003 were extraordinarily successful, and the market capitalization of some of these companies soared to around US$1 billion shortly afterward.[36]

Yet the companies with the highest market capitalization did not have products to sell.[37] Compared to the US, the Japanese valuations were inflated. In particular, US bioventures having IPOs were only a couple of years older on average than their Japanese IPO counterparts, but they had raised over US$100 million before the IPO on average and most had drugs in phase 2 or 3 clinical trials.[38] (This is consistent with the previous discussion that it is easier for bioventures to obtain funding from angels and pharmaceutical companies in the USA. Also, they can hire more people and develop products more quickly.) But despite having raised more private investment and being more advanced, their post-IPO valuations were lower than the Japanese companies.[39]

The air began to go out of the bubble in 2004. Four Japanese biotechnology companies had IPOs that year but share prices did not surge the way they had the two previous years. Then the stock prices of most of the publicly traded bioventures began to decline.[40] There was a sense that too much of the risk for financing the growth of new companies was being shifted to public investors, who cannot be expected to have as much in-depth awareness of risks as private investors. In order to prevent shares of inherently weak companies being sold on public markets, the major exchanges tightened listing requirements. In particular, for therapeutic-oriented ventures, a company now ought to have two drugs in clinical trials and these drugs should be covered by a licensing or cooperative research agreement with one or more pharmaceutical companies.[41] Throughout 2005 and 2006, there were no IPOs for Japanese bioventures on any of the three major exchanges.[42] Instead of turning to public markets, they have had to struggle to raise funds from private sources. Based on discussions with a number of companies, this has not been easy, although some companies with promising technologies are managing to raise funds.[43] In other words, the Japanese system seems to be making necessary, although difficult adjustments.

Nevertheless, the current situation also gives rise to concerns. First, most second, third and later round funding is being raised from traditional venture capital, whose level of investment sophistication and ability to help companies grow is increasing only slowly. Pharmaceutical companies and pension funds are still standing on the sidelines.

Second, even though some companies manage to raise later round financing, it may not be sufficient to keep pace with overseas rivals. Also, because of social and demographic factors and less reliance on stock options, the same amount of funding probably employs fewer skilled researchers and managers in Japan than overseas.

Third, it may be unwise for equity markets to apply strictly their guidelines that alliances with pharmaceutical companies cover a venture's main drugs. Similar requirements exist for listing on Nasdaq. But in the USA, large companies are generally open to partnering with new high technology companies, and a bioventure with promising technology is likely to find several interested prospective partners. But in Japan, a de facto requirement that the main drugs of bioventures be covered by alliances with pharmaceutical companies places the former in even weaker bargaining positions. In view of large Japanese companies' typical stance of malignant neglect toward domestic ventures, forced partnerships are likely to be on the terms of the large companies.[44] There is evidence that when biotechnology companies are forced into alliances with pharmaceutical companies under which they lose control over their lead drugs, the likelihood that those drugs will be developed decreases.[45] Thus strict implementation of the current listing guidelines may harm pharmaceutical innovation in Japan.

Fourth, the immaturity of the pre- and post-IPO financial markets is increasing financial pressure on ventures. Aside from the founders and other individuals (often family members), most pre-IPO investors in bioventures have been VC funds affiliated with banks and insurance and securities companies, as well as a small number of traditional VC companies.[46] Lacking specialized knowledge about the science and business of their portfolio companies and incentives to help them over the long-term, these traditional VC investors tend to push for early exits so that they can recoup their investments early with some profit.[47] Nonetheless, VC must shoulder a greater proportion of the funding burden than in the USA and recently must do so over a longer period, beginning soon after formation of the venture. Thus investor pressure for early IPOs is intense.

However, institutional investors, particularly pension funds, have been hesitant to invest in post-IPO high technology ventures that do not have a strong revenue stream.[48] In the case of bioventures, this exacerbates the cooling of post-IPO-bubble investor sentiment and declining share prices.[49] One positive development is that insurance companies are beginning to increase investments,[50] in part because they are hiring gatekeepers who can assess the technical and business potential of young high technology companies.[51]

If astute post-IPO institutional investment grows, the system may self-correct and fuel the growth of more high technology ventures. If not, a vicious cycle may become more entrenched. The stringent listing requirements for bioventures that have led to a continuing suspension of biotechnology IPOs on the major markets, may restore the confidence of post-IPO investors. But at least in the life sciences it may choke off the only source of funding that has so far sustained most high technology ventures.

THE IPO LANDSCAPE

Examination of the list in Appendix Table 5A.1 of technology-focused[52] companies that have had IPOs in Japan between 2000 and 2004, suggests the following points.

1. Only about 19 percent of IPOs (147 of 791) are for technology-focused companies, only 12% if software companies are excluded.

2. Software companies account for the largest number of high technology IPOs (38%), although the numbers of IPOs for software companies has been falling. Numbers of IPOs for IT hardware[53] and other engineering

companies were steady. Life Science, including companies that manage research and clinical trials, is the only category that showed an upward trend, but numbers are small compared to engineering and software. Furthermore, IPOs for pharmaceutical and biotechnology companies fell sharply after 2004. Surprisingly for a country that is investing heavily in nanotechnology and other fields of materials research, the number of IPOs in these fields is small.

3. Most of the technology-focused IPOs (121 of 147 or 82%) are taking place on the concessionary markets where listing requirements are less stringent.

4. Perhaps most surprisingly, except for software and life sciences, most of the companies having IPOs are old. Even among the IT hardware companies, more than half were incorporated more than twenty years before their IPO, and only 30 percent were incorporated within ten years of the IPO. All of the IPO companies specializing in high technology materials (including nano-materials) were either old at the time of their IPOs or were subsidiaries of large established companies. In other words, most of the money that is being invested in newly listed high technology companies is supporting old companies. Also, VC investment in these old companies is substantial.[54] A few of these old companies have evolved quickly and creatively into new fields of technology,[55] a process that access to investment funds probably promotes. Also IPOs are probably allowing some of the older companies to break out of subcontractor status and begin innovating in fields that were outside their scope of business as subcontractors. Nevertheless, my working hypothesis is that when old companies change to more 'modern' fields of technology, they are probably following the leads of earlier pioneers of these technologies, rather than being pioneers, themselves.[56] This raises the question whether scarce investment funds and human resources are being used optimally when so many are being allocated to old companies.

5. The only companies that had IPOs between 2000–2004 and are founded on university discoveries are bioventures.[57] In part, this is a reflection of the low number of promising university startups in IT hardware, other engineering, and materials compared to biomedicine and software.[58] As shown in Chapter 3, although much university research occurs in these fields and a substantial proportion of university patent applications are in these fields, university discoveries in these areas are being preempted by established companies.

APPENDIX 1

Table 5A.1 (pp. 178–9) Constitutes the main item in this appendix. Immediately following are notes concerning methodology:

1. Particularly in the case of software companies, I had to decide on the basis of brief descriptions of the companies on Tokyo IPO (one of the web sites providing comprehensive online lists of companies that have had IPOs) and information on the companies' home pages, whether they were significantly focused on new product R&D or more oriented toward sales, services, consultation, etc.

2. Some companies could have been assigned to more than one technology category, and in these cases I made a judgment call assigning the company to only one field. For example, Intec Web and Genome Informatics Corp. (IPO 2000) is developing data and internet systems with both biomedical and nonbiomedical applications. Because the bioinformatics applications are a significant part of their business, I assigned it to 'biotech bioinformatics', but I could also have assigned it to 'software and information management.'

3. In the case of Tokai Introduction Systems' IPO in 2002 and JMNet's IPO in 2004, I considered Nagoya and Fukuoka to be concessionary exchanges.

APPENDIX 2: ANALYSIS OF MAIN PRE- AND POST-IPO INVESTORS IN JAPANESE BIOVENTURES

To better understand who are the investors in biomedical ventures before and after IPOs,[59] I have examined lists of stockholders for most of the biomedical ventures that had IPOs in 2002 and 2003. The pre-IPO lists were submitted as part of each company's application to list its stocks on public exchanges. The ten investors holding the largest value of post-IPO shares are available from the exchange web sites, and I used the lists current as of mid-2004.[60]

In the case of *Precision Systems Science*, financial corporations and traditional VC funds were prominent among the top ten holders of pre-IPO stock. Precision Systems had its IPO on Japan Nasdaq in Feb. 2001. By June 2004, all these corporate investors had exited, except JAFCO, which had reduced its shareholdings by about 80 percent. In their places were two individuals who were not previously investors, Osaka Securities Finance Co., and companies holding the stock of previous inventors in trust.

In the case of *Anges MG* which had its IPO on TSE Mothers in Sept. 2002, the leading investors were the same. They are all individuals except for one independent VC fund focusing on bioventures.

In the case of *TransGenic* which had its IPO on TSE Mothers in Dec. 2002, insurance and traditional VC companies were prominent among the leading pre-IPO investors. By March 2004, they had all exited. Two life insurance companies that had not sold

Appendix 1, Table 5A.1. Manufacturing or R&D-focused Japanese companies with IPOs 2000–2004

Software & information management

2000	2001	2002	2003	2004
Met's 1988 M	CIJ 1976 J	B Map 1998 N	Nippon Information Development 1967 J	Software Service 1969 H
Toukei Computer 1970 T2	Access 1984 M	Symplex Tech. 1997 J	Yasukawa Information Systems 1976 T2	10art-ni 1997 M
Kozo Keikaku Engin. 1959 J	Open Loop 1997 N	Tokai Info Sys. Consultation 1974 Nagoya	CNA 1996 M	
Showa Sys. Engin. 1966 J	Dream Technologies 1995 N	TCC 1969 J	SunJapan 1989 J	
Digital Design 1996 N	Intelligent Wave 1984 J	Dawn N 1991	Wacom 1983 J	
Information Creative 1978 J	Planex Communications 1995 J	XCat 1973 J	Associant Tech. 1994 M	
NEC Soft 1975 T1	*Matsushita Info Systems* 1999 J	SystemPro 1983 N	*NEC System Tech.* 1977 T1	
A&I System 1987 N	Solxyz 1981 J	SoftFront 1997 N	Aplix 1986 M	
Cyboz 1997 M	Artiza Networks 1990 M	Digital Arts 1995 N		
Nippon Comp. Dyn. 1967 J	JIEC 1985 T2	NetVillage 1997 N		
Softscience J†	Nippon Time Share 1993 J	NS Solutions 1980 T1		
ImageOne 1984 N	Open Interface 1992 N	Quest 1965 J		
Fujitsu BSC 1963 J	CSI 1996 M	Pixela 1982 M		
Info. Serv. Int'l Dentsu 1975 T1	e-System 1994 N			
Real Vision 1996 M	Nippon System Tech. 1972 J			
	Intage 1960 J			
	Works Applications 1996 J			
	Touhou System Science 1971 J			

IT hardware

2000	2001	2002	2003	2004
Tesec 1959 J	MEC 1969 N	*NEC Mobiling* 1972 T2	Seiko Epson 1942 T1	Raytex 1988 M
Seikoh Giken 1972 J	Interaction 1992 M	NewTech 1982 J	*NEC Electronics* 2002 T1	Arm Electronics 1980 J
Yozan 1990 J	Kubotec 1985 M	North 1990 M	Wintest 1993 M	Tazmo 1972 J
NEC Machinery 1972 O	Santec 1979 N	Axel 1996 J		*Elpida* 2004 T1
Meiko Electronics 1975 J	Thine 1992 J			JMNet 1995 Fukuoka
V Technology 1997 M				Taiyo Industrial 1960 J

Electronics, machinery, other engineering

2000	2001	2002	2003	2004
Nichidai 1967 J	Kokusai Co., Ltd. 1969 J	Takeuchi Mfg 1963 J	A&D 1977 J	Ulvac 1952 T1
Tohoku Pioneer 1966 T2	Takamatsu Machines 1961 J	Phoenix Elec 1976 J	A&T 1978 J	Hephaist Seiko 1962 J
Takatori 1950 O	Photonics 1982 N		Techno Medica 1987 J	CCS 1992 J
Patlite 1963 O	Samco Int'l 1979 J		General Packer 1961 M	ADTEC Plasma Tech. 1985 M
Hongo 1961 J				Shicoh Engineering 1976 M

Materials & nanotech

Nakanishi 1930 J
Nidec-Read Corp. 1991 O
Cimeo Precision 1963 J
Tanaka Precision Eng. 1957 J
Showa Shinku 1958 J

Nihon Micro Coating 1941 J

Taisei Lamick 1966 T2

Nihon Ceratec 1987 J
Tocalo 1951 T2
Okamoto Glass 1928 J

Myotoku 1951 J
Kawamura 1995 M
Optoelectronics 1976 J
NS Tool 1954 J
Kawaden 1940 J
GMB 1943 O
Asahi Intec 1976 J

Materials & nanotech

Toyo Gosei Co. Ltd. 1954 J

River Eletec 1949 J

Chemicals & cosmetics

Tanaka Chemical 1957 J
Sun A Kaken 1942 J
Hagihara Industries 1962 T2

Japan Pure Chemical 1971 J

Dr. C Lab 1999 J (cosmetics)
Haba 1983 J (cosmetics)
Matsumura Oil Res. 1958 J

Daiichi Kigenso 1956 T2
Kuriyama 1940 T2

Biotechnology & bioinformatics

Intec Web & Genome Informatics 1989 M

Precision Systems Science 1985 N

Anges MG 1999 M
TransGenic 1998 M

MediNet 1995 M
OncoTherapy 2001 M
MediBic 2000 M

DNA Chip 1999 M
Sosei 1990 M
LTT Bio Pharma 2003 M
Takara Bio 2002 M

Clinical trials & health information management

EPS 1991 J

I'rom 1997 J
Soiken 2001 M
Site Support Inst. 1999 M

Shin Nippon Biomed. Labs 1973 M

Energy

Japan Wind Dev. 1999 M
Japan Petroleum Exp. 1955 T1

J-Power 1952 T1

Key for each entry: **Company name**, incorporation year, market* on which IPO occurred. *Italics* = company is a subsidiary of or spin-off from an established company.

Source: Tokyo IPO and information on specific companies.

* *Abbreviations for markets with standard listing requirements*: T1 = Tokyo Stock Exchange (TSE), 1st Section; T2 = TSE, 2nd Section; O = Osaka. *Abbreviations for concessionary equity markets*: M = Mothers (part of Tokyo Stock Exchange), J = Jasdaq (over the counter), N = Nasdaq Japan (in business only 2000–2003), H = Hercules (part of Osaka Stock Exchange). Hercules has taken over many of the companies listed by Nasdaq Japan.

† merged with iNet 2001.

their shares moved into the ranks of the top ten investors, as did a local government regional investment fund, a public utility (Japan Power), a mutual fund/securities firm (Matsui Shouken), and a company holding shareholders' equity in trust.

In the case of *MediNet*, which had its IPO in June 2003 on TSE Mothers, lead pre-IPO investors consisted of individuals, a government regional investment company, banks, traditional VC companies, and insurance companies. Fifteen months after the IPO, the lead bank and the two lead insurance companies had either exited or significantly reduced their shareholdings. Only one traditional VC and one insurance company (Daimond Capital and Dai-ichi Life Insurance) remained among the lead investors. The only significant new investor was a diversified Osaka manufacturing company, Kaneka.

In the case of *Soiken*, which had its IPO in Dec. 2003 on TSE Mothers, the pre- and post-IPO leading shareholders were all individuals whose shareholdings remained little changed six months following the IPO.

In the case of *OncoTherapy Science*, which also had its IPO in Dec. 2003 on TSE Mothers, its lead investors had changed little three months after its IPO. These included the lead investor, a dedicated bioinvestment fund formed within a VC sub-sidiary of a software company (CSK), and the local government's regional investment corporation. Two individuals and the university TLO had sold or reduced their shares, and as a result, individual pre-IPO investors moved up in the rankings.

NOTES

1. Willoughby (2005) found that 64% of financing for 184 bioscience firms in New York and Utah was from revenue (sales and contracts) in the late 1990s, and that the mean annual revenue per firm was US$35 million. Willoughby's sample consisted of most of the bioscience firms in these two states—old and new, large and small firms in pharmaceuticals, biotechnology, medical devices, and biological systems (e.g. environment and energy). Nevertheless most of the firms were young (47% were no more than 10 years old) and small (60% had no more than 25 employees). Even firms less than 6 years old obtained on average about 46% of financing from revenue, as opposed to about 21% from equity (almost entirely angel) investments.

   However, Willoughby's findings do suggest that life science firms that rely on VC or angel investments for a larger than average proportion of financing tend to have more revenue, faster revenue growth, more R&D spending, and more R&D spending per employee, but also lower profits than other life science firms. In other words, VC and angel equity financing does enable R&D intensive firms to grow rapidly beyond the limits that revenue, grants and the founders' personal funds, would otherwise place on R&D.

   Dibner's survey findings (1988) are consistent. 45% of US biotechnology companies said that over half their total financing came from private sources (presumably personal assets, loans, sales, and maybe also angel investment). Only 15% said that VC accounted for more than half their financing, and about the same percentage said public equity accounted for at least half their financing.

These analyses apply only to life science companies. But, because ventures focusing on drug development rely disproportionately on VC during their early years (because sales are low and R&D expenditures high), ventures in engineering, materials, and software probably rely even less on VC and other forms of equity financing.

2. In 2003, VC investments in private (pre-IPO) US biotech companies was nearly US$3 billion. In 2003, pre-IPO VC biotech investment was recovering from its 2001–2 slump, and this level was second only to the US$3.3 billion invested in 2000 (SBLSEM 2004).

For 2003, 156 alliances related to drug R&D or marketing rights and involving payments to biotechnology companies were recorded in the Recap rdna.com database of biotechnology and pharmaceutical alliances. The average payment per alliance was US$77 million, implying that total payments were about US$12 billion (McCully and van Brunt 2004). Having used this database previously (see Kneller 2003 and Chapter 2), I know that some of the bioventures receiving funding under deals recorded in this database are non-US biotechs (20% being a rough estimate). Among the remaining biotechs, some have already had IPOs, usually on Nasdaq (70% being a rough estimate). Therefore, I roughly estimate that about US$3 billion of the total US$12 billion in alliance payments went to pre-IPO US bioventures, approximately equal to the amount of pre-IPO VC investment in all US biotechs, including those not focused on therapeutics.

However, some of these Nasdaq-listed companies probably had IPOs that followed closely on the conclusion of the alliances. (In other words, these alliances set the stage the IPO, and thus 70% may have been too high a discount factor to apply in order to exclude post-IPO alliances.) Also some (perhaps many) alliances involving pre-IPO companies are not included in the rdna.com database. Alliances come to the attention of the database managers when one of the partners notifies the US SEC (which may happen only if the recipient of funding has had or is planning an IPO), or if one of the alliance partners issues a press release. The Recap database recorded 738 therapeutic alliances for 2003 but had financial data on only 158 of these (of course many of the remainder may not have involved financial payments). Therefore, the US$3 billion estimate for the amount of pre-IPO alliance financing is conservative.

In summary, the amount of alliance funding paid by pharmaceutical companies to pre-IPO US bioventures in 2003 in the form of R&D contract payments, license fees, other milestone payments, cash for equity, etc. was probably more than US$3 billion.

3. See the analyses in Chapter 4 of the companies featured in the case studies, employment in Japanese therapeutic-oriented bioventures, the twenty-eight

bioventures on which I have sales and investment data, and ventures from the University of Tokyo, Keio University, and AIST. The one sector not covered in depth by these analyses is non-life science, non-startup new companies, although the case studies suggest even these face the same problem.

There are exceptions. These include the large pharmaceutical company that has supported Gene Angiogenesis, Mitsubishi Pharma's sponsorship of clinical trials for LTT Bio-Pharma and Lequio Pharma, Internal Search Engine's parent's support for ISE and other spin-offs, Toyota's support of Lattice Technology in 1998, and most recently Fuji Film's US$8.4 million (22% equity) investment in Perseus Proteomics. In the latter case, Fuji Film was probably motivated by its desire to expand into medical diagnostics and the complementarities between its photo-optical expertise and Perseus's expertise in diagnostic and therapeutic antibodies, as well as Perseus's clinical genomics system for identifying diagnostic markers and drug targets. However, these cases are probably exceptions that prove the rule.

4. Willoughby's surveys in the late 1990s of bioscience firms in New York and Utah (see n. 1) showed that 7.6% of financing was from angels, almost equal to the contribution of grants (8%) and the founders' personal funds (7%), but substantially less than revenue (64%). However, when the analysis was limited to companies less than 6 years old, angel investment accounted for 21% of financing, substantially higher than any other form except revenue (46%).

5. Gershwiler, May, and Hudson (2006); CVR (2006).

6. A considerable portion of which is later stage VC investment, i.e. beyond the funding at startup and the first round of VC financing following startup (later stage making up 45% of total VC investments in 2005 according to NVCA 2006).

7. NVCA (2006); Van Osnabrugge and Robinson (2000: 68).

8. Freear, Sohl and Wetzel (2002).

9. Judging from company reports, GCR's angel probably invested between US$100,000 and US$140,000 in 1999, while MBV's could have invested as much as US$600,000 in 1985. Nanofilm Applicator and Big Crystal received investments from US-based angels.

10. In 2005, 49,500 US entrepreneurial ventures received a total of $23.1 billion in investment from angels—of which 70% ($16.2 billion) were first time investments (CVR 2006). This suggests the mean first-time investment per company in 2005 was higher than $327,000. ($327,000 is an underestimate because I still used 49,500 ventures as the denominator.)

Survey responses from 56 Japanese ventures that had received angel investment up to 2003 (the majority of which were probably made after 1994) indicate that the median initial investment by the lead angel investor was just over 10 million yen (approximately US$85,000). Since these are median averages and may not include coinvestments by other angels, these averages also should be increased to be comparable with the US averages. The average investment by any individual angel at any stage (initial and follow on, lead or coinvestor) was between 1 and 5 million yen (US$8,500 to US$45,000) and the average angel respondent (of which there were about 50) had made 5 investments (JASMEC 2002).

11. The main government body collecting data on SMEs tried to identify all angels and all their target companies by identifying persons who had applied for special tax deductions for angels (see below); screening newspapers, business publications and the internet; and making inquiries to angel networks, individual venture businesses, securities companies, organizers of venture business conferences, and regional METI bureaus. 19% of the provisionally identified angels gave usable responses to the survey and of these 78% (54) were judged to be actual business angels. 9% of the provisionally identified target companies gave usable responses, but only 36% of these were judged to actually have received angel investment (JASMEC 2002).

Consistent with this, the 2005 Global Entrepreneurship Monitor indicates that the percentage of Japanese adults who invest in new companies is among the lowest in the world (0.6%) compared to about 4.6% of US adults.

12. See NIF News (2002, under Chapter 4 references), describing a 2002 amendment to the 1997 Angle Tax Incentive law that extends similar tax benefits to angel investors investing through investment cooperatives. This report notes that in the five years since enactment of the law (1997–2002), only 226 angels, who had invested a total of 400 million yen (US$3.5 million) in only 15 companies, had applied for tax benefits under this law.

13. Surveys of university startups support the observations in the Chapter 4 case studies that many ventures rely on government grants and contracts for financing. Tsukuba University (2006: 41) indicates that about 27% of initial funding for startups comes from their universities (probably mainly as government sponsored research).

METI (20006: 34, 53) indicates that on average, about 22% of startup R&D expenses (about US$200,000 per startup in 2005) were financed from government sources, mostly through grants or contracts (*hojokin*, in Japanese; estimates based upon 30% response rate to a questionnaire sent to all identified startups). However, fewer than half of all startups received such assistance. For those startups that actually received such assistance, the average amount was about US$800,000 in 2005.

14. In 1988, about 80% of VC investments were in the form of loans and listed stocks. By 2002, pre-IPO equity investments accounted for about 70% of total Japanese VC investments, compared to eighteen, four and a negligible percentage for bonds, loans and listed stocks, respectively (VEC 2002).

In 1997, only 25% of Japanese VC investment went to companies less than 5 years old, and only 42% went to companies less than 10 years old. In contrast 83% of US VC investment was in companies less than 10 years old (Nakagawa 1999). However, by 2003, about 64% of Japanese investments went to companies within five years of their establishment, with about 72% going to companies within ten years of their establishment (VEC 2003).

In 1997, less than 40% of Japanese VC investment was in high technology industries compared with about 80% in the USA. Only 23% of Japanese VC investment in 1997 was in IT related fields compared to about 60% in the USA (Nakagawa 1999). By 2004, 48% of new venture investments were in IT, 33% in biomedicine or health and 15% in manufacturing or energy (VEC 2005).

Overseas VC investments accounted for about 50% of total in 1997 and 1998, but only 20% in 2003 and 2004.

15. Japanese A̲ssociation of S̲ecurities D̲ealers A̲utomated Q̲uotations.

16. M̲arket o̲f High-Grow̲th and E̲merging S̲tocks.

17. Between 1997 and 1999, requirements in the Public Welfare Pension and Insurance Law (section 130.2) that the majority of a company's pension funds be managed by designated trust banks or insurance companies, and that the majority of pension funds be invested in government bonds and less than 30% in joint stock companies, were abolished (PFA 1999).

18. US corporate pensions invest about 7% of their funds, in private equity investments. In contrast in 2003 only 1–7% of Japanese corporate pensions funds were investing at all in private equities, and some of the most active among these were investing only 2% of their assets in private equities (Ueda 2003).

19. In 2006, some public pension funds were discussing with ministries the possibility of investing in private equities as well as increasing investments in publicly trade equities (Shimizu 2006).

20. See *Nikkei Financial Daily* (2006) regarding formation on new funds by JAFCO and SBI Holdings to invest corporate pension assets in venture companies. However, these investments are being made into companies that are anticipating IPOs in the near future. Based upon recent IPO experience, these include few new companies developing new technologies.

21. In 1979, the US Department of Labor declared that pension fund managers could take into account portfolio diversification in determining whether their investments met the 'prudent man' standard set by the Employee Retirement Income Security Act (ERISA) of 1974. Prior to this ruling, most pension funds avoided investing in venture businesses because they thought the risk might expose them to liability under the 'prudent man' standard. New funds raised annually by US VC companies rose from US$482 million in 1978 to US$6.2 billion five years later, and to US$69.7 billion in 2000 (all amounts in year 2000 US$). Between 1996 and 2000 pension assets have accounted for at least 40% of VC funds annually (Gompers and Lerner 2001: 92–4). In contrast, in 2003 only 2.3% of money under management by Japanese VC funds was from pensions (VEC 2003). However, insurance companies and trust banks still manage corporate pensions, thus the actual Japanese percentage is higher.

22. VEC (2005).

23. By the end of 2003, at least 25 VC companies or VC consortia had established funds that had raised over US$400 million in capital for investing in university or GRI startups. This money will be invested over several years and some will be held in reserve. NEDO contributed to many of these funds. SMRJ has also

contributed to various VC funds to encourage them to investment in startups. In addition, traditional VC funds managed by banks, insurance companies, and securities companies also invest in university startups.

24. Nakagawa (1999). Since 1999 some independent and some public VC companies have been formed, but my impression is that, aside from a few established independent companies that were in existence at the time of Nakagawa's survey, most are still subsidiaries of large financial or manufacturing corporations.

25. At least in biomedicine, some of the VC companies have hired MS or Ph.D. scientists or persons with pharmaceutical industry experience. The level of scientific and business expertise is rising, but in general it is still quite thin throughout the VC industry.

26. They have few career incentives to focus on a fairly small set of high risk companies and to try to ensure that some of these succeed. The following paraphrased statement illustrates this phenomenon, which I have heard at several financial institutions:

> The amount of money that I would invest in [ABC] company is much smaller than investments I normally make. If ABC Company happens to have a successful IPO within a year which doubles my investment, I will have made only one million profit for my institution. This will hardly show up on our overall profit and loss tabulations. It will not bring me any special credit. But if the investment turns out badly, and my group shows a million dollar, or even a half million dollar loss, that will look bad on my record. My superiors will notice this loss and say I exercised bad judgment in making the decision to invest.

27. In 1997, the average new Japanese VC investment was 40 million yen (Nakagawa 1999). In 2004, the average VC equity investment by the lead VC company was 46 million yen (about US$400,000) and the average pooled coinvestment by other VC companies was 76 million yen (about US$700,000) (VEC 2005). In contrast, the average VC investment in the USA is 1–5 million US$.

   The bioventures with which I am familiar that were formed 2000–3, typically began with startup funding in the range of a few hundreds of thousands of US dollars from several VC investors. The first formal round of VC funding then raised about US$1 million, contributed by five or more financial institutions. Subsequent funding rounds would raise larger amounts. Over the past few years, the amount that each institution contributes on average per round probably has increased, but only marginally. Contributions are still dispersed among a large number of investors.

28. Each of the life-science-focused VC companies has at least one manager or general partner with biomedical experience.

29. Cancer Answers and Reverse Targeting owe their competent managers in part to efforts by the dedicated biomedical VC funds that had invested in them and helped to recruit those managers. As described in Chapter 4, Chip Detect and Internal Search Engine were also helped by small VC companies focused on IT investments.

30. A large proportion of this VC company's investments are overseas. One of the smaller hands-on VC companies also has a substantial proportion of its portfolio companies overseas (mainly in California).

31. In the case of the five biomedicine-focused VC companies that I know best, the one that has raised the most money does not receive government investments. All the others do. Between one-third and two-thirds of their most recent funds' capital comes from government organizations.

    The investments by insurance companies are of some note because they may involve pension funds. Trading companies are also important investors in many of the hands-on VC companies.

32. See Appendix 2. Nevertheless, the hands-on VC companies can claim successes. One of these was an early investor in OncoTherapy, that had one of the most successful IPOs world wide in 2003. The head of another of these VC companies invested in Anges MG nearly 3 years before its successful IPO in 2002.

33. See e.g., 'Accelerating the formation of new businesses from university technologies' [Daigaku hatsu gijutsu kigyou ni hazumi]. *Nikkei Business Daily*, March 4, 2003, p. 3, which describes a special relationship between Osaka University and Nippon Venture capital.

34. See the companies listed under biotech and bioinformatics in Appendix 1 of this chapter. These nine do not include the companies I have classified under 'clinical trials and health information management'. Among these latter companies is Soiken, a startup from two Osaka universities, that also had a highly successful IPO in 2003, and like the other companies has seen the value of its stock erode.

35. Medical Biological Laboratories in 1996, Intec Web and Genome Science in 2000, and Precision Systems Science in 2001.

36. According to SBLSEM (2004), Transgenic and Anges MG had each raised about US$15 million prior to their IPOs, Medinet about 8 million, OncoTherapy and Medibic about 4 million and Soiken barely 1 million. In May 2004, five months after Soiken's and OncoTherapy's nearly simultaneous IPOs, each had market capitalizations of about US$1 billion, while the six previously mentioned bioventures together had a mean market capitalization of nearly US$600 million. Yet both Soiken and OncoTherapy marketed only services (by 2006, OncoTherapy would have one drug in phase 1 clinical trials). Anges MG had a pioneering drug in early stage clinical trials and its market capitalization was about US$550 million. Two companies that were selling concrete products, TransGenic (knockout mice) and Medinet (cells and related services for patients undergoing immunotherapy), had relatively low-market capitalizations, 100 million and 300 million US$, respectively.

37. See previous note.

38. Three stages of clinical trials are needed to obtain approval to market drugs. Each stage requires more patients. Phases 1 and 2 focus on dosing and safety, phase 3 on efficacy.

39. In May 2004, the average market capitalization of 16 US bioventures that had IPOs between Oct. 2003 and April 2004 was 350 million, compared to 550 million for the 6 Japanese bioventures (including Soiken) that had IPOs between 2002 and 2003 (SBLSEM, 2004).

40. Although, with some exceptions, the prices did not collapse. Also the market capitalization of Anges MG has tended to increase, while that of Takara Bio increased or remained steady until well into 2005.

41. TSE (2006).

42. In 2005, Effector Cell Institute (a University of Tokyo startup) had an IPO on the Nagoya exchange, and MediciNova had an IPO on the Hercules (Osaka) exchange. MediciNova is a spin-off from Tanabe Pharmaceuticals developing compounds in-licensed from Angiogene (UK) and Kissei, Kyorin and Mitsubishi Pharma. (Japan), but since its founding in 2000 its operations have been based in San Diego.

43. For example, a few bioventures embarking on clinical trials have raised over US$40 million in second and later round financing. Nevertheless, between 2005 and 2006 such later round financing declined for Japanese bioventures as a whole (Tsujimoto, Kenji (2006), 'Sector review and introduction', presentation in Japanese at the 13th Nomura Bio Conference (Nov. 20, Tokyo)).

44. The listing requirements of the Mothers exchange, in theory, allow exceptions, but the conditions for such exceptions are strict. A therapeutic venture that wants to raise funds on public markets but does not have alliances covering its lead pipeline drugs must make clear its development strategy. More importantly, it must be able to confirm on the basis of clinical trial results the likely effectiveness of its drugs, and it must be able to explain the marketability of its pipeline drugs (TSE 2006). These conditions probably mean that the venture must have completed at least phase 2 trials. They could easily mean that the venture must have some positive phase 3 data, plus evidence it has feasible arrangements (access to sales staff, etc.) to market the drugs. These are quite stringent conditions. In effect, they require the venture to obtain a considerable amount of private VC financing if it wants to avoid unfavorable terms offered by pharmaceutical companies.

    Some drug-focused ventures have drugs in phase 1 or 2 clinical trials and are trying hard to establish partnerships, knowing that their ability to list on public exchanges is at stake (as is their ability to raise private equity funding from a large number of investors who want near term exit opportunities). But so far, few have been successful in establishing partnerships on terms they consider reasonable.

45. Lerner and Tsai (2000) have studied alliance agreements between large corporations and small bioventures, which usually involve a pharmaceutical company giving the venture financial support in return for equity and/or rights to control the development, clinical trials, marketing etc. of one or more of the venture's technologies. They showed that contracts signed at downturns in VC funding cycles (i.e. when it is difficult for venture companies to raise VC) and that assign

substantial control over the venture's technology to the large company have significantly *worse* outcomes (in terms of progression to the next development phase or reaching the market) than contracts signed when VC funding is plentiful or contracts that let the venture retain control over the development of its technologies.

46. Examination of lists of pre-IPO investors in bioventures (see Appendix 2) shows that the most frequent pre-IPO corporate investors are VC funds affiliated with insurance companies, followed by funds affiliated with banks and then a few independent traditional VC companies. Rarely do any of these companies hold more than 5% of pre-IPO stock, suggesting their investments usually were not large. According to publicly available lists of pre-IPO stockholders, the only cases of a traditional VC company holding more than 5% of pre-IPO stock at the time of IPO are Japan Asia Investment Corp. and Tokyo Fire and Marine Insurance, which held slightly over 5% equity in TransGenic and Sosei, respectively. Also no particular type of investor has emerged as having made consistently good bets on what companies will have IPOs. The pattern of investment seems haphazard with little sign that investors are developing expertise in selecting good investments.

47. See the descriptions in Appendix 2 of the major pre-IPO shareholders in the bioventures that had IPOs in 2002 and 2003, and which of these exited within a few months of the IPOs. These suggest that most traditional financial institutions take advantage of the IPO to exit, but with a few exceptions, other private corporations do not invest substantially following the IPO.

   On the other hand, the hands-on VC companies tend to hold much of their equity post-IPO, suggesting a continuing commitment to these companies and a desire to have continuing influence over their operations. Appendix 2 also shows that substantial public funds remain invested in these companies even after IPO.

48. See Appendix 2, and previous note.

49. The largest purchasers of IPO or post-IPO shares are probably wealthy individual clients of the large securities firms that underwrite most of the IPOs, according to discussions with officials in several of these firms. Most of the individual client-investors are not savvy about the companies whose stocks they buy, but rather rely on the advice of the securities companies. The pool of such investors is limited, and by the end of 2004, falling share prices and lackluster IPOs suggested it was being tapped out.

50. See Appendix 2 which indicates insurance companies, practically alone among major corporations, appear fairly frequently among the leading pre- and post-IPO investors. Some insurance company investments may represent pension funds the insurance companies are managing.

51. Sometimes these gatekeepers are specialized capital management companies that will pool money from insurance companies for investment. These companies are just beginning to become a presence in VC investment in Japan.

52. *Technology focused* means companies whose primary business is focused on manufacturing or R&D (including companies offering support for clinical drug trials and laboratory research).

53. e.g. IC chips, flat display panels, and printed circuit boards, and machines to make or test such products.

54. I obtained lists of pre-IPO stock holders for over half the engineering and materials companies listed in Appendix Table 5A.1. In over half of these cases, VC companies (including SMRJ-affiliated public investment corporations) are listed among the top ten holders of pre-IPO stock. Foreign VC funds also appear both among the top ten pre-IPO and (even more frequently) post-IPO equity investors in these old companies.

55. e.g. Taiyo Industrial Co. (IPO on Jasdaq in 2004) started business in 1960, manufacturing textile printing rollers. It expanded into flat and rotary screen engraving, then into computer aided design, and since the 1990s into computer graphics and the design and manufacture of PCBs and related testing equipment.

56. Based on review of the technologies of the companies that had IPOs and media reports.

57. Based on examination of the web pages of the companies in Table 5A.1. Most companies would mention close links to universities if they had them. Many of the web pages of life science companies mention cooperative research activities with universities and some mention university affiliations of their top managers. But no web page of any non-bio company in this cohort of IPO companies mentions a university as the source of key technology, nor a founder having come from a university.

58. See Chapter 4, Appendix 2 dealing with startups from the University of Tokyo, Keio University and AIST.

59. And perhaps by extension nonbiomedical high technology ventures that may be planning to have IPOs.

60. One limitation of this analysis is that I only know the names of the top ten post-IPO investors.

REFERENCES

Center for Venture Research (CVR) (2006). 'The Angel Investor Market in 2005: the Angle Market Exhibits Modest Growth'.

Dibner, Mark D. (1988). *Biotechnology Guide U.S.A.*, 1st edn. New York: Stockton Press.

Freear, John, Sohl, Jeffrey E., and Wetzel, William (2002). 'Angel on Angels; Financing Technology-Based Ventures—A Historical Perspective', *Venture Capital*, 4(4): 275–87.

Gershwiler, James, May, John, and Hudson, Marianne (2006). 'State of Angel Group: A Report on ACA [Angel Capital Association] and ACEF [Angel Capital Education

Foundation]', PowerPoint presentation of Marianne Hudson of the Kauffman Foundation at the ACEF Summit in April, 2006.

Gompers, Paul A. and Lerner, Josh (2001). *The Money of Innovation: How Venture Capital Creates New Wealth*. Boston, MA: Harvard Business School Press.

Japan Small and Medium Enterprise Corporation (JASMEC) [Chuushou kigyou sougou jigyoudan] (2002). Survey report on business angels [Bijinesu enjeru no jittai chousa; houkokusho].

Kneller, Robert (2003). 'Autarkic Drug Discovery in Japanese Pharmaceutical Companies: Insights into National Differences in Industrial Innovation', *Research Policy*, 32: 1805–27.

Lerner, Josh and Tsai, Alexander (2000). 'Do Equity Financing Cycles Matter? Evidence from Biotechnology Alliances', Working Paper Series No. 7464. National Bureau of Economic Research (NBER). Available at www.nber.org/papers/w7464

McCully, Michael G. and van Brunt, Jennifer (2004). 'Late Stage Deals Have Lots of Appeal', *Signals Magazine* at www.signalsmag.com (published online, Oct. 27, 2004).

METI (2006). 'Basic Survey Report on University Ventures' [Daigaku hastu benchyaa inkansuru kiso chousa: jisshi houkokusho] (May, in Japanese).

Nakagawa, Katsuhiro (Dec. 1999). 'Japanese Entrepreneurship: Can the Silicon Valley Model be Applied to Japan?', adaptation of presentation given on May 19, 1999 at the Asia Pacific Research Center, Stanford University.

National Venture Capital Association (NVCA) (2006). 'Venture Capital Investing Steady at $21.7 Billion in 2005, Holding on to 2004's Gain', (Jan. 24). Available at www.nvca.org

Nikkei Financial Daily (2006). 'Nenkin, VC fundo e ryuunyuu: tsuzuki IPO meigara ninki (shigunaru)' [Pension funds pumping more money into VC funds: a signal of the continuing popularity of IPO stocks], Jan. 25, 5.

Pension Fund Association (PFA) [Kousei nenkin kikin rengoukai] (1999). *Management of Pension Fund Assets in the Age of Management Freedom* [Unyou jiyuuka jidai no nenkin kikin no shisan unyou]. Tokyo: Touyou Keizai Shimbun Sha [Toyo Keizai Co.], pp. 44–5.

Shimizu, Takashi (2006). 'Civil Servant Pension Funds to Boost Stock Investments', *Nikkei Weekly*, May 29, 24.

SoftBank Life Science Equity Management (SBLSEM), LLC (2004). Japan Country Seminar. 'Japan Partner Focus. Venture Capital'. Presentation at Bio 2004, June 6, 2004 in San Francisco. Available at http://www.biojapan.org/fs/presentations/Nakada_Final.pdf

Tokyo Stock Exchange (TSE) (2006). 'Mothers Listing Guide', available in Japanese at http://www.tse.or.jp/mothers/guide/5.pdf, pp. 80–1, and in English at http://www.tse.or.jp/english/listing/mothers/05.pdf, pp. 70–4.

Tsukuba University, Industry–University Liaison Joint Research Center (2006). 'Survey of University Ventures and Promotion Strategies' [Daigaku nado hatsu benchaa no kadai to suishin housaku ni kansuru chousa kenkyuu], issued March, in Japanese.

Ueda, Takashi (2003). 'More Corp Pension Funds Eye Private Equity Investments', *Nikkei Financial Daily*, April 15 (translation from Japanese available at www.nni. nikkei.co.jp)

Van Osnabrugge, Mark and Robinson, Robert J. (2000). *Angel Investing: Matching Start-up Funds with Start-up Companies*. San Francisco: Jossey-Bass a Wiley Company.

Venture Enterprise Center (VEC) (2003). 'Survey of Venture Capital Investment 2003' (also issued for other years), available at www.vec.or.jp/touchi.html

Willoughby, Kelvin W. (2005). 'How Do Entrepreneurial Technology Firms Really Get Financed and What Difference Does it Make?', Paper for the US Association for Small Business and Entrepreneurship, Technology Entrepreneurship Division, Annual Conference, Tucson, AZ, January 12–15, 2006.

6

Amoeba Innovation:
The Alternative to Ventures

Introduction

In view of the problems facing independent ventures in Japan, might not the best hope for Japanese industry to remain at the forefront of innovation rest with its large, established companies, or else spin-offs from such companies? In contrast to the USA, where new companies often pioneer successive generations of technologies and compete successfully with large incumbents, in Japan large established companies often remain dominant in their industries even when technologies evolve substantially. This cross-national difference has been documented in the case of personal computers, integrated circuits, photolithography, hard disk drives, and other technologies.[1] At least until recently, the scarcity of independent high technology ventures has implied that Japan has no choice but to rely on its established companies for innovation in new fields of technology.

There are many examples of established newcomers succeeding spectacularly when they moved into new fields related to areas in which they already had expertise. Examples include Toyota's move from weaving machines into automobiles and Honda's from motorcycles into automobiles. NEC built on its expertise in computing and image recognition to develop an automated fingerprint identification system that became the choice of police departments in both Japan and the USA.[2] Sharp developed an early LCD calculator which was a commercial failure, but then went on to make breakthroughs in large active matrix LCD displays and became the world's largest manufacturer of LCDs. It incorporated LCD technology into its core television manufacturing business to become the leading manufacturer of large LCD screen televisions.[3] Canon's expertise in photo-optics helped it to become a leading manufacturer of photocopiers and second generation mask aligners and steppers for manufacturing IC chips.[4]

This process of successful technical diversification has been described by others.[5] It is beyond the scope of this chapter even to summarize the possible reasons for the historical success of Japanese companies relative to established

US counterparts except to mention the following advantages attributed to Japanese companies: Among the advantages over which there seems to be little debate are close communication with customers, close communication within work units and resulting close attention to detail and quality, and a high level of tacit knowledge shared by company employees as a result of lifetime employment and frequent rotations within the company.[6] Reasons that have been debated, or that may have been applicable only when Japanese companies were in a catch-up phase, include: controlled competition mediated either directly by the government or large government controlled corporations such as NTT,[7] a policy of weak protection for IP that encouraged sharing of new technologies especially those originating abroad, a close follower strategy involving rapidly refining or improving technologies that others had pioneered,[8] access to long-term funding from main banks, and a de-emphasis on vertical integration as a means of corporate management and reliance instead on alliances between semi-independent companies.

This latter reason refers not only to the system of manufacturing keiretsu, under which a number of companies are linked to a large manufacturing company. It also refers to diversification by forming spin-offs that the parent companies support by various means and over which the parents maintain partial ownership and control.

Hereinafter, I refer to these as *tethered spin-offs*. I refer to established companies that have moved into new field of technologies that are significantly different from their current core businesses as *established newcomers*.

Compared to independent venture companies, established newcomers and tethered spin-offs often have greater access to complementary assets such as skilled researchers and managers, financing, manufacturing facilities and networks of suppliers and customers. Established companies may also have greater access to complementary technologies, including in-house expertise[9] and technologies of other companies that have been obtained as a result of in-licensing, collaborative research, or other means.[10] In addition, established companies do not face the appropriability problems that small independent ventures often face. If intellectual property rights or first-to-market advantages[11] are weak, innovative ventures run the risk that their hard won technical developments will be quickly copied by rivals. By providing complementary assets, a large company can provide innovators the resources they need to develop new technologies shielded from the awareness of rivals. They can also provide manufacturing facilities and distribution and customer networks that help to maintain a first-to-market advantage.[12] In view of the difficulties Japanese ventures still face in recruiting skilled researchers and managers and in establishing networks of customers and collaborators, and considering also the access of the large, diversified, high-technology Japanese manufacturers to a range of complementary technologies, it might seem natural to expect that established companies will remain the engines of innovation in most new fields of technology in Japan.

Yet at least in the case of biomedical technologies,[13] established newcomers and tethered spin-offs generally are not particularly successful. The case of nonbiomedical technologies is less clear. Nevertheless there is also evidence that, when established companies enter nonbiomedical fields that are far from their core businesses their innovative competitiveness also diminishes.

PART I: ESTABLISHED NEWCOMERS

Established Newcomers in Biomedicine

A Tale of Two Breweries

Kirin and *Suntory* are two of Japan's most respected makers of alcoholic beverages. Kirin was incorporated in 1907, Suntory in 1921. With the establishment of the Suntory Institute for Biomedical Research in 1979 and Kirin's Pharmaceutical Division in 1982, both companies entered pharmaceutical R&D just when the potential of biotechnology was beginning to be apparent and revenues from the sale of alcohol and other beverages were flat. These were not steps into completely alien territory. By emphasizing the development of drugs based on antibodies and naturally occurring proteins that could be mass produced by commonly used microorganisms utilizing new recombinant DNA techniques,[14] these companies could build on their expertise in fermentation. This expertise could be used both to manipulate microorganisms during early stage drug discovery and later to scale up manufacturing to commercial levels. Both were driven to enter the nascent biotechnology field by CEOs with strong personalities and executive powers who perceived the potential of the new science and thought that it would be an avenue to expand their companies' business and visibility. At the time, both companies had positive balance sheets and abundant cash, Japanese companies being under little pressure to return profit to shareholders.[15]

Both companies recruited young scientists as well as some experienced team leaders for their new pharmaceutical operations. The ratio of researchers with Ph.D.s may have been somewhat higher than for mainstream pharmaceutical corporations.[16] However, although they recruited some young researchers from leading Japanese universities, they felt they were not able to recruit enough good scientists and luring good people away from established pharmaceutical companies was nearly impossible.[17] Both companies had close ties with university researchers in Japan and overseas.[18] Both laboratories were well equipped, researchers seemed competent and enthusiastic, and some volunteered that they had considerable freedom to pursue their own drug discovery projects.[19]

Suntory's first and most successful drug was pilsicainide. Launched in 1991 to treat fibrillation and other arrythmias of the upper chamber of the heart, it is synthetic version of a compound naturally secreted by the kidneys. It is a small molecule, not a typical biotechnology therapeutic. In 2001, Suntory earned about US$90 million in sales of pilsicainide—far from *blockbuster* status. Its next best selling drugs were the combined penicillin–cephalosporin antibiotic, Farom® and recombinant carperitide to treat heart failure.[20] All of these drugs have been marketed only in Japan.

In 2000, the Biomedical Research Institute was incorporated as Suntory Biomedical Research, Ltd. (SBR). At the end of 2002, Daiichi Pharmaceuticals bought two-thirds of SBR's stock. Conversations with industry insiders suggest that the reasons Suntory divested its pharmaceutical business relate to the death of the chairman who had championed the pharmaceuticals division,[21] a general retrenchment from several areas remote from Suntory's core business, a realization that substantial investments would be needed to make the pharmaceutical operations competitive, and a difference in corporate culture between the individualistic and free-thinking pharmaceutical researchers and the traditionally minded employees in the rest of the company. Daiichi reportedly pledged that it would keep the Suntory researcher teams intact in the near term, and it continues to operate SBR under the name Daiichi Suntory Pharma Co., Ltd. Clearly, however, control has passed to Daiichi.

In the case of Kirin's Pharmaceutical Division, one strategic decision overshadowed all others in its early years, its 1984 partnership with Amgen under which Kirin helped to bankroll Amgen's development of bioengineered erythropoietin (EPO) and granulocyte colony stimulating factor (G-CSF).[22] In return, Kirin received the right to market these drugs in Asia and comarket EPO in Europe through a joint venture with Amgen. This partnership has paid off financially. EPO and G-CSF are the world's two best-selling biotechnology drugs,[23] and they account for the bulk of Amgen's and almost all of Kirin's pharmaceutical revenue.[24] As of the end of 2003, all of the drugs Kirin was marketing originated in outside laboratories.[25]

Research on EPO and G-CSF was left primarily to Amgen. Although Kirin received licenses to Amgen's technology, to this day it appears that Kirin is not competing with Amgen in drugs that are similar to EPO or G-CSF. For Kirin, the future of its pharmaceutical operations lies with a technology to genetically engineer mice or other animals to produce purely human polyclonal antibodies that can be used as drugs to treat cancer, infectious diseases, and autoimmune diseases.[26]

The point of this comparison is not to pinpoint various corporate strategies that made the difference between failure in Suntory's case and the prospect of continuing success in Kirin's. Luck probably played as great a role as corporate management. What does seem clear, however, is that despite being able to ramp up pharmaceutical operations quickly and to build on their fermentation expertise, the road for both these new entrants was perilous.

Ajinomoto was founded in 1908 by a professor of physical science at the University of Tokyo who had isolated glutamic acid from the broth of cooked seaweed (konbu) and identified it as the source of the savory taste in traditional Japanese dishes made with konbu. He formed the company to manufacture and market glutamic acid as a flavor enhancer. Ajinomoto has grown into a diversified food products company with its core technology centered on the production and utilization of amino acids. In the 1980s, it began to use genetic engineering to produce proteins. Its pharmaceutical R&D also dates from this time, with its genetic engineering of *Escherichia coli* to produce purified interleukin 2 and 6. However, it has ceased development of interleukins, probably because of competition from US biotechnology companies,[27] and is now focusing on the discovery of small molecule drugs using genomic and proteomic technologies. Pharmaceuticals account for only about 8 percent of Ajinomoto's total net sales—a higher percentage than for Kirin.[28]

Ajinomoto can claim at least one partial success. Its researchers were among the first to make a new type of drug to treat adult onset diabetes, a drug that acts quickly to stimulate secretion of insulin by the pancreas and decrease the damaging surges in blood glucose levels that occur during mealtime.[29] Lacking a sales force, Ajinomoto licensed marketing rights outside of Japan to Norvatis, which markets the drug under the brand name Starlix®.[30] FDA-approved Starlix® for use in the USA at the end of 2000. Ajinomoto's other main drug in 2004 was Actonel® for osteoporosis, which is in-licensed from Procter and Gamble.[31]

A new experimental small molecule drug originating in Ajinomoto's laboratories is AVE-8062, the leading drug in a new class of compounds that disrupt both existing and newly developing blood vessels in tumors. Ajinomoto licensed worldwide rights to the drug to Aventis in 2001. As of 2006, early stage clinical trials were still in progress.

In 2000 only about 5 percent of Ajinomoto's researchers had doctoral degrees. It has long-standing relationships with US universities, including MIT, to which it regularly sends researchers. It also has collaborations with German, French, and Dutch universities. All told, it sends about fifteen researchers annually to overseas universities for training and about the same number to Japanese universities. These numbers are high in comparison to the largest Japanese pharmaceutical companies and indicate a considerable commitment of human resources. Ajinomoto provides research support to many Japanese university laboratories.

These close relations with universities may have paid off. The basic active compound of Starlix® was coinvented with Keio University researchers.[32] Among all the Japan-origin drugs approved by the US FDA from 1998 to 2002, only Starlix® had university researchers listed as inventors on the underlying patents. Along with the university input to Kirin's new antibody technology, this suggests that established newcomers are making use of collaborations with universities to develop innovative products in a way that established incumbents usually do not.[33]

Takara Shuzo was incorporated in 1925 and is best known as a sake brewer. In 1986, a Japanese biochemist who had been a research director at Centocor in the USA joined Takara and began to build its biotechnology operations. Beginning with manufacturing of various enzymes, reagents and test kits for genetic engineering laboratories; these operations expanded to include genome and protein analysis using technologies in-licensed from abroad; then large scale genome sequencing to discover links between genes and diseases and sensitivity to drugs in Asian populations; then gene-therapy using technologies in-licensed from abroad. In 2005, Talara Bio's revenue was mainly from the sale of protein synthesis systems and services based largely upon overseas technologies, supplemented by sales of health food products.[34] Although the company's biomedical operations do involve research at the forefront of science, available information suggests that most of its business activities are based on standard or in-licensed technologies.[35]

Since 1993 Takara Shuzo has been organizing its biomedical operations into tethered spin-offs: Takara Biotechnology in Dalien, China (1993) to produce genetic engineering reagents and to process samples collected in China for genetic analysis; Dragon Genomics near Nagoya (2000) as the gene sequencing center; and Takara Bio near Kyoto (2002) as the main R&D center. Unable to sustain the burgeoning biomedical research budget,[36] Takara Shuzo arranged for Takara Bio to have an IPO on Mothers at the end of 2004. This strategy may have paid off as the market capitalization of Takara Bio three months after the IPO was about US$1.5 billion, higher than for any other new life science company. As of mid-2006, market capitalization was still over US$1 billion.

Japan Tobacco (JT) was incorporated in 1985 as a wholly owned government corporation, continuing the government's monopoly over the sale of domestically produced tobacco products. Beginning in 1994, the government began to sell some of its shares in JT, but as of 2004 it still owned 50 percent of the company. As part of JT's diversification, it established a pharmaceutical research center in 1993 and bought Torii Pharmaceutical Company in 1998 to be the main marketer for its drugs. Currently marketed pharmaceuticals consist of drugs developed by Torii prior to the merger and drugs in-licensed from overseas. The ratio of pharmaceutical R&D expenditures to sales has been very high,[37] indicating a substantial investment of overall corporate revenue, mostly from the sale of cigarettes in Japan and overseas.[38] For its pharmaceutical division, losses as a percentage of sales have also been high.[39] Although total revenue has continued to increase, JT has scaled back its pharmaceutical R&D. In 2002 it had ten in-house origin drugs in early clinical trials, but by 2005, it had pared this number to six.[40] Some of the bioventures I interviewed said that they had recently hired researchers from JT.

However, one of these drugs is among the first in a promising new class that increases high density lipoprotein and thereby reduces the risk of heart disease in persons with high cholesterol.[41] Late in 2004, JT licensed worldwide rights to this drug to Roche, although it retained marketing rights in Japan and Korea.

Several *chemical and foodstuffs companies* have discovered drugs that have subsequently been developed by major Japanese pharmaceutical companies. They play a similar role in relation to the major companies as US biotechnologies play vis-à-vis multinational pharmaceutical companies, although they are the source of a much smaller proportion of the pharmaceutical companies' pipelines than are the US biotechnology companies. However, the drugs discovered by chemical and foodstuffs companies generally are not groundbreaking drugs. Rather they are variations on classes of drugs that have been pioneered by other companies.[42] The pharmaceutical operations of the chemical and foodstuffs companies are generally small. They receive modest funding from the parent, and they employ small numbers of researchers. Other than Ajinomoto, I know of no cases where R&D in such companies has been ramped up, either by generous funding from the parent or by substantial revenues from successful products, to sustain pharmaceutical R&D on a scale that can produce a continuing sequence of drug candidates entering clinical trials.

Asahi Glass produces purified recombinant (genetically engineered) proteins for bioventures and some major pharmaceutical companies to meet their research needs and also for pilot-scale (precommercial) production. Asahi's system uses a yeast isolated from east African beer to synthesize proteins.

However, this basic system was developed in the early 1990s, largely by scientists in the New York State Department of Health, and it appears that other companies are using this system for similar purposes.[43]

Hitachi, Toshiba, Canon, NEC, and Fujitsu all have entered the field of biochips and bioinformatics. *Toshiba* patented and in 2005 was nearing the end of prototype testing of a new type of DNA chip that relies on electrochemical signals rather than fluorescence to detect binding of unknown gene sequences to known strands of DNA.[44] Toshiba claims the system is quicker and more accurate than the 'industry standard' Affymetrix chips, and more suitable for large-scale use in molecular diagnostic laboratories. However, it holds fewer DNA probes than conventional chips and therefore one chip can detect fewer types of DNA or fewer types of mutations. Nevertheless, it might be useful in clinical settings where patient samples are being tested for a limited number of genetic mutations or genetic variations.

Canon, drawing on its expertise in ink jet printers, is developing a new way to make DNA chips by spraying DNA solution onto glass slides. The new chips will be used to diagnose cancer (or cancer susceptibility) and infectious diseases.[45]

Hitachi's activities related to DNA chips are described in one of the case studies in Chapter 4. But Hitachi's involvement in the biological aspects of life science[46] goes far beyond DNA chips.

The development of gene sequencing machines is an intriguing side story to the race to sequence the human genome between the public international consortium[47] and the private sequencing effort of Celera and its lead scientist and CEO, Craig Venter.[48]

Applied Biosystems Incorporated (ABI) was founded by venture capitalists in 1981 to commercialize DNA sequencing technology largely pioneered by Dr Hunkapillar and his research team at Caltech.[49] Among this team's key inventions were methods to attach fluorescent dyes to each of the four nucleic acids that make up DNA sequences, thus enabling their identification when exposed to laser light. By 1987 ABI had a sequencer on the market, although its speed was too slow to meet the original goal of sequencing the human genome by 2005. In 1993 ABI was bought by the mainline scientific instrument company, Perkin Elmer (PE), which soon began to reorient its entire business toward the life sciences.[50]

However, by the mid-1990s ABI had competition. A team at UC Berkeley had also developed sequencing technologies that became the basis for founding another venture company, Molecular Dynamics. In order to gain access to marketing resources and appropriate dye technology, Molecular Dynamics entered into a strategic alliance with Amersham, which had obtained access to ABI's fluorescent labeling technology through a series of technology swaps with ABI.[51] By 1997 the MegaBace sequencer being developed by Molecular Dynamics–Amersham was faster than any ABI machine.

Meanwhile in 1981, approximately three years before the idea of sequencing the human genome began to crystallize in the minds of US scientists, the Japanese Science and Technology Agency (STA) had launched a project to involve various companies, universities, and GRIs in the development of automated DNA sequencing technologies. This project and a follow-on project from 1984–7 were brainchildren of Akiyoshi Wada, professor of physics at the University of Tokyo. Although Seiko Instruments was originally designated as the lead developer of an automated sequencing system under the first of the Dr Wada–STA projects, Hitachi was to make the greatest contribution to genome sequencing technology. As a participant in the second of these projects, Hitachi began development of its own DNA sequencer. A team headed by Dr Hideki Kambara (a former student of Dr Wada) developed new ways to configure the array of capillaries carrying fluorescently labeled gene sequences and the laser beam that would illuminate these sequences. By 1993 this team had developed the *sheath flow capillary array method* that greatly improved the speed and reading accuracy of the sequences, although Hitachi's complete machines were only in prototype stage.

In order to counter the Molecular Dyamics–Amersham threat, ABI licensed this technology from Hitachi around 1997. This enabled ABI to build its new 3700 model sequencer, which became the workhorse for sequencing of the human genome as well as the genomes of other organisms. This machine and later models allowed ABI to continue to hold over 70 percent of the world market for sequencers. Dr Kambara's team made further improvements, simplifying the sheath flow mechanism and adjusting the optical characteristics of the capillaries to improve laser beam focusing. The result is a more compact, lower maintenance system incorporated in the latest DNA sequences sold by ABI designed particularly for clinical use. Hitachi currently markets only a few gene sequencers annually under its own brand name. But ABI brand sequencers sold since the late 1990s contain key technologies from Hitachi.

However, the relationship between Hitachi and ABI has not been entirely cordial. ABI and Hitachi never agreed on terms under which ABI would license its fluorescent tagging patents to Hitachi. Hitachi scientists maintain that this technology was not necessarily crucial, and Hitachi's in-house-originating technology was sufficiently unique and comprehensive that Hitachi could have manufactured and marketed sequencers on its own and probably survived a patent infringement suit by ABI. However, Hitachi felt it would be at a disadvantage marketing its own brand name sequencers internationally and thus its senior management felt a partnership was necessary. At one time Hitachi and Amersham explored a development partnership.

Ironically, Hitachi ended up partnering with the company that some Hitachi scientists regarded as its arch rival.

This story is relevant for this book because it illustrates two different approaches to innovation in new technical fields: (*a*) the US approach where very early prototypes are made in universities, then developed by newly formed ventures which finally partner with large companies to assist in marketing; and (*b*) the Japanese approach where almost all R&D from conception to final product is done in large established companies.

However, one problem for the Japanese effort was the relatively low level of company and government funding devoted to the project. During the peak years of Hitachi's sequencer development efforts, about twenty to thirty Hitachi researchers were working on this project, about ten of whom were under Dr Kambara developing the sheath flow technology. Hitachi funded all the R&D that led to its sheath flow capillary array breakthrough.[52]

In all probability, Molecular Dynamics and ABI each had larger numbers of researchers working on sequencer development. ABI and Molecular Dyanamics both benefited initially from access to substantial VC funding and later from the support of their large partners, Perkin Elmer and Amersham, respectively. Japanese government funding never compensated for the reluctance of Hitachi executives to devote large resources to the project. Over the seven years duration of the two Dr Wada–STA projects, STA contributed a total of only about US$13 million to the genome sequencing projects.[53] In contrast, the US NIH and DOE each began contributing over US$100 million per year beginning around 1989, while the UK government through the MRC and the Wellcome Trust also contributed significant funding.

The Japanese project from the beginning had a strong focus on developing sequencing technology. The US/UK project was more focused on the scientific and medical benefits from sequencing the genome. But ironically the scientifically and medically focused US/UK government funding ultimately provided a greater incentive for the development of sequencer technology than the instrument-focused Japanese project.

Separate from sequencer operations, the Hitachi Life Science Group and other groups within Hitachi offer various genomics and proteomics services that build on Hitachi's experience in gene sequencing and analysis.[54] Hitachi Life Science has been the main provider of data analysis and hardware for two government organized consortia.[55] However, the genomics and proteomics services that Hitachi Life Sciences offers seem duplicative of services available elsewhere in Japan and abroad. Conversations with Japanese researchers suggest that Japanese laboratories sometimes do much of this analysis themselves. Takara Bio offers similar services as do Affymetrix, Gene Logic, Celera, or Roche Diagnostics, to name just a few examples. In other words, although

Hitachi has become a major provider of genomics and proteomics services in Japan, outside its core competence of developing electronic instruments, it has not developed new technologies that have given it a competitive advantage. It does not appear to be playing the same role in innovation that venture companies play in the USA.[56]

Nonbiomedical Cases

Although I am more familiar with biomedical technologies, the following example of a consortium research project in optical communications bears resemblance to the case of DNA sequencers just described. Based on my knowledge of other large scale collaborative projects and conversations with government and business officials, it is probably fairly representative in terms of priority setting, funding, and organization of high-priority government-initiated collaborative R&D projects. As noted in Chapter 7, because these projects are so numerous, they probably constitute one of the main mechanisms by which established companies enter new fields of technology.

This particular project was inspired by the success of US companies such as Cisco Systems and Juniper Networks[57] in creating systems for transmitting large amounts of data efficiently and securely over fiber optic networks. It aims to have each member of a consortium of well-known Japanese companies develop cutting-edge expertise in specific components of broadband optical communication—expertise that can then be integrated into a commercially viable system that each consortium member would contribute to and profit from. Thus, for example, a major electronics company is responsible for R&D in optical switches, a major manufacturer of fiber optic cables for packaging, a major telecommunications company and university researchers for integration, and another major electronics company for cables, splitters, couplers, and tunable lasers.

Japanese government funding averages roughly US$1 million per corporate project participant, each of whom is expected to devote some of its own resources to the project. The corporate participants are expected to develop at least a prototype of the equipment or system assigned to them.[58] However, it is difficult for them to convince higher corporate management to commit the additional resources to refine the prototype and scale up manufacturing for a viable commercial product. Usually much more funding is required for such translational research following development of a prototype. But in this case and others, senior managers of large corporations, whose attention is directed mainly to existing product lines and to customers who each account for hundreds of millions of dollars of annual sales, are reluctant to commit scarce resources to an uncertain technology, the current market for which is only a few million dollars.[59]

No venture companies or other SMEs are taking part in this consortium. Some of these technologies are capital intensive and thus perhaps not suitable for venture companies. However, in some technical fields of optical communications, the acknowledged industry leaders are US venture companies. For example, the consortium member responsible for tunable lasers[60] perceives its main rivals to be Agility,[61] Iolon,[62] and Santur,[63] all of which are new VC backed companies. In 2005 all three companies were manufacturing tunable lasers for commercial sale. All had patent portfolios.[64] As of the end of 2004, Agility had raised over US$200 million in venture financing, Iolon approximately US$85 million, and Santur US$60 million. Any of these amounts is probably greater than the combined investment of the consortium member responsible for tunable lasers and the Japanese government in this technology.

This situation appears analogous to Hitachi's and STA's investment in DNA sequencers compared to that of ABI or Molecular Dynamics and their VC investors and large company partners. By the time large market size became apparent, US ventures already had a substantial development lead.[65]

The ministry promoting the consortium would like to have Japanese SME participants, but no eligible companies could be found. One reason is that private VC funding for new companies in IT and materials fields is difficult to obtain for companies without a revenue stream. But a related reason, discussed in the final chapter, is that the government's policy of cobbling together consortia of large companies and major universities to pursue R&D in new fields of technology leaves few high-growth-potential niches for venture companies or entrepreneurial faculty to exploit. To the extent SMEs are involved at all in high priority, cutting-edge projects, their role is usually limited.[66]

Concluding Observations on Established Newcomers

These cases suggest that when established companies move into new fields of technology that are relatively distant from current areas of expertise, the road is difficult. Although there have been some successes in terms of new products, the newcomers in drug development have not shown a distinct advantage over established Japanese pharmaceutical companies.

Even though they offer the assurance of large reputable corporations, many of the established newcomers have had difficulty recruiting skilled persons for their new operations. Even though, in theory, they had an opportunity to embark on new lines of research unencumbered by prior business goals, often they ended up pursuing lines of R&D that are not new. Some, such as Ajinomoto and Japan Tobacco, did eventually pursue innovative projects, but it took time and they devoted considerable resources to less innovative projects. Often it seems as if the decision to move into a new area was made before specific new projects were clearly conceived. Thus there was a tendency to focus on tried and true technologies that offered the prospect of a relatively quick although modest return.

The case of Hitachi's gene sequencer R&D is different. The project was close to Hitachi's prior core operations, building on experience in engineering and medical instruments. Moreover, from the outset, an experienced scientist in Hitachi had a clear vision of the project, including its importance and the technical challenges that needed to be overcome. The decision of senior management to move into a new field seems to have been matched[67] by the desire of experienced research scientists and their realistic confidence in the company's ability to carry forward the project.

Are the challenges of the established newcomers less if they are entering an uncrowded field? In the 1980s and early 1990s, genetic engineering and the related fields of protein and antibody science were new to most pharmaceutical companies. Kirin, Suntory, and Ajinomoto all initially targeted these biotechnology fields. Nevertheless, they ended up at a disadvantage with respect to US bioventures. Whatever advantages the Japanese established newcomers possessed with respect to access to complementary assets did not compensate for the greater ability of US biotechnology companies to assemble and concentrate resources on promising new fields of drug discovery—and perhaps also to benefit from the in-depth academic knowledge base and plentiful supply of skilled researchers resulting from generous, astutely allocated, NIH funding for basic research.

Is the picture different with respect to nonbiomedical fields? This chapter began with numerous examples of relatively large Japanese automotive and electronics companies outcompeting even larger US or European companies in fields pioneered by the latter. But what about fields that are very new, where successful commercial applications are still few and where the competitors also include venture companies?

Nanotechnology may soon provide another test case. Since commercial applications are still few, all companies are newcomers. Progress is probably most advanced in the US and Japan. Again, the US companies in the forefront of R&D include new as well as established companies, while the Japanese leaders are almost all large established companies. For example, among nineteen companies identified in a 2005 survey as leaders in the electronic application of carbon nanotubes, six are large Japanese companies, five are large US companies, six are US ventures (four of which are university startups), one is a major Korean electronics company, and one is a large spin-off from a large European electronics company.[68] All told, approximately 500 US companies are developing commercial applications of nanoscale technologies compared to about 50 in Japan.[69]

Japanese and US government spending for nanotechnology R&D were both close to US$1 billion in 2004.[70] In the Japanese case, much of this was for collaborative university–industry projects.[71]

PART II: TETHERED SPIN-OFFS AND KEIRETSU

Background: The Case for Spin-Offs as Engines of Innovation

Many of Japan's best known companies, including some of its leading high technology and financial investment companies such as Toyota, Fujitsu, Mitsubishi Electric, Mitsubishi Motors, Nomura Securities, and JAFCO, orig-inated as spin-offs from established parents.[72] The Hitachi group has over 650 companies, most of which are spin-offs. The Matsushita Electric industrial group[73] has over 150 companies, many of which are spin-offs.[74] Spin-offs from established companies may be the most common way that new companies in high technology industries are formed in Japan.[75]

Spin-offs may be initially 100 percent owned by their parents. As time goes on and particularly if the spin-off is successful, the parent's ownership share will likely diminish. Sometimes the spin-off grows to be much larger than the parent and may own more of the parent than vice versa. For example, by 1990 Toyota Motors's sales were twenty times larger than that of its parent, Toyoda Automatic Loom, and it owned 25 percent of its parent, while the parent owned only 4.3 percent of Toyota Motors.

Are tethered spin-offs, that is, spin-offs partially controlled by their par-ents, likely to succeed better in new technology fields than independent ven-tures? Can they combine the advantages of independent ventures[76] with the advantages of large established companies?[77] To answer these questions, I first explore reasons established companies form spin-offs and then the degree of control they exercise over their spin-offs, before considering information about actual spin-offs.

Formation of spin-offs can be a means to reduce labor costs[78] or to increase the number of high level management positions for senior employees who might otherwise have to retire at age 60.[79] It can be a means to outsource the manufacturing of component parts, allowing the parent to maintain some control but also increasing incentives to market to outside companies.[80] It can be a step in preparation for obtaining outside investment, or for divestiture.[81] But most importantly from the perspective of spin-offs as engines of innovation, it is a means to let a promising new line of business flourish on its own, to give scope to entrepreneurship among new managers, and to relieve the parent of the burden and complexity of having to manage the operations of the spin-off internally so that both the parent and spin-off can focus on their core competencies.[82] In other words, spin-off formation is a growth and adaptation strategy for Japanese companies where primary value is placed not on a single corporate entity but on a family or loose federation of firms.

Those that emphasize this pro-entrepreneurship rationale for spin-offs, along with the rationales of management efficiency and maintaining focus on core competence, often contrast the Japanese style of growth and diversification through spin-offs with the tendency of large US companies to be more diversified and vertically integrated.[83] They note that vertically integrated, diversified firms often encounter problems related to coordination, inappropriate incentives, and hierarchical control that deadens initiative.[84] Operations that are not internalized have to be managed by arm's-length market transactions.

For example, Dyer (1996) analyzed manufacturer–supplier relations in the automobile industry and showed that coordination and integration, including sharing of information valuable for productivity improvements, between Toyota and Nissan and their partially owned suppliers were better than between GM and Ford and even their internal parts divisions—and substantially better than between Ford and GM and their arm's-length suppliers. Coordination between Toyota and Nissan and their independent suppliers was closer than between GM and Ford and their affiliated suppliers, and of course much closer than between the US automakers and their independent suppliers.[85] Dyer concluded that the Japanese system of production based on close alliances between each of the main automakers and their networks of supplier companies (some partially owned, some independent) resulted in greater overall value chain specialization. This in turn allowed for gains in productivity which could not be matched by their US competitors whose parts suppliers were either internal parts divisions or independent companies whose relationship with the main manufacturer was defined by arm's-length contracts.[86] Others have noted the innovative capabilities of Japanese spin-offs and affiliated suppliers, particularly in the auto industry, and the extent to which main manufacturers rely on their supplier affiliates for important product innovations.[87]

In other words, at least in the automobile industry, there is evidence that a system of manufacturing and innovation based on close long-term coordination between a 'family' of companies led by a main manufacturer[88] can be more effective than a system where manufacturing and innovation occur either within a single, large, hierarchical company or in independent suppliers dealing at arm's length with the main company. So both the genealogy of various leading high technology companies and Dyer's case study of the automobile industry show that spin-off formation in Japan can enhance entrepreneurship, management efficiency, and improvement of core technical competence.

Control versus Flexibility

But to what extent does parental assistance and control compromise spin-offs' ability to be competitive innovators in new fields of technology?

It is common for parents to provide spin-offs with management support, especially on launch or if the spin-off runs into trouble. A substantial proportion of a parent's managerial effort may be devoted to cooperation with or supporting spin-offs.[89]

In the case of a spin-off that is supplying components or services to its parent, incentives do exist for the spin-off to upgrade its technical capacity, for example from production of components designed by the parent to components it designs on its own. Apportionment of risks and benefits in the supplier–buyer relationship is relatively equal in the case of parents and their subsidiaries.[90] If a spin-off is successful, that is if it generates growing revenue, the parent typically reduces its ownership share over time, often considerably below the 33 percent that constitutes veto power over major corporate decisions.[91] Even spin-offs over which the parent maintains a substantial ownership interest are often able to sell to competitors of the parent, although parents will discourage their selling products that may leak key technologies to the parent's main competitors.[92]

So far the picture is of a relatively benign, if somewhat, paternal relationship where both spin-off and parent usually operate under a mutual-benefit obligation.[93] But in the case of ventures that are developing new technologies requiring large investments in R&D, where substantial sales may be years away, does this system allow for the benefits associated with independent ventures?

First let us consider the seven University of Tokyo, Keio, and AIST startups that are at least one-third owned by another company.[94] Only two of these had annual revenues greater than US$1 million,[95] approximately the same proportion as for all startups from these institutions.[96] In other words, being closely tied to a larger company does not seem to increase the chance of rising above the low average indices of success that characterize the start-ups from these three institutions.

Next let us consider the case of UP Science, which was spun off from Sumitomo Electric Industries (SEI) in 1999 to commercialize the achievements of a biomedical R&D group within SEI related to a class of enzymes linked to cancer and autoimmune and neurological diseases.[97] Backed by Sumitomo Pharmaceuticals and JAFCO, UP Science would continue the development of assay systems to screen candidate compounds to correct or mitigate the effects of the defective enzymes, optimize candidate drugs, and finally take the lead candidates into clinical trials. It would also screen compounds submitted by pharmaceutical companies and possibly enter into joint drug development partnerships with pharmaceutical companies. It aimed for an IPO in 2004 or 2005.

UP Science called itself Japan's first satellite bioventure.[98] It was praised by knowledgeable independent observers of the Japanese biotechnology scene as the harbinger of the future for Japanese ventures. It was to have independence and to be subject to good corporate governance procedures.[99] Yet it would also have the backing of one of Japan's largest electrical equipment and engineering companies and a midsize pharmaceutical company also within the Sumitomo group. JAFCO would provide not only funding but also advice on business development. In other words, it had at its disposal a wide range of complementary assets of the type that constitute the main advantage of incumbent companies over independent ventures. The head of the biomedical research laboratory in SEI was given leave to be the CEO. Recruitment of other skilled personnel would not be a problem. Most would simply transfer to the new company from the parent. They would do so without the fear of the company failing because UP Science had strong backing from large companies.

Fail it did. By mid-2004, UP Science had ceased operations. Of potential interest, it appears that some, perhaps most, of the key staff did not return to SEI and but instead had to find jobs in universities and other companies. In other words, the project was not reconstituted back in the parent, and there was no safety net for the employees.[100] The reasons for UP Science's failure are not completely clear, but evidence suggests that the control exercised by the parent was an important factor. My requests to interview the company in 2001 and 2002 (before I had any idea it was in trouble) were refused. But the reason given for the refusals, that management was busy preparing reports for the parent, is consistent with management being preoccupied with relations with the parent. After it failed, sources familiar with the Japanese biotechnology industry said concerns had arisen about inaccurate reporting of scientific data, and that pressure from the parent to meet development milestones probably were at the root of these reporting irregularities.

Have similar problems arisen between US ventures the private VC funds that have invested in them? Undoubtedly—so perhaps not too much should be made of this case. However, it was launched with some fanfare and it appears to have been well planned. Its failure suggests that the traditional way parents manage relations with tethered spin-offs may not be appropriate for fields of considerable technical uncertainty, where sustained sales revenue is remote and flexibility is necessary to respond to changing risks and opportunities.

There are other tethered bioventure spin-offs or affiliates of large Japanese companies. As a group, they account for a small proportion of bioventure drugs under development and sales of bioventure products and services.[101] Some of these feel their development is constrained by limited funding from their parents and the parents' unwillingness to yield control to outside investors. Investment analysts now tend to be skeptical about tethered spin-offs and express concerns that they lack the independence necessary to adjust their businesses quickly in order to grow and meet the challenge of competitors. In the USA, the situation appears similar with independent bioventures tending to outperform those owned by established firms.[102]

What about tethered spin-offs in nonbiomedical fields? The former director of Sony's computer science laboratory is reported to have said that all sixty spin-offs based on business plans submitted by employees were unsuccessful.[103]

A person who has discussed spin-offs with executives in leading Japanese electronics companies notes that spin-offs face unique challenges because of their relationship with the parent corporation. These challenges relate to personnel, organization, strategy, resource availability, and general decision-making. He writes:[104]

In the late-1990s, at the height of the telecommunications boom, corporations such as Sony and Toyota announced that they would spin out dozens, perhaps a hundred, venture companies. After the dust settled, there were in fact very few viable spin-outs, at least among those pursuing pioneering R&D. A review of the reasons reveals two general problems: (1) lack of consensus throughout management regarding the priority for spinning out companies and (2) unproductive intrusion of the parent corporation into the operation of the venture company which often handicapped substantially the growth of the venture.

The success of venture companies typically stems from their ability to exploit speed and focus. However, spin-off ventures must deal with their parent corporations, which usually adds inefficiency and frustration. I know of spin-outs that are being smothered by the need for their CEOs to spend a great deal of time responding to inquiries and directives from the parents. This takes away large chunks of valuable time from the needed task of running the venture.

Issues that arise include the following:

Personnel. The parent company often wants to dispatch the CEO—who may have no venture business skills—as well as the staff—who may be 'second tier' staff that the corporation's personnel department is desperate to rotate.

Decision making. The parent corporation often wants to participate or control key business decisions of the venture. The parent usually does not know the details of the venture's needs and changing environment it faces. The need to negotiate with the parent usually slows the venture. The venture usually loses if disagreements arise.

Strategic focus and strategic flexibility. Venture businesses have limited staff. Thus management attention and staff time have to be focused on moving forward toward clear targets. At the same time, most ventures change direction as they develop their business lines. So an ability to make quick, well defined strategic adjustments is also essential. Large corporations are not flexible once directions have been established.

Process orientation versus outcome orientation. Large corporate organizations typically emphasize processes. This often becomes ingrained in the habits of workers and managers. Venture businesses focus on outcomes.

Resource availability and financing. It is common for the large corporation to want to control the spending of the venture, requiring time consuming justifications for everyday operating expenses. Typically, spin-offs feel they need more resources than the parent corporations allow. Allowing for outside investment is an important asset for venture businesses.

Strategic partnerships hindered when they go beyond corporate group. While the parent generally agrees that the spin-off can sell products and services widely, strategic partnerships that might involve transfer of core proprietary technologies or joint investments in high priority projects are often restricted to companies within the parent's group. Conversely, companies outside the group may not trust a venture associated with a different corporate group.[105]

Limitations on acting against the interests of companies in the same group. Competition against companies in the same group is discouraged, and thus vigorous growth is often handicapped.[106]

An example was recounted to me of a project by an internationally respected high technology company to establish a spin-off venture in a new field of technology. Although the technical field was new, the company had world leading expertise in related technologies that could be applied to the new field. The complementarities between the company and its outside partners were good in terms of technical experience and other resources. The company assigned a senior manager with good business and technical expertise to head the venture. The researchers were competent, and lines of communication between the various parties seemed good.

However, the head of the spin-off had to spend much time seeking permission from the parent for a wide range of decisions related to the collaboration and in reporting back to the parent. He had to spend almost as much time managing the parent as managing the venture. Particularly time consuming were decisions related to funding by the parent and expenditures by the venture. Another problem was the parent's involvement in personnel decisions, both in recommending/insisting that particular persons from the parent be transferred to the venture even though they are not suited and in questioning the venture head's attempts to obtain persons with skills needed by the venture. Moreover, the venture head had to justify his requests and actions to multiple hierarchical levels within the parent. The delays caused by this internal oversight delayed the entire project and ultimately turned the venture toward more conservative technologies.[107]

The reasons UP Science and the venture described above ran into trouble seem similar to the problems that beset 19 internal ventures initiated by Exxon in the late 1970s and early 1980s. None managed to reach a break-even point or to have an IPO or merger. By 1986 Exxon had terminated and written off all of them.[108] More generally, the points raised above echo some of the problems besetting US spin-offs, particularly in competing for resources, markets, and the dedication of managers transferred from the parent. So long as the primary motivation for spin-off creation is strategic (to contribute to the growth of the parent and its affiliates), as opposed to financial (to increase revenue and profits), then providing the spin-off greater autonomy increases the potential for conflict with the parent's established business.[109]

The opposite problem can also occur. A former director of international licensing at IBM commented that when large companies form spin-offs, they usually have to hire someone from the outside who understands the particular business to head the spin-off. But if they adopt a hands-off policy, they often end up spending lots of money at the behest of the outside manager. Because monitoring is ineffective, resources are often wasted on projects that should have been terminated or redirected earlier.[110]

As a more promising example, let us consider the New Ventures Group (NVG) that Lucent Technologies established internally in 1997 to form spin-offs to commercialize some of Bell Laboratory's technologies. By December 2001 when Lucent sold most of its interest in NVG, the fund had launched thirty-five ventures. Eight had had IPOs or mergers. Together with the US$100 million that Lucent received from the sale of most of its ownership stake in NVG, these 'liquidation events' gave Lucent a 46 percent gross annual internal rate of return on its investments from 1996 through 2001, a rate that constitutes a financial success. Some of these were internal ventures, where NVG provided almost all the investment. However, a larger number were syndicated investments involving NVG coinvesting with outside VC companies that took the lead in forming the management team and overseeing the businesses.

The program was probably also a success in terms of business and technology development. Chesbrough attributes the success to NVGs combining the benefits of traditional private VC investment (including insulating the ventures from having decisions reviewed, delayed, and possibly overturned by Lucent executives)[111] with benefits associated with internal corporate investments.[112]

Nevertheless, when the downturn in the IT industry struck in 2000 and Lucent's year-on revenues fell over 26 percent from 2000 to 2001, Lucent had to divest all but its core business activities in order to survive. It sold all but 20 percent of its interest in NVG and its portfolio companies to a group led by Coller Capital, which specialized in secondary equity investments. The NVG team, renamed New Venture Partners, became the general partner of the fund and now manages the portfolio for the new investors.

Does Japan provide a better environment for a corporate spin-off program modeled on Lucent's?[113] On the positive side, large Japanese manufacturing companies have not yet faced the life or death situation that Lucent faced in 2001, so they have not been forced to shed noncore businesses. Furthermore, these large companies are less constrained by short-term earnings targets. Because of the uncertain and lumpy nature of revenue from venture businesses, Lucent may not have been able to build revenues from venture investments into such targets, which were nevertheless key to maintaining the support of investors and creditors.

But on the negative side, Lucent's experience suggests that a successful corporate spin-off program requires either a long-term commitment to building and maintaining a professional in-house venture business team or a willingness to let outside VC companies manage the spin-off process and the resulting portfolio companies. To my knowledge, very few Japanese manufacturing companies have taken either of these steps. More fundamentally, the underlying objective of spinning off companies is usually diversification and growth of the parent and its affiliates, rather than financial returns.[114] For this reason, there appears to be a strong tendency for parents to maintain control over spin-offs, particularly those requiring support from the parents, rather than to give them the autonomy to maximize growth and thus financial returns to the parents.

There are exceptions. Fujitsu seems to have the reputation not only of creating many spin-offs, but also giving them considerable autonomy—regarding them more as sources of profits and perhaps part of a voluntary commonwealth than as part of a mutual support alliance dominated by the core manufacturer.[115] One of the nonbiomedical venture case studies in Chapter 4 concerns a Fujitsu spin-off. Information from this venture, as well as a few other Fujitsu spin-offs, tends to be consistent with this reputation. However, I know of no other large manufacturing company that has adopted this approach to spin-offs. The other nonbiomedical spin-off profiled in Chapter 4, Chip Detect, also has autonomy (at least for the time being), but only after a less-than-cordial separation from its parent made possible by unique circumstances enabling it to garner outside support.

It may seem that some of the recent high profile spin-offs of major operating divisions from large electronics companies are also exceptions. But on closer examination, at least some of these have been plagued by interference from the parents. Elpida was formed in 1999 as a joint venture between Hitachi and NEC to absorb the loss making DRAM operations of both those companies. In the fiscal year ending March 2005, it registered its first operating profit and sales growth, although its share of the global DRAM market was only about 5 percent. However, when its current CEO, Yukio Sakamoto, took over in 2002, the company was in crisis. Morale among engineers was low due to downsizing and also because work teams combined employees from both parents who often did not communicate well with each other. Hitachi and NEC bickered about issues such as where to site the new manufacturing plant, from which parent to source purchases, and who should fill executive positions. Sakamoto had to demand that NEC and Hitachi cede him investment authority. He forced an end to the practice of appointing executives alternately from the parents.[116] Sakamoto was not affiliated with either parent, having previously headed the Japan operations of UMC.[117] His leadership helped to turn Elpida around and reduce interference from the parents. He was aided in 2003 by investments from about thirty outside companies, including Intel.[118] Intel's influence as a major investor was important in helping Hitachi and NEC to tone down their squabbling and to give the company real autonomy. Following a late 2004 IPO on the Tokyo Stock Exchange, Hitachi and NEC each held about 24 percent of Elpida's stock and Intel held about 6 percent.

Even mergers between members of the same group can be problematic and time consuming, as was the case of the 1994 merger of Mitsubishi Kasei and Mitsubishi Petrochemical.[119]

Fanuc, the robotics spin-off from Fujitsu which now is one of the world's largest manufacturers of computer controllers for machine tools, is sometimes cited as a successful spin-off in a new field of technology.[120] However, at the time of Fanuc's founding in 1972, it had approximately 300 employees and its first year operating revenue was approximately US$20 million.[121] Six years after founding, those revenues began to increase dramatically. In other words, when it was spun off, Fujitsu's robotics division already was fairly large, and it had sales revenue which, although not particularly large, had the potential to realize substantial increases within a relatively short period.[122] This size and access to revenue may have given it a degree of independence that spin-offs aiming to develop more early stage technologies with more distant market prospects lack.

The above examples do not prove that tethered spin-offs are less effective than independent ventures as sources of innovation. But they do suggest that across a wide range of industries, tethered spin-offs face obstacles to becoming successful pioneers of new technologies, and perhaps the most significant of these relate to control by the parents.

Conclusion

This chapter has endeavored to provide representative case studies of established companies attempting to innovate in new fields of technology that are removed from their core expertise. These cases suggest that these attempts usually do not lead to internationally competitive operations. Some of the reasons may be unique to Japan, for example, the prevalence of lifetime employment that prevents established newcomers from hiring experienced researchers and managers, and the deference paid to the welfare of a corporate family that may prevent established newcomers and spin-offs from competing vigorously. However, other reasons are not unique to Japan.[123] Thus, in other countries as well, established newcomers and tethered spin-offs may face similar difficulties.

Chapter 7 considers the evidence (and also the circumstances) under which independent ventures can, in many industries, be superior innovators in new fields of technology. It also examines the remaining possible strategies for established Japanese companies to become innovation leaders in new fields.

NOTES

1. See overviews of this issue in Chesbrough (1999) and Rtischev and Cole (2003). Chesbrough (1999) has documented this phenomenon with respect to hard disk drives (HDD). Henderson (1996) and Henderson and Clark (1990) have documented it with respect to photolithography and IC chipmaking technologies, and Fransman (1995) with respect to manufacturers of mainframe computers entering into the manufacture of PCs.
2. This latter example is described in Fransman (1995). Another reason NEC developed a better system than its main rival, Rockwell International's Printrak

system, was because it worked closely with retired Tokyo police officials to determine the most important aspects of fingerprint identification, aspects that the police officials themselves sometimes had not consciously conceptualized (e.g. the importance of ridge counts in differentiating between sets of fingerprints, particularly when prints had been degraded).

3. See Johnstone (1999) and Jim Frederick, 'A Sharper Focus', *Time Magazine*, May 9, 2005, 36–37.

4. See Suzuki and Kodama (2004), which analyzes the subject classification of Canon's patents over time pertaining to cameras, copiers, and semiconductor manufacturing equipment. This analysis shows the flow of technology from cameras into copiers and also into mask aligners and steppers (see also Henderson, 1996).

5. See, e.g. Friedman (1988), Henderson and Clark (1990), Henderson (1996), Kodama (1991), Aoki and Dore (1994, 1992), Odagiri and Goto (1993), Fransman (1995, 1998), Nonaka and Takeuchi (1995), Goto and Odagiri (1997), and Johnstone (1999).

6. See Chapter 7.

7. Fransman (1995).

8. e.g. Canon with respect to lithography (Henderson and Clark 1990, Henderson 1996), and Hitachi with respect to hard disk drives (Christensen 1993).

9. Sometimes formally protected as intellectual property, sometimes simply uncodified, tacit knowledge.

10. See Chesbrough (1999) for a discussion of access to *complementary assets*.

11. Also known as first mover advantages, i.e. the ability to maintain market share by being first to produce and market a new product.

12. The flip side of this argument, that strong IP protection enables innovators to organize their efforts in independent ventures and to obtain the complementary assets they need (sometimes even more so than they could if restricted by a large corporate bureaucracy), while maintaining entrepreneurial drive, flexibility, and responsiveness to customers, is discussed in Chapter 7.

13. Biomedicine is the one field where I have a fairly comprehensive picture of the main Japanese innovators, and where I feel can make comparisons between Japanese and overseas companies.

14. As opposed to small molecules usually made by synthetic chemical processes.

15. To date, Suntory remains a privately held company. Kirin's stock has been publicly traded on the Tokyo Stock Exchange (First section) since 1949.

16. One of Kirin's main laboratories, the Central Laboratory for Key Technology, was home to thirty-seven researchers, eighteen of whom (just under half) had Ph.D.s. Within this laboratory, the protein engineering group (which would have been considered to be working on cutting edge technologies) consisted of two scientists with doctoral degrees and three with master degrees (Protein Engineering in Japan 1992). In comparison, about 20–30% of researchers

in the large mainstream pharmaceutical companies have doctoral degrees (Kneller 2003).

17. It is interesting to compare the 1992 observations of the Protein Engineering study team with discussions with one of the senior scientists in one of these companies in 2004. What was not apparent to the foreign observers in 1992 was that the new companies were having recruitment problems and often had to rely on researchers from their brewery divisions.

18. A professor at the University of Tokyo provided Suntory researchers with the amino acid sequence of a bacterial protein that catalyzes the breakdown of penicillin. Using this information and information about the structure of a similar bacterial enzyme, beta-lactamase, Suntory researchers were able to design a combined penicillin–cephalosporin antibiotic, Farom®, that overcame bacterial resistance and has been marketed in Japan since 1997. This professor and another professor at the Kyoto University would create mutations at specific points in the DNA sequence of the gene coding for this resistance protein. The resulting changes in protein structure and function would give Suntory researchers clues as to good drug targets. Suntory was one of the founding members of the Protein Engineering Research Institute (PERI), a consortium of companies and academic researchers organized by MITI to pursue protein research. Suntory also had collaborative research agreements with Rockefeller University and the University of California at Irvine.

Kirin collaborated on government sponsored projects with several Japanese universities and GRIs. It was also one of the founding members of the PERI consortium. It still has an important collaborative relationship with the La Jolla Institute for Allergy and Immunology dating from 1988. It regularly sends its researchers there and also to the University of Oregon. It also has collaborated in genetic engineering with Konstantz University (Protein Engineering in Japan 1992).

19. Protein Engineering in Japan (1992). However senior scientists did voice frustration that young scientists, whom the company had sent to PERI and perhaps to other academic institutions, became enamored with basic research, and on their return experienced difficulty readjusting to applied corporate research. Also, as noted above, they were in fact concerned about the quality of the researchers they were able to recruit.

20. 2001 sales about US$50 million and US$40 million, respectively. Farom® is described in note 18 above.

21. He was succeeded by his son, a graduate of an American business school.

22. EPO is a naturally occurring hormone that prompts the body to increase red cell production. It is used to treat anemia resulting from chronic renal failure (e.g. in dialysis patients) and cancer radiation and chemotherapy. G-CSF stimulates the production of white cells and is used in the treatment of neutropenia and some malignancies and also in bone marrow transplantation. Amgen scientists cloned the genes coding for these two naturally occurring substances [US patents 4,703,008 (EPO) and 4,810,643 (G-CSF)]. The patent record suggests this was a competitive field with several US biotechnology companies

and several Japanese pharmaceutical companies pursuing drug discovery research related to these two compounds in the early 1980s.

23. Edwards, Murray and Yu (2003).

24. Most of Kirin's pharmaceutical revenue has come from domestic sales (55.3 of 57.5 billion yen in 2003). Of its domestic sales, 96% came from sales of EPO and G-CSF.

25. Aside from EPO and G-CSF originating from Amgen, its other two marketed products, Rocalcitrol® to treat hyperparathyroidism and Phosblock® to treat hyperphosphatemia, are in-licensed from Roche and Genzyme, respectively.

26. By virtue of being fully human, these antibodies are less likely to generate adverse immune reactions than many currently marketed antibody drugs that are either pure mouse antibodies or antibodies that are partly mouse and partly human. By virtue of being polyclonal rather than monoclonal, they are not directed against a single molecular structure (antigen) on a tumor cell or an invading infectious particle, but rather an array of such antigens. The basic technique for transferring complete human genes coding for completely human antibodies into mice was invented by a Japanese professor of medicine in a university distant from Japan's major urban centers. As in the case of many US biotechnology companies, Kirin is building its future pipeline on discoveries made in universities.

27. Interleukin 2 (IL-2) is used to stimulate the immune system of patients with cancer and some infectious diseases. Interleukin 6 is used to stimulate the production of immune cells following bone marrow transplantation or chemotherapy. Ajinomoto holds several US patents covering the gene for IL-2 and methods for producing IL-2 using genetically engineered cells. However, in 2000 company officials said Ajinomoto is no longer developing IL-2 or IL-6. To my knowledge, most IL-2 sold commercially in Japan is manufactured either by Chiron, which has FDA approval to market the drug in the US under the tradename Proleukin®, or Shionogi under license from Biogen.

28. Ajinomoto's 2003 net sales were about 1 trillion yen (just under US$10 billion), while Kirin's were 1.2 trillion yen. Kirin's pharmaceutical sales accounted for 5% of the company's net sales.

29. Antidiabetic drugs in this class are known as meglitinides. Ajinomoto's drug was second in its class after Novo Nordsk's repaglinide approved by the US FDA in 1997.

30. However, worldwide sales (US$96 million in drug 2002, about half the level of Novo Nordsk's drug) did not meet expectations at least in the first years after launch.

31. Procter & Gamble (P&G) holds the basic patent covering this drug. Aventis comarkets Actonel with P&G outside Japan. In Japan, Takeda also markets this drug under the brand name Benet®.

Ajinomoto's pharmaceutical products also include oral amino acid supplements for patients with liver disease (Livact®) and other medical nutritional supplements.

32. See US Patent No. 4,816,484.

33. See Kneller (2003) and Chapter 2.

34. Takara Bio's total revenue for the fiscal year ending March 2006 was 16 billion yen or about US$140 million, 85% from biotechnology systems and services and 12% from health food products (Accounting summary [kessan tanshin], May 2005, available at www.takara-bio.co.jp/news/pdfs/05051302.pdf.)

35. Takara Bio in-licensed genome and protein analysis technologies from Affymetrix, Lynx Therapeutics and other overseas biotechnology companies. It has been testing gene therapy technologies in-licensed from an Italian biotechnology company and Indiana University to treat leukemia, solid tumors and HIV/AIDS. It has been conducting clinical trials with the gene for vascular endothelial growth factor (VEGF) to produce new blood vessels in legs disabled by circulatory problems—an effort that closely parallels that of Gene Angiogenesis (Chapter 4). Most Variants of VEGF, the genes coding for it, and even therapeutic methods were discovered by researchers in the US and Europe, and related patents are held by US and European universities and companies. Among the key technologies underlying its current genetic engineering/protein synthesis core business, mRNA interferase was invented by New Jersey Medical and Dental University researchers, and the cold-shock expression vector system was jointly invented by researchers at the same university and Takara Bio. (Various public sources including www.takara-bio.co.jp and the USPTO patent data base.)

36. Hayakawa (2003). In 2002, biotech R&D accounted for 3.1 billion yen (~US$28 million) compared with 500 million yen (~US$4.5 million) related to brewing. Beginning in 2004, Takara Bio was planning to increase R&D expenditures to almost US$50 million annually.

37. About 56% in 2002.

38. Since 1999 when JT purchased the international tobacco operations of RJR Nabisco, JT has marketed Camel, Salem, and Winston cigarettes outside the US. Non-US sales of these brands substantially exceed US sales. In 2003, tobacco accounted for over 81% of JT's net sales, food products 12%, and pharmaceuticals 5%.

39. About 20% in 2002.

40. For the fiscal year ending March 2004, pharmaceutical R&D expenditures were down by 34% compared to their peak of 35 billion yen (about US$330 million) in the fiscal year ending March 2002.

41. This new class of drugs is known as cholesteryl ester transfer protein (CETP) inhibitors. Pfizer is also developing a drug in this class.

42. One exception is sorivudine, discovered by scientists at Yamasa, a soya sauce maker and a new type of treatment for shingles and some types of herpes rash. Unfortunately, when given to patients receiving 5-fluorouracil (5FU), a common chemotherapy for cancer, it sometimes resulted in 5FU rising to toxic levels. Therefore it is no longer marketed as a systemic medication although it is now being developed as a topical medicine.

None of the Japanese drugs approved between 1998 and 2003 on a high priority basis by the US FDA originated in small foodstuffs or chemical companies, although one, the anticancer drug oxaliplatin, originated in a metals company.

43. The yeast is *Schizosaccharomyces* pombe. See Frederick D. Ziegler, Trent R. Gemmill, and Robert B. Trimble (1994). 'Glycoprotein Synthesis in Yeast', *Journal of Biological Chemistry* 269 (no. 17, April 29): 12527–5; and Michael A. Romanos, Carol A. Scorer, and Jeffrey J. Clare (1992). 'Foreign Gene Expression in Yeast: A Review', *Yeast* 8: 423–88. See also http://www.stratagene.com/Newsletter/vol10_2/p72-74.htm

44. Toshiba is cooperating with universities, such as Osaka University and Tokyo Women's Medical University, on development of this system.

45. 'Canon to enter pharmaceutical business, focus on DNA chip'. *Nihon Keizai Shimbun*, March, 29, 2005.

46. As opposed to diagnostic and therapeutic machines of which it has long been a leading developer and manufacturer—e.g. MRI and CT scanners, ultrasound machines, automated biochemical analyzers, and electron beam accelerators for radiation therapy.

47. Funded primarily by US NIH, US DOE, UK Medical Research Council (MRC), and the Wellcome Trust.

48. The following account is based on the following sources: Kambara and Takahashi (1993), Kishi (2004), Pollack (2000), Takeda Foundation (2001), Yoshikawa (1987), various materials from Hitachi, various Japanese and US government memoranda and reports, and interviews with industry officials and scientists.

49. Hunkapillar's team was working under the direction of Lloyd Smith and Leroy Hood, two well-known geneticists.

50. Later in 1998, PE and ABI were to bankroll the formation of Celera as ABI's sister company and Celera's entry into the genome sequencing race.

51. Amersham was to buy out Molecular Dynamics in 1999.

52. According to Hitachi scientists, no government funds were used, at least for the sheath flow R&D.

53. 'Concensus elusive on Japan's genome plans, 1998'. *Science* 243, March 31, 1998: 1656–7.

54. e.g. gene sequencing, functional genomics (predicting protein function from information about the genes that code for the proteins), protein structure determination, drug target identification, and prediction of adverse reactions to drugs.

55. One initiated around 2001 was intended to allow pharmaceutical companies to match their data on various proteins with the data on gene sequences held by a METI-affiliated research company, Helix, to help the pharmaceutical companies decide which of the proteins might be useful drugs or drug targets ('Firms to set up genome laboratory', *Asahi Shimbun*, May 5–6, 2001).

Another project initiated in 2004 involved Riken, Hitachi Life Science, the University of Tokyo, the National Center for Genomics Research, and fifteen companies in research to elucidate the genetic basis of various diseases and to construct a related database that would initially be open only to the consortium members and paying companies. (Riken, GSC shutai no genomu netto 9 gatsu shidou, bousai na bunshikan saiyou o DB ka [Beginning Sept. 2004 Riken's Genome Science Center to lead research into genome networks and construction of data base on multiple molecular interactions] (*Kagaku Kougyou Nippou* [Chemical Engineering News], Aug. 23, 2004).

56. This assessment is based on published reports in the general and trade-oriented press, but it is generally confirmed by discussions with biomedical researchers. It is possible, of course, that some members of the Hitachi group have made life science breakthroughs unrelated to relatively large-scale instruments that are not yet apparent. In 2002 Hitachi Life Science was planning to increase its number of employees to approximately 120 (Nikkei Shimbun Sept. 10, 2002, 8). It declined to release sales data.

57. Founded, respectively, by Stanford researchers in 1984, and by a researcher from Xerox PARC's Computer Science Laboratory in 1996.

58. In some projects of this nature, companies dispatch their researchers to the participating university laboratory where they may work alongside researchers from other companies involved in the project. In other cases, most of the R&D occurs in the corporate laboratories, and the university laboratory plays more of a coordinating and synthesizing role, perhaps hosting regular meetings where all the corporate and university researchers can discuss progress.

59. This echoes Christensen's (1993) observations how existing product lines and customers confined the perceived technology development options of large US manufacturers of computer hard disks, even though venture companies later developed small drives that captured most of the market from the large companies (see Chapter 7).

60. Tunable lasers help to distribute the load of broadband Internet communication (e.g. enabling just in time capacity) and also allow optical communication infrastructure companies such as Lucent and Nortel to handle long and short distance communications in the same system.

61. Founded in 1998 by researchers at the University of California at Santa Barbara.

62. Spun off from Seagate in 2000.

63. Santur was founded in 2000 by researchers and managers from SDL, who were soon joined by key personnel from Nortel/Xros. SDL was a manufacturer of semiconductor diode lasers for fiber optic data transmission. It was formed in 1983 as a joint venture between Xerox and Spectra Physics, but it was bought out by a competitior, JDS Uniphase for US$41 million in 2000, the same year that some of its staff left to form Santur.

64. As of May 2005, Agility had eighteen issued US patents and ten published pending patents, Iolon twenty-one issued patents and two published pending patents, and Santur seven issued patents and one published pending patent.

65. This is not, however, to say the position of the US ventures is secure. In 2004, reports were circulating that Agility was in trouble, although in early 2005 it was still releasing new products. See Agility's February 28, 2005 press release available at www.agility.com. See also the August 6, 2004 report, Iolon's Alright, available at www.lightreading.com

66. See the discussion of consortia research in ch. 7 and the example of Phoenix Wireless in Chapter 4.

67. Perhaps, in this case, overmatched.

68. See the summary of carbon nanotube electronics under NanoMarkets. Market Report: Semiconductors/Electronics (May 4, 2005) available at www. nanomarkets.net. The companies identified in this summary are:

> Large Japanese: Fujitsu, Hitachi, Mitsubishi, NEC, Noritake and NTT
> Large US: DuPont, General Electric, IBM, Intel and Motorola/Freescale
> US ventures with founding date and university affiliation (if university source of core founding technology): Eikos (1996), Molecular Nanosystems (2001, Stanford), Nanomix (~2001, U California at Berkeley), Nano-Proprietary (1989), Nantero (~2000, Harvard), and Xintek/Applied Nanotech (2000, U North Carolina)
> Large Korean: Samsung
> Large European: Infineon, the 1999 spin-off of Siemens's semiconductor operations

69. See NanoInvestorNews.com. Nanotech Company Distribution (Nov. 15, 2004) available at www.nanoinvestornews.com under Facts and Figures.

70. Regarding government budgets, the Tokyo Office of the NSF estimates that Japanese government support for nanotechnology R&D amounted to 94.6 billion yen in 2003 and 93.5 billion yen in 2004. (See Report Memorandum 05-02, Japanese Government Budgets for Nanotechnology JFY, 2005 available at www.nsftokyo.org.) Nano Investor News estimated that in 2003 US, Japanese, and EU government spending for nanotechnology R&D amounted to 800, 780, and US$660 million, respectively (available at www.nanoinvestornews.com under Nanotechnology Facts and Figures). Since nanotechnology encompasses many fields including biology, materials, and electronics, estimates may vary according to the definition of *nanotechnology*.

71. See Chapter 7.

72. Toyota Motors was spun off from Toyoda Automatic Loom in 1937, two years after the latter began to produce trucks for the army. Toyoda Automatic Loom was itself a spin-off from Toyoda Boshoku, established in 1895 by Sakichi Toyoda who invented high quality automatic looms.

Fujitsu was spun off from Fuji Electric in 1935 in order to concentrate on automatic exchange equipment and telephone sets. Fuji Electric was established in 1923 as an electrical machinery joint venture between Furukawa Electric and Siemens. Furukawa Electric was spun off in 1883 from Furukawa Co., a copper mining company, in order to concentrate on wire making. Fujitsu itself spun off a high technology robotics subsidiary, Fanuc in 1972.

Mitsubishi Electric was spun off in 1931 from Mitsubishi Shipbuilding (now Mitsubishi Heavy Industries) so that the former could pursue its own growth and be more independent with respect to manufacturing equipment, components, administrative resources, engineers, and sales. Mitsubishi Heavy Industries also spun off its automobile division in 1970 to form Mitsubishi Motors.

Nomura Securities was spun off by Osaka Nomura Bank in 1925. Managers of the bank thought it was necessary to separate the capital of the banking and securities operations as the securities industry grew, and thus formed a separate securities company that could expand into this growth area. In 1926 the parent bank then changed its name to Nomura Bank and then in 1948 to Daiwa Bank (no connection with Daiwa Securities, Nomura Securities's largest competitor). In 2003 Daiwa Bank merged with Asahi Bank to become Resona Bank. But Resona, beset by financial troubles in 2003, is less prominent in the banking industry than its child, Nomura Securities, is in the securities industry.

JAFCO was established in 1974 with the backing of Nomura Securities and other financial institutions such as Nippon Life Insurance and Sanwa Bank, and it has remained Japan's largest VC company in terms of invested capital.

Sources: Ito (1995) and Ito and Rose (1994). Also Odagiri and Goto (1993) with respect to the entrepreneurship of Sakichi Toyoda, and various corporate histories with respect to Nomura and JAFCO.

73. Matsushita sells under brands such as Panasonic, National, Technics, and Quasar. The Matsushita group includes Japan Victor Corporation, which originated the VHS standard and which is majority owned by Matsushita.

74. The notion of *Hitachi* or *Matsushita group* brings up the complicated issue of terminology. These membership totals are from Ito (1995), whose definition of *group* probably means companies affiliated by familial (spin-off) relationships or companies in which one of the main members has acquired a significant ownership interest, usually by acquisition or joint venture.

As used by Odagiri, however, the notion of *keiretsu* (literally *related linkage* or simply *linked companies*) focuses on supplier–assembler relations within a particular industry. So under this definition, keiretsu would include a dominant manufacturing firm, its related spin-offs and acquired subsidiaries, and companies linked by long-standing subcontracting relationships (*shita-uke* and more independent subcontractors). In other words, it could be narrower than the notion of corporate group (because it is limited to a particular industry or even a particular line of manufacturing) or broader (because it includes companies linked only by subcontracting relationships).

Another definition of *keiretsu*, which Odagiri (1992) calls *kinyu-keiretsu* (financially linked companies), refers to a group consisting of a bank and the companies for which it is the main supplier of funds. Prior to the consolidation of Japanese banks around 2002, the most influential kinyu-keiretsu were those organized around the most influential banks, Mitsubishi (Tokyo-Mitsubishi after its merger with the Bank of Tokyo), Mitsui (now merged with Sumitomo to form Sumitomo-Mitsui), Sumitomo (merged with Mitsui), Fuji (Yasuda before World War II, now merged with First Industrial Bank and Industrial Bank of Japan to form Mizuho), Sanwa (now part of UFJ which merged in 2005 with Tokyo Mitsubishi to form Mitsubishi-UFJ), and First Industrial Bank (Daiichi Kangyo, now part of Mizuho). Gerlach (1992), Odagiri (1992), and Gilson and Roe (1993) provide helpful descriptions of these kinyu-keiretsu. Odagiri emphasizes the independence of the kinyu-keiretsu members, noting that rates of cross shareholdings and shareholdings exclusive to group members are not particularly high—which should be even more so today as banks and companies have sold many of their cross held shares. He concludes the main benefits they offer to their members are information exchange, some degree of mutual insurance, and reduction of the risk of hostile take over.

75. See Odagiri (1992), Ito (1995), Dyer (1996), and Gerlach (1992), all of which describe the role of spin-offs in the Japanese economy. None of these estimate the proportion of new, technology-oriented companies accounted for by spin-offs. But according to Ito, 17.5% of the companies listed on the Tokyo Stock Exchange were spin-offs, in contrast to only 20 (1.3%) of the companies listed on the New York Stock Exchange (17 of which were established as the result of antitrust actions).

76. e.g. entrepreneurship, motivated staff, flexibility, and access to private capital.

77. Especially access to supply and distribution networks, manufacturing resources, funding, and other complementary assets.

78. Until very recently, and still within many organizations, wages within a company have been primarily seniority based. Transferring workers into a start-up allows them to be paid according to different salary scales and also to have different retirement benefits (Odagiri 1992). Also Chesbrough (1999), quoting from T. Tatsura and S. Adachi (interview at the Tokyo office of FDK, a Fujitsu spin-off), Tokyo, March 19, 1998).

79. Ito (1995) and Odagiri (1992). Ito notes that in order for spin-offs to provide greater employment opportunities, the spun-out operations need to grow faster in the spin-offs than they would have if they had remained within the parent.

80. Odagiri (1992) describes examples of spin-offs from NEC to manufacture relatively low technology components. Chesbrough (1999) describes examples of

spin-offs from Hitachi, Fujitsu, NEC, Toshiba, and Matsushita, some of which engaged in sophisticated development of disk drives (HDD). For example, NFL, a joint venture formed by Hitachi and Fujitsu, was the first Japanese company to reverse engineer the IBM 3340 and 3350 HDDs and to develop 8″ and 5.25″ HDDs. Other spin-offs or partially owned subsidiaries also manufactured HDDs: NEI for NEC; Fuji Electric (Fujitsu's parent) for Fujitsu; and JVC, KME, MKE, and MCI for Matsushita. Other spin-offs developed HDD testing equipment (e.g. Hitachi DECO), manufactured HDD heads (e.g. Hitachi Metal and FDK from Fujitsu), or provided maintenance for such heads (e.g. Hitachi Electric Service).

81. Suntory's spinning off its pharmaceutical operations just before their sale to Daiichi may be an example of preparation for divestiture, while the spinning off of Takara Bio from Takara Shuzo and UP Science from Sumitomo Electronics (see below) may be examples of preparation for raising outside funds.

82. Some of the HDD spin-offs mentioned in note 80 above are probably examples of high technology spin-offs formed to give the new companies greater operational flexibility than they would have had as branches within their parents.

83. Ito (1995) and Dyer (1996). See Odagiri (1992) for evidence that the proportion of diversified firms is smaller in Japan than the US.

84. In economists' terms, specialized assets are vital to the productivity of any firm. Centralized management of a variety of specialized assets can create economies of scope. Sometimes, however, these economies of scope are not achieved and the transaction costs associated with management, e.g. difficulties in coordination, bureaucracy, loss of individual initiative, shirking, low morale, etc., outweigh the benefits from accumulation and central management of specialized assets. (Ito 1995, Dyer 1996).

85. Interestingly, these unaffiliated Japanese suppliers relied to a lesser extent than either the American affiliated or unaffiliated suppliers on one auto manufacturer for their sales (19% of sales to one manufacturer for the Japanese independent suppliers, on average, vs. 34% for both the US affiliated and independent suppliers). So the greater cooperation among the Japanese firms was not due to the Japanese suppliers being more beholden to the main manufacturer than their American counterparts.

86. Of some interest, the same advantages that Dyer attributes to Japanese manufacturing families compared to US integrated companies, particularly the disincentives to innovation faced by internal suppliers, could also be attributed to independent ventures with respect to manufacturing groups. In other words, taking Dyer's analysis at face value and extending it to its logical conclusion, the ideal supplier–manufacturer relationship would be independent small companies as suppliers, provided they could communicate effectively with the manufacturer. The subsequent discussion in the text deals with some of the problems that arise when a main manufacturer tries to exert too much control over spin-offs and other subsidiaries.

87. See e.g. Asanuma (1992), Odagiri (1992), and Nishiguchi and Ikeda (1996) with respect to the automobile industry. See also Friedman (1988) and

Whittaker (1994) with respect to innovation in machine tools and other fields of manufacturing.

88. Odagiri (1992) and Hikino, Harada, Tokuhisa and Yoshida (1998) use the term *manufacturing keiretsu* to refer to such families or groups of firms centered around a dominant manufacturer. See also n. 74.

89. Overall about 17% of parent companies' management staff is on loan to spin-offs—including about 12% of director level executives. Overall, about 7% of total employees are on loan to spin-offs (Odagiri 1992 citing 1989 Ministry of Labor statistics). (Some of these may be to non-spin-off affiliates.)

90. Odagiri (1992).

91. Ito (1995).

92. Among the 171 companies in Toyota's supplier association in 1985, Toyota owned at least 20% of the stock of only 36 of these companies (21%), indicating that the glue that holds together Toyota's family is probably not shareholding but simply long-term involvement in Toyota's production process. (One of these is Nippon Denso, a 1949 spin-off from Toyota which is now the largest producer of electronic automobile components in Japan, and which was 30% owned by Toyota in 1990.) Even among these 36 firms, 16 (44%, including Nippon Denso) were members of the supplier association of another automobile manufacturer. *But only two of these were members of Nissan's supplier associations.* In other words, membership in one manufacturing *family*, even to the extent of being a spin-off, often does not preclude membership in that of a competitor. Nevertheless, there are limits to the freedom of family members to diversify their markets, and it seems that Toyota would not tolerate suppliers working with an arch-rival if there were danger of technology leakage. Also other main manufacturers may typically exert more control over their spin-offs and other partner companies (Odagiri 1992).

93. Although as the economic slump that began around 1990 grew longer and as large manufacturers began to outsource more of their operations to China and other Asian countries, the limits of these mutual obligations have often been reached, and many subsidiaries have had to diversify their customer and technology base. It is not clear the extent to which spin-offs receive preferential treatment compared to independent subcontractors (*shita-uke* companies) in hard economic circumstances.

94. CMD Research (Keio U, 47% owned by Simplex), InternetNode (Keio U, 50% owned by Yokogawa Electric), EcoPower (Keio U, 82% owned by Ebara), GenoFunction (AIST, 95% owned by Hisamitsu Pharmaceuticals), Summit GlycoResearch (UT, >33% owned by Sumitomo Pharmaceuticals), Fluidwave Technologies (UT, >33% owned by Pentax), StarLabo (UT, 40% owned by Sumitomo Electric). Probably most of these university-related spin-offs arose by the 'parent' backing formation of a new company based on discoveries from UT, Keio, or AIST. See the analysis of the startup from these three institutions in Chapter 4, Appendix 2.

95. GenoFunction and EcoPower.

96. See Chapter 4, Appendix 2 Table 4A2.4.
97. These enzymes are involved in the intracellular decomposition of proteins, and dysfunctions of these enzymes are linked to various diseases.
98. Sumitomo Electric News Release, Nov. 2, 1999, announcing the formation of UP Science.
99. JAFCO and Sumitomo Pharma would have substantial minority shareholdings and be represented on the board of directors.
100. The fact that the employees ended up without a bridge back seems somewhat unusual for Japanese spin-offs. It suggests that one of SEI's main motives in forming the company was cost savings, i.e. it wanted to shed its biomedical research division, and SEI's willingness to support UP Science through rough business periods may have been limited.
101. Two have had IPOs Takura Bio and DNA Chip. However, of 13 Japanese bioventures identified at the end of 2006 by an organization that follows the biotech sector closely as having drugs in (or about to start) formal clinical trials, only one Takura Bio, is a tethered spin-off (Tsujimoto, Kenji, 2006. 'Sector view and introduction', presentation at the 13th Nomura Bio Conference (Tokyo, Nov, 20). This spin-off's therapy in human trials is based upon technology in-licensed from abroad. (I happen to know of two other tethered spin-offs that together have three drugs in (or about to start) formal clinical trials. Both of these companies have the same, mid-size pharmaceutical parent. Two of the drugs are licensed from the parent. The third was discovered in a university.) Among the other tethered bioventures with which I am familiar, two had annual revenues over US$1 million in 2005—mostly from contract research.
102. Zahra (1996) surveyed US biotechs in the early 1990s and found that those started by independent entrepreneurs and those owned by established companies had introduced similar numbers of new products, but the independent biotechs had more pioneering products and higher sales.
103. These sixty were competitively selected from among the employee business plans, and financed by an internal fund to start new ventures based on such business plans (Rtischev and Cole 2003).
104. From a 2005 communication.
105. Conversations I have had with venture companies substantiate this, although most of these conversations have been with independent ventures assessing the conditions of tethered spin-offs. It is not so much that there is an absolute prohibition against strategic alliances with outside group companies, but rather that such alliances need special approval from the parent and important decisions related to the alliance require frequent back and forth communication with the parent.

    Lincoln and Gerlach (2004) tabulated reports between 1992 and 1997 from the *Nikkei Shimbun* and similar newspapers about alliances involving 128 large publicly traded Japanese which they classified as belonging either to one of ten vertical networks (Hitachi, Toshiba, NEC, Fujitsu, Sony, Matsushita, Oki, Kobe Heavy Industry, Sumitomo Electric, and Yasukawa Electric) or to none of these networks. They found that the likelihood of R&D alliances between firms in different networks was not different from the likelihood of alliances between firms in the same network. However, considering the sources of these reports, these alliances probably dealt with mature or downstream technologies. Moreover, the analysis does not consider what was at stake for the parent in the alliance and how it turned out. The case described below of Elpida, the joint venture spin-off of Hitachi's and NEC's DRAM operations, is an example of a strategic alliance with a strong R&D component between two large companies from different families. However, it involved protracted negotiations between NEC and Hitachi.

106. This constraint on growth is also suggested by Odagiri (1992: 196, Chapter 7), who notes that when the markets a new firm wishes to enter are already served by other members of the same group, entry is often discouraged because it is likely to create intragroup competition, which will threaten the group's harmony and cohesion. See also Hikino et al. (1998: 117) who describe the same process of inhibited competition among chemical companies that are members of the same bank keiretsu.

107. These observations echo descriptions of the Japanese chemical industry, where construction and operation of large petrochemical complexes required the participation of many companies, usually from the same manufacturing or bank keiretsu. Hikino et al. (1998) observe, 'The complexity of the ownership, transactional and operational ties among firms forming a petrochemical complex became a significant structural rigidity. [D]ownsizing often meant the liquidation of those enterprises, a strategy that their management (and some of their parent companies) vigorously opposed.'

108. Chesbrough (2000), summarizing the first-hand account of Hollister Sykes in 'The Anatomy of a Corporate Venturing Program: Factors Influencing Success,' *Journal of Business Venturing* 1 (1986): 275–93.

109. Chesbrough (2000) summarizing studies by Eric von Hippel, Norman Fast, Kenneth Rind, R. Siegel, E. Siegel, and I. MacMillan.

110. Discussion with Robert Myers, Fairfield Resources International and Adjunct Professor, Columbia School of Business, Nov. 3, 2004.

111. Other such benefits including hiring outsiders as CEOs and financing in staged milestone-dependent increments (Chesbrough 2000, Chesbrough and Socolof 2000).

112. These include flexibility regarding life of fund (no end-of-fund drive for liquidity), ability to rejoin company, the associated retention of group learning, and the potential for large-scale funding for capital intensive businesses.

113. The following builds on the analysis in Chesbrough and Socolof (2003).

114. Ito (1995) makes this point most clearly, but it also appears to be the consensus among other observers of Japanese industry.

115. See Takahiro Shibuya (2003), 'Fujitsu fosters spin-off system'. *Nikkei Weekly*, November 24, 2003, 37. In contrast, Mr Yuji Mizuno, senior staff writer of the *Nihon Keizai Shimbun*, considered Hitachi's plan to launch more spin-offs 'hardly different from organizational reforms of a smaller scale,

where authority is delegated to departments or sections'. Mr Mizuno added:

> Concerned about keeping various operations within the group, Japanese companies tend to transfer assets from one company to another, thereby failing to improve business efficiency. ... If Japanese companies want to be truly accountable to shareholders, giving up their stake in spinoffs is worth considering. ... In contrast [to US corporations such as ATT and Hewlett Packard], Japanese companies resist spinning off some [independent] businesses, citing the importance of maintaining synergy. But the practice of expanding a corporate group at the expense of business efficiency does not really benefit shareholders regardless of what it does for synergy. ('Time is ripe for spinoffs in true sense'. *Nikkei Weekly*, October 15, 2001, 9).

116. See the following three articles on Elpida in the *Nikkei Weekly*: Shuhei Yamada, 'Sole DRAM maker sets the bar high', Nov. 24, 2003: 12; Hiroyuki Shioda, 'DRAM maker Elpida ready for big time', June 20, 2004: 30; and 'Elpida goes solo with daring scheme', Nov. 22, 2004: 3. I am also grateful to Professor Yoshitaka Okada of Sophia University for information related to Epilda, some of which will be contained in his forthcoming book, *Struggles for Survival: Institutional and Organizational Changes in Japan's High-Tech Industries* (Springer-Verlag). Any misrepresentation of Professor Okada's perspective is my responsibility.

117. UMC is a Taiwanese chip foundry.

118. These 2003 outside investments totaled about US$1.6 billion, of which Intel contributed US$110 million.

119. Hikino et al. (1998: 118).

120. Ito (1995).

121. Six billion yen, 1 US$ being equivalent to about 300 yen in 1972.

122. Early revenue and employment data at www.fanuc.co.jp/ja/profile/ir/index. htm

123. e.g. the inability of large corporate bureaucracies to anticipate market demand and to allocate adequate resources in a timely manner to R&D projects in new fields, and excessive control over spin-offs. Moreover, lifetime employment may be prevalent in other countries including those of Continental Europe and Korea.

REFERENCES

Aoki, Masahiko and Dore, Ronald (1994). *The Japanese Firm: Sources of Competitive Strength*. Oxford: Oxford University Press.

Asanuma, Banri (1992). 'Japanese Manufacturer–Supplier Relationships in International Perspective: The Automobile Case', in Paul Sheard (ed.), *International Adjustment and the Japanese Firm*. St. Leonarts, NSW, Australia: Allen & Unwin.

Chesbrough, Henry W. (1999). 'The Organizational Impact of Technological Change: A Comparative Theory of National Institutional Factors', *Industrial and Corporate Change*, 8(3): 447–85.

—— (2000). 'Designing Corporate Ventures in the Shadow of Private Venture Capital', *California Management Review*, 42(2): 31–49.

—— and Socolof, Stephen J. (2000). 'Creating New Ventures from Bell Labs Technologies', *Research-Technology Management*, March–April: 13–17.

—— —— (2003). 'Sustaining Venture Creation from Industrial Laboratories', *Research-Technology Management*, July–August: 16–19.

Christensen, Clayton M. (1993). 'The Rigid Disk Drive Industry: A History of Commercial and Technological Turbulence', *Business History Review*, 67: 531–88.

Dyer, Jeffrey H. (1996). 'Does Governance Matter? Keiretsu Alliances and Asset Specificity as Sources of Japanese Competitive Advantage', *Organization Science*, 7(6): 649–66.

Edwards, Mark G., Murray, Fiona, and Yu, Robert (2003). 'Value Creation and Sharing among Universities, Biotechnology, and Pharma', *Nature Biotechnology*, 21(6): 618–24.

Fransman, Martin (1995). *Japan's Computer and Communications Industry*. New York: Oxford.

—— (1998). *Visions of Innovation: The Firm and Japan*. Oxford: Oxford University Press.

Friedman, David (1988). *The Misunderstood Miracle: Industrial Development and Political Chang in Japan*. Ithaca NY: Cornell University Press.

Gerlach, Michael L. (1992). *Alliance Capitalism, the Social Organization of Japanese Business*. Berkeley CA: University of California Press, especially chapter VI.

Gilson, Ronald J. and Roe, Mark J. (1993). 'Understanding the Japanese Keiretsu: Overlaps between Corporate Governance and Industrial Organization', *Yale Law Journal*, 102: 871–906.

Goto, Akira and Odagiri, Hiroyuki (eds.) (1997). *Innovation in Japan*. Oxford: Oxford University Press.

Hayakawa, Akira (2003). 'Takara Research Unit Shifts Focus to In-House Drugs', *Nikkei Weekly*, Sept. 8, 16.

Henderson, Rebecca (1996). 'Product Development Capability as a Strategic Weapon: Canon's Experience in the Photolithographic Alignment Equipment Industry', in Nishiguchi Toshihiro (ed.), *Managing Product Development*. New York: Oxford University Press.

Henderson, Rebecca M. and Clark, Kim B. (1990). 'Architectural Innovation: The Reconfiguration of Existing Product Technologies and the Failure of Established Firms', *Administrative Science Quarterly*, 35: 9–30.

Hikino, Takashi, Harada, Tsutomu, Tokuhisa, Yoshio, and Yoshida, James A. (1998). 'The Japanese Puzzle: Rapid Catch-Up and Long Struggle', in Ashish Arora, Ralph Landau, and Nathan Rosenberg (eds.), *Chemicals and Long-Term Economic Growth*. New York: John Wiley & Sons, pp. 103–35.

Ito, Kiyohiko (1995). 'Japanese Spinoffs: Unexplored Survival Strategies', *Strategic Management Journal*, 16: 431–46.

——and Rose, Elizabeth L. (1994). 'The Genealogical Structure of Japanese Firms: Parent-Child Subsidiary Relationships', *Strategic Management Journal*, 15: 35–51.

Johnstone, Bob (1999). *We Were Burning: Japanese Entrepreneurs and the Forging of the Electronic Age*. New York: Basic Books.

Kambara, Hideki and Takahashi, Satoshi (1993). 'Multiple-Sheathflow Capillary Array DNA Analyzer', Nature, 361(Feb. 11): 565–6.

Kishi, Yoshinobu (2004). *Genome Haiboku [Genome Defeat]*. Tokyo: Diamond Press.

Kneller, R. W. (2003). 'Autarkic Drug Discovery in Japanese Pharmaceutical Companies: Insights into National Differences in Industrial Innovation', *Research Policy*, 32: 1805–27.

Kodama, Fumio (1991). *Emerging Patterns of Innovation: Sources of Japan's Technological Edge*. Boston, MA: Harvard Business School Press.

Lincoln, James R. and Gerlach, Michael L. (2004). *Japan's Network Economy: Structure, Persistence and Change*. Cambridge, UK: Cambridge University Press, especially pp. 353–66.

Nishiguchi, Toshihiro and Ikeda, Masayoshi (1996). 'Suppliers' Process Innovation: Understated Aspects of Japanese Industrial Sourcing', in Toshihiro Nishiguchi (ed.), *Managing Product Development*. New York: Oxford University Press.

Nonaka, Ijujiro and Takeuchi, Hirotaka (1995). *The Knowledge Creating Company: How Japanese Companies Create the Dynamics of Innovation*. New York: Oxford University Press.

Odagiri, Hiroyuki (1992). *Growth Through Competition, Competition Through Growth: Strategic Management and the Economy in Japan*. Oxford: Clarendon Press, especially Chapters 6 and 7.

—— and Goto, Akira (1993). 'The Japanese System of Innovation; Past, Present, and Future', in Richard Nelson (ed.), *National Innovation Systems: A Comparative Analysis*. Oxford: Oxford University Press, pp. 76–114.

Pollack, Andrew (2000). 'The Microsoft (and Gates) of the Genome Industry', *New York Times*, July 23.

Protein Engineering in Japan: Report of a U.S. Protein Engineering Study Team on its Visit to Japan (1992). US Government limited circulation report.

Rtischev, Dimitry and Cole, Robert E. (2003). 'Social and Structural Barriers to the IT Revolution in High-Tech Industries', in Jane M. Bachhnik (ed.), *Roadblocks on the Information Highway*. Lanham, MD: Lexington Books, pp. 127–53.

Suzuki, Jun and Kodama, Fumio (2004). 'Technology Diversity of Persistent Innovators in Japan: Two Case Studies of Large Japanese Firms', *Research Policy*, 33: 531–49.

Takeda Foundation (2001). Takeda Award 2001 Achievement Fact (official announcement and explanation of award to Michael Hunkapillar and J. Craig Venter).

Whittaker, D. Hugh (1994). 'SMEs, Entry Barriers, and "Strategic Alliances"', in Masahiko Aoki and Ronald Dore (eds.), *The Japanese Firm: Sources of Competitive Strength*. Oxford: Oxford University Press.

Yoshikawa, Akihiro (1987). 'In Search of the "Ultimate Map" of the Human Genome: The Japanese Efforts', Document prepared for the Office of Technology Assessment of the US Congress, and presented at the Berkeley Roundtable on the International Economy.

Zahra, S. (1996). 'Technology Strategy and New Venture Performance; A Study of Corporate Sponsored and Independent Biotechnology Ventures', *Journal of Business Venturing*, 11: 289–321.

7

Innovation Across Time and Space: Advantage New Companies

INTRODUCTION

This book began with the assertion that innovation in Japan depends primarily on large companies, while, in the USA, universities, venture companies, and established companies all play major roles in the discovery and early improvement of technologies with commercial potential. The most striking difference between the two countries is the role of venture companies. The middle chapters described the challenges and opportunities facing ventures in Japan.

Now we come face to face with the issue of which innovation system is better, mindful that innovation systems are part of larger social and institutional systems and that effectiveness depends on the specific technology and industry at issue.

This in turn raises the question of the extent to which these differences have influenced the changing fortunes of high technology industries in Japan, the USA, and other countries over the last decades of the twentieth century and the initial years of the twenty-first century. What were the roots of Japan's economic miracle from the 1960s to 1980s? Why did many US and European industries seem to lag behind Japanese counterparts in terms of innovation capabilities, and why did Japan seem to falter and the USA recover beginning around 1990? Finally, what are prospects for the future, and the lessons various countries can learn from this experience? These are complex issues which this book can only address in part.

To lay my cards on the table, I think one of the key factors concerns the vitality of high technology venture companies. This vitality is based on their ability to draw liberally on and to develop quickly university discoveries, to provide a professional rebirth for scientists and engineers whose ideas are not being developed in large companies, and to find receptive customers and development partners among established large companies. In Chapters 2–6, I have tried to show how Japan's innovation system has problems with respect to all these factors. These problems, juxtaposed against the more nourishing US environment for ventures, probably explain to a significant degree the varying fortunes of the high technology industries in Japan and the USA over the past half century.

However, because of deeply rooted social factors, the environment for ventures in Japan and many countries of Continental Europe is unlikely to become similarly supportive in the foreseeable future. Thus, these countries will continue to rely on their large companies for innovation, while the USA will rely to a large extent on new companies to be innovation leaders.

Whether one agrees with these assessments (or thinks their obviousness precludes the need for further analysis), because the role of ventures in US and Japanese high technology industries differs so markedly, these two countries provide a natural experiment shedding light on the types of environments ventures need to flourish, the strengths of shortcomings of the two types of innovation systems, and how government policies, university–industry relations, and social factors, influence the balance between ventures and established companies as innovation leaders.

Part I compares the innovative capabilities of ventures and established companies. The work of others suggests that in a fairly large number of UK and US industries, independent venture companies can be superior early stage innovators. The data presented at the end of Part 1 show, in the case of pharmaceuticals, that on a world wide basis venture companies are generally better innovators than established pharmaceutical companies, and that America's innovation leadership is due to its venture companies. Part II discusses whether these findings are likely to be applicable to other industries. Whether strong intellectual property (IP) rights can protect ventures in fields other than pharmaceuticals from encroachment by competitors is a related issue. The conclusion is that ventures can be innovation leaders in many fields besides pharmaceuticals, and that IP is vital for them, too.

Parts III and IV explore the sustainability of an innovation system based on venture companies. Part III explores whether venture companies undercut the innovation capabilities of large companies, particularly by poaching employees. The conclusion is that, while losing employees does hurt companies, probably the advantages of labor mobility outweigh the disadvantages. Part IV discusses whether ventures are capable of sustained innovation even though many companies are narrowly focused and lack broad integrated learning and organizational capabilities. The conclusion is that in an entrepreneurial region such as Silicon Valley, integrated learning and organizational capabilities that foster continued innovation are built into the disaggregated yet networked system of independent companies. However, there are questions whether the Silicon Valley model can be replicated in other regions. There are also questions whether ventures need a large supply of skilled, young researchers and whether US demographics and immigration policies can meet this need.

Applying these perspectives to Japan, Part V suggests that Japan's dearth of venture companies probably is hurting its innovation capabilities in a number of fields. Nevertheless, Japan's manufacturing industry has unique strengths that contributed to Japan's economic miracle and to its continuing strength in some industries. Part VI analyzes some of these strengths: the behind-the-scenes role of small companies and personal management policies based on a system of lifetime employment, as well as the purported benefits of government-organized R&D consortia. Some of these factors are also sources of weakness for Japanese ventures. Part VII examines whether the system of lifetime employment, autarkic innovation in large companies, and preemption of university research is likely to change in the near future, and concludes that the likelihood is low. In part this is because these factors underlie the strength of large companies, and in part because of deeply rooted social factors, including Japan's system of education. However, another reason is the new strategy of large Japanese companies to cooperate more closely with universities—to rely on them for basic research and potentially valuable early stage discoveries—which will leave ventures with even less space to grow.

In the concluding part, I suggest that this strategy probably will not allow established companies to remain in the forefront of early stage innovation. In other words, without vibrant new companies, Japan's prospects for long-term strength in new fields of technology are dim. On the other hand, some of the key factors that make the USA a fertile environment for ventures are also factors that undermine the innovative strength of its large companies, in particular labor mobility and its corollary, low job security. Thus, the USA has no choice but to rely on venture companies if it is to remain a leader in most new fields of technology. Moreover, although the US system is robust, the continued existence of all the various conditions required for its success cannot be taken for granted. Thus the USA and Japan are each locked into dependence on a different innovation system. The sources of each country's strength with respect to one system are the sources of its weakness with respect to the other. Yet the continued success of each country's system is not assured.

This phenomenon has parallels in previous comparisons of the institutional frameworks of innovation systems, in particular the contrast between liberal market economies, such as the USA and UK, and coordinated market economies, such as Japan and Germany. However, this chapter suggests that the different impact of various institutional systems are largely mediated by whether they create environments that favor early stage innovation by new companies or confine such innovation to established companies.

In order to distill the conclusions of this book into specific policies that can be the focus for productive debate, this chapter closes with suggested reforms to improve the environment for high technology ventures in Japan, the USA, and developing countries.

PART I: ARE NEW-OR-SMALL OR ESTABLISHED-AND-LARGE COMPANIES BETTER INNOVATORS?

Of course the answer is, 'It depends'. The ability to realize advantages attributed to either large or small size (or longevity vs. youth)[1] often depends on the particular technology and industry, as well as the overall innovation system and social context. Nevertheless advantages attributed to new/small companies include the following:[2]

1. Many *people want to be their own boss*, and when they are, they generally are more satisfied with their work and with life in general.[3] Assuming job satisfaction and motivation translates into more productive effort on the part of the entrepreneur and other key employees than would be the case if they worked in large companies, then the productivity of these small businesses should be higher.

2. Small companies are *more able to focus and experiment on a new product or new class of products*—even though these products may seem incremental or risky and the market may seem small.

3. With a large number of small companies pursuing many different approaches, the chances of developing *optimal approaches* will be higher than if only a small number of large companies are involved.

4. *Vision and goals regarding the application of innovative ideas are less constrained in new companies* than they are in large organizations, where past successes and current customers may limit the applications of a new idea. An established company benefiting from steady relationships with other companies may be disinclined to push a new technology that might upset those relationships. Lacking such constraints, it might be easier for new companies to aim for a global market, which their new products might create.[4]

5. *Less bureaucracy*. Internal communication is better, decisions can be made quickly. Thus the young-or-small company can respond quickly to technical and market opportunities.

6. Young companies are more likely to *start out with people who are right for the job*, both in terms of management and technical experience, and their personalities and motivations. Moreover, a system that fosters the

formation of new companies and the movement of people between companies helps to *redeploy human resources to where they are most productive.*

7. *Motivation and morale can be very high.* The feeling that 'We are all in the same boat together', is widespread. Minimal hierarchy, good communication, shared goals, and shared expectation that all who contribute will be rewarded, all make it easier to sustain a high level of focused dedication from employees.[5]

8. Because of this well-fitting, focused grouping of personalities, and talents, the company's *ability to learn and grow quickly is great*, at least in its initial area of expertise.

9. Provided capital markets function well, promising new small companies may be able to raise funds more quickly than established companies (even large companies) that have to fund projects internally. *Venture capital may provide more* funding to develop a new technology than internal funding which may face various internal budgetary constraints and bureaucratic hurdles.[6]

10. Small companies focusing on developing new technologies *may be less vulnerable* than large companies, whose revenue is derived mainly from manufacturing or licensing of previously invented technologies, *to having their resource base undercut by lower-cost foreign competitors.*

Advantages attributed to large companies include the following:[7]

1. *Capital* to start and build a new operation, and to achieve economies of scale: Usually the single greatest hurdle new companies face is raising capital. The resources of a large company can be used to purchase capital equipment and hire skilled managers, scientists and engineers, and support personnel. A large company can fund the initial costs of a new operation with income from another.

2. *Complementary assets*: These include supplier, manufacturing and distribution networks; brand name and customer or supplier loyalty; as well as ability to tap into various sources of advice related to management and regulatory issues.

3. *Ability to invest in training* to increase know-how that is specific to the company and tailored to the company's goals.

4. *Ability to integrate* the products and services of many suppliers, which is especially helpful in the case of products requiring complementary technologies.

5. Ability to develop an *integrated learning base and integrated organizational capabilities*, enabling a company to build on previous experience and to tailor and integrate all the steps in the value creation process, from research to marketing, with appropriate feedback mechanisms—in other words, smart, responsive vertical integration.

6. Supplier and customer *brand recognition and trust.*

7. *Ability to appropriate returns* from its investment in innovation: Because of the need to disclose their technology to attract investment, human resources and suppliers, and the need to outsource manufacturing and distribution, it is harder for new companies to prevent their technology from becoming known to potential competitors, particularly larger companies that could pursue rapid development once a new company has developed its technology to proof of concept stage. These power imbalances can be exacerbated if small firms lack skilled management. Conversely it is easier for a large company to keep its new technologies under wraps until they are close to marketing, and thus to establish leads over potential competitors in terms of manufacturing and marketing capabilities, and brand recognition. (These are known as first mover advantages.)

8. Ability to *undercut smaller competitors*, either through price competition or by engineering around (or infringing) their technologies.

9. Strength at *process and incremental* innovations: Because large companies often have manufacturing facilities and development staff, while new companies need to focus all their efforts on developing one or two initial products, the large companies are better able to invest in incremental and process innovations. Such innovations can rely on existing suppliers and marketing capabilities.

Of course the goal of most new companies is to grow into large companies. Almost by definition, young rapidly growing companies are among the most successful and have contributed much to innovation. Writing in the middle of World War II before the advent of VC backed ventures, Joseph Schumpeter noted the contribution of rapidly growing companies that manage to establish market dominance. Because of their efficiency and control over many complementary productive assets, they could often innovate better than small firms. But, as part of a renewal process built into the capitalist, free market system, Schumpeter felt that most large companies were destined to be replaced by successive gales of creative destruction, as innovation driven by entrepreneurship caused old technologies and skills to become obsolete. However, he warned that this system, although advancing human well-being, contained the seeds of its own destruction. Schumpeter predicted that as large organizations became more prevalent, most innovation would be routinized and individual entrepreneurship would be replaced by work of a team. Individual entrepreneurs would be replaced by organization men, with less entrepreneurial drive and less commitment to capitalism itself. This warning may be more prescient in the case of large Japanese manufacturers, because of their very success in terms of technology and management compared to many large US and European counterparts.

In any event, what evidence is there that small or large firms are more innovative? Throughout this book, I have tried to assess innovation in terms of actual products or processes, rather than indirectly using proxy measurements such as R&D expenditures, academic publications, or patent applications.[8] There are a few data sets of actual innovations by specific companies enabling analysis according to firm size. These involve mainly US inventions compiled from reports in trade journals[9] or UK inventions compiled by industry experts.[10] The UK data show that from 1945 to 1983, the share of innovations contributed by firms with less than 500 employees rose to nearly 40 percent of the total, while that of firms with 500–9,999 employees declined and that of firms with at least 10,000 employees stayed constant at 42 percent.[11] The ratio of inventions per US$1 million of R&D expenditure is highest for small firms and lowest for large firms.[12]

However, both the US and UK data show considerable variation by industry. Among the thirty-five most innovative US industries, the number of innovations by large firms was higher in twenty-one industries while the number of innovations by small firms[13] was higher in fourteen industries, including the two with the most innovations.[14] The innovations per employee ratio for large versus small companies also varies greatly among US industries. It is considerably higher for large companies in the tire, agricultural chemical, general industrial machinery, and food products industries; and for small companies in the computing, instrument, and synthetic rubber industries.[15]

Not surprisingly, industries with relatively high capital costs tend to be those where large firms are more innovative, while industries where skilled labor is relatively important tend to be those where small firms are more innovative.[16] Also as firm size increases and the focus of activity includes more manufacturing and product refinement, a greater proportion of R&D is probably devoted to process as opposed to new product R&D.[17]

Although innovation decreases as industry concentration increases, the most innovative industries have large, influential firms.[18] However, in these industries, innovations tend to emanate more from small than large firms.[19] Also the balance between small and large firm innovations has changed, with small companies having become prominent innovators in IT hardware and in biomedicine.[20] Finally, average R&D costs per innovation do not appear to be increasing, in other words high costs have not closed out innovation possibilities for small companies.[21]

In summary, these data sets of actual innovations suggest that, at least in industries where capital costs do not pose high barriers to entry, the innovative potential of small firms is strong. In some of the industries with relatively low capital entry barriers, small firms do indeed seem to be the main founts of innovation.[22] Furthermore, there is little evidence that the cost of innovation in general is rising beyond the capabilities of small companies. If anything, over time independent small firms are probably accounting for a greater share of total innovations, at least in the USA and UK. However, the most innovative industries appear to be those with a mixture of large and small companies. Large companies tend to have a higher proportion of process inventions compared with small companies. These findings in turn suggest that the most conducive environment for innovation involves symbiotic relationships between large and small companies, where small companies carry on much of the discovery and early stage product development, while large companies rely to a large extent on these companies for new discoveries and concentrate much of their efforts on assimilation of outside discoveries, later stage development, manufacturing, and marketing.[23]

The Importance of Ventures in Drug Discovery

Fortunately, in one industry, pharmaceuticals, we can have a fairly clear picture of where each new product originated, its degree of innovativeness, and the contribution of ventures, large companies and universities to its discovery and early development. From 1998 through 2003, the US FDA approved 169 new prescription drugs for the US market.[24] Most of the drugs were covered by patents. I obtained copies of the key patents covering the therapeutic compounds underlying each drug or, if compound patents did not exist, the main methods of manufacturing or using the drugs.[25] I identified the inventors listed on the key patents, and then confirmed their place of work and the identity of their employers at the time of the original patent applications.[26] Thus, I attributed a type of work organization (pharmaceutical company, biotech.[27] or university/GRI) and a national location of work place to each inventor. Weighting each inventor equally for each key patent, but weighting key patents according to the criteria in note 25, I attributed the proportionate origins of each drug according both to type of discovering organization and to country.[28]

Overall, about 40 percent of these drugs originated outside of large pharmaceutical companies, in either universities (22%) or biotechs (19%). Nearly three-quarters of university-origin drugs were licensed to biotechs, confirming that biotechs are the main initial development partners for drugs originating in universities.[29] Of the seventy drugs whose origins were at least 50 percent attributable to universities or biotechs, nearly one-third were transferred to pharmacentical companies that completed development and obtained FDA marketing approval, and over half were ultimately marketed in the USA or Europe by pharmacentical companies.[30]

However, national variations in this regard are striking. Drugs discovered by biotechs, or discovered in universities and transferred to biotechs, account for 62 percent of US origin drugs, 50 percent of Canadian drugs, 14 percent of UK drugs, and less than ten percent of Japanese, Germans Swiss and French drugs. This shows the importance of biotechs in drug discovery and early stage development in North America in contrast to their negligible role in Japan and Continental Europe. Of drugs discovered in biotechs, or in universities and licensed to biotechs, 84 percent were discovered in the USA or Canada.[31]

The FDA approval process embodies an indicator of innovativeness. New molecular entities (NMEs, small molecule drugs that accounted for 144 of the 169 newly approved drugs) are accorded *priority* review status and fast-tracked for approval, if they offer substantial benefit over currently marketed drugs. All FDA applicants want the priority designation, because it enables them to market their drugs more quickly. Priority NMEs account for just under half of all NMEs.[32] The proportion of priority NMEs discovered in biotechs or in universities and licensed to biotechs is over two fold that of nonpriority NMEs.[33] *In other words, the drugs that originate in the laboratories of established pharmaceutical companies tend to be less innovative than those discovered in biotechs or in universities and then licensed to biotechs.* The USA dominates as the origin of priority NMEs, accounting for just over half followed by the UK and Germany accounting for 9 percent each.[34] Biotechs, or universities licensing to biotechs, are the origin of over half of US-origin priority NMEs. In other words, *the US pharmaceutical industry now relies on biotechs to discover its most innovative drugs.*

These differences are still more striking with respect to new therapeutic biologics (NTBs).[35] Genetic engineering techniques using microorganisms to produce human-type NTBs were pioneered in universities and then, in a parallel fashion, in industry.[36] The narrow definition of *biotechnology* refers to these techniques and the underlying science. However, within industry, biotechs rather than pharmaceutical companies took the lead and have maintained it. In general, the discovery and development of NTBs takes longer and is more costly than that of NMEs, and the diseases they address are more challenging.[37] Only 12 percent of the NTBs originated in the in-house laboratories of established pharmaceutical companies. A full 84 percent of NTBs originated in biotechs, or in universities and were then licensed to biotechs. All but two of these biotechs were US companies.[38] Over 85 percent of NTBs are of US origin, and of these over 90 percent were discovered by biotechs, or in universities and then licensed to biotechs.[39]

Data on drugs still in clinical trials also indicate that the center of gravity of new drug discovery and early stage development is in biotechs rather than established pharmaceutical companies, with biotechs predominating as the sponsors of clinical trials for drugs in a variety of disease fields.[40]

Outside the USA, biotechs and universities that license to biotechs also discover a greater proportion of the priority compared to nonpriority NMEs than do pharmaceutical companies.[41] However, in the case of Japan, Germany, Switzerland, and France, no NTBs and only two priority NMEs originated in a biotech or in a university and was then licensed to a biotech.[42] Thus, although the USA may have the most fertile or most mature environment for biotechs, they also are innovation leaders in other countries—except in Japan and the large countries of Continental Europe, where drugs discovery and early stage development relies almost exclusively on established pharmaceutical companies.[43]

But to reinforce an earlier point, a large proportion of the university and biotech originating drugs are licensed to established pharmaceutical companies that complete clinical trials or market the drugs.[44] Thus, these data describe a symbiotic relationship (except in countries such as Japan, Germany, Switzerland, and France) where the center of gravity of drug discovery is shifting to new companies with close university ties, but where cooperation between biotechs and established pharmaceutical companies is essential for final commercialization. The biotechs and the universities that license to them may be leaders in discovery, but much of the heavy lifting in terms of clinical trials and marketing depends on large pharmaceutical companies. Thus there is a suggestion of a rational division of effort, mediated by much back and forth, and sometimes tense, negotiations—and also by entrepreneurship in universities and their startups.

But to observe that bioventures are now more innovative in terms of drug discovery than large pharmaceutical companies does not automatically mean that they are inherently better at early stage innovation. What if bioventures never existed? Would pharmaceutical companies have been able to produce a similar number of innovative new drugs? Prior to the advent of most biotechs, large pharmaceutical companies were developing at least some of the university discoveries that pointed toward important drugs. University or GRI discoveries pointed the way to the discovery of a majority of the most therapeutically important drugs introduced between 1965 and 1992—although, in only one case was the actual active compound synthesized in a university or GRI.[45] Large pharmaceutical companies developed and commercialized all but one of these.[46] If this method of drug discovery and development based solely on established companies and universities had continued, would we have as many new drugs as we do today?

It has been suggested that the playing field for large companies versus ventures is not level with respect to training and movements of personnel—more specifically, that large US companies have curtailed R&D, in part because VC financing lures key R&D employees to venture companies.[47] Key personnel have indeed moved from large pharmaceutical companies to biotechs. For example, George Rathmann resigned as Abbott's Vice President for Research and Development to become the first CEO of Amgen in 1980. However, this was apparently an amicable parting, and within half a year, at Rathmann's urging, Abbott became one of the largest investors in Amgen's initial round of financing.[48] Genentech, formed in 1976, recruited most of its key scientists from universities, as have some other biotechs.[49] To my understanding, however, poaching of lead researchers and managers by biotech companies has generally not been a great concern for the pharmaceutical industry. As for the ability to hire star university researchers away from academia, large pharmaceutical companies are probably more able to do so than new biotechs.[50] Thus, at least in drug discovery, the evidence seems weak that biotechs' innovation advantage is due to their poaching personnel from pharmaceutical companies or being able to out-compete pharmaceutical companies in hiring away academic researchers from universities.[51]

Also, there is no clear evidence that biotechs' innovation advantage stems from an inherent tendency by US universities to license to startups rather than to pharmaceutical companies. Large pharmaceutical companies often decline to license early stage pharmaceutical candidates or drug targets from universities, which end up licensing these to biotechs as a last resort.[52] Pharmaceutical executives have said that if they have the option to develop university and in-house-originating candidate drugs for the same disease, they prefer to develop the in-house discovered compounds, because they can retain greater control.[53] In other words, if big pharma wanted to license more early stage university pharmaceutical compounds, it probably could. Similarly, if it wanted to fund more research in universities and have the right to exclusively license compounds arising from such research it could do so, too. The fact that big pharma has instead chosen to let biotechs and their investors bear much of the risk for developing early stage discoveries, suggests an element of economic rationality in the present system. In other words, at least in North America, big pharma seems to have made a calculation that it makes more economic and scientific sense to let universities and biotechs discover most of the compounds that eventually become new drugs, even though they may pay substantial license fees when they eventually in-license the partially developed compounds.[54] This trend has continued despite university entrepreneurialism, suggesting that such entrepreneurialism, rather than hindering innovation, may be an important part of an environment that fosters it.[55]

Finally, the data on drugs recently approved by the FDA provide clues as to the pace and innovativeness of drug discovery in an environment without biotechs. For practical purposes, Japan, Germany, Switzerland, and France together constitute such a region.[56] Each country's share of the world pharmaceutical market in 2003 serves as a benchmark against which to assess its output of newly discovered drugs. The proportion of new drugs originating in these four countries is higher than their proportion of world pharmaceutical sales—although without Switzerland, the three remaining countries' share of new drugs equals their world market share. The US proportion of new drugs equals its share of the pharmaceutical market, while the combined share of the USA, the UK, and Canada—three countries where biotechs are active in drug discovery, exceeds their combined market share.

However, considering only NTBs and *priority* NMEs, the share of Japan, Germany, Switzerland, and France combined is less than their combined share of the world market. Leaving out Switzerland, the share of the three remaining countries is considerably below their combined share of the world pharmaceutical market. In contrast, the share of the USA, the UK, and Canada combined is greater than their world market share. The US share alone is also considerably greater than its market share. Although these results do not show conclusively what the pace of drug discovery would be in the absence of biotechs, *they are consistent with the hypothesis that a pharmaceutical innovation system comprised of vibrant biotechs linked both to universities and large pharmaceutical companies produces more innovative drugs than a system with weak biotechs.* Moreover, the involvement of biotechs, not drug discovery in universities per se, seems to be the most important condition for innovative drug discovery.[57] The case of Switzerland shows it is possible for a country to have large pharmaceutical companies that discover a disproportionate share of the world's drugs. But perhaps this is possible only in a small country where pharmaceuticals has become the premier high technology industry.

PART II. CAN VENTURES BE INNOVATION LEADERS ONLY IN PHARMACEUTICALS? REQUISITES FOR A FAVORABLE ENVIRONMENT

Can venture companies also be the lead innovators in other industries? In software and some of the engineering and applied physics fields of IT, they can—witness the growth of Google, Netscape, Lycos, Sun Microsystems, Cisco

Systems, and Qualcomm to name just a few. But what about other fields? Or
are pharmaceuticals and IT so unique that it is unlikely that venture companies in other fields can assume the same leading role in innovation? Aside from
the data sets mentioned at the beginning of this chapter and a study of the
patented technologies underlying 3G mobile telecommunications[58], I know of
no systematic data on how various companies (large and small, old and young)
contributed to the development of final products in a particular industry.

However, let us examine the possible reasons pharmaceuticals are so unique
that the situation described above probably does not apply in other industries.
This examination suggests that although ventures in the life sciences, and particularly pharmaceuticals, have some advantages over those in other new high
technology fields, the barriers to US ventures in these other fields probably are
not prohibitively high.

1. *When it comes to drugs, diagnostics and medical devices, universities and
their startups have a clear advantage, access to patients, that does not apply
in other industries.* The access is ideal because conditions can be controlled
and monitored and because well trained research-oriented physicians, nurses,
and technicians are close at hand. With the possible exception of some fields
of software, there is probably no other industry where universities provide
ideal laboratories for development up to the point of commercialization. Thus
there is probably no other industry where startups closely linked to universities have a similar advantage in terms of access to resources for late-stage
development.[59] Nevertheless, major pharmaceutical companies have developed strong clinical development capabilities that are largely independent of
universities. Sometimes they manage clinical trials on their own. Sometimes
they rely on contract research organizations (CROs) to conduct trials. They
and the CROs have their own networks of hospitals and physicians that they
can use as sites for clinical trials and to recruit patients, respectively. Sometimes these are university hospitals and medical school faculty, but probably
in the majority of cases they are not. Similarly, in fields such as medical
devices and diagnostics, large companies routinely cooperate with universities
in R&D.[60]

In summary, access by university biomedical startups to patients and clinical research resources probably gives them an advantage relative to startups
in other fields. But these are probably not decisive advantages vis-à-vis large
companies. Independent spin-offs from large companies may benefit in a
similar way from knowledge related to product development and marketing
that their founders gained in the parent companies. Of course, this raises
the question, 'What about ventures that are neither university biomedical
startups nor spin-offs from large companies, for example many of the
materials science/nanotechnology start-ups in existence in 2005?' Is one of

the reasons few nanotech university startups seem to have taken off due to this absence of development facilities in universities? As of 2005, the main uncertainty among investors with respect to nanotech in general was that most practical applications seemed still in the future, and still in need more development work. So perhaps this is a key advantage of biomedical ventures. The answers are not yet clear.

2. *Particularly in the USA, university R&D is heavily weighted toward the life sciences. Thus, it is to be expected that there will be more startups in the life sciences than in other fields.* Life science accounts for approximately 59 percent of total US university R&D expenditures in all S&E fields, engineering for about 15 percent, the physical sciences (astronomy, physics and chemistry) for 9 percent, and earth atmospheric and ocean sciences for 6 percent.[61] Nevertheless, engineering alone accounts for about US$5 billion in annual US university R&D. Thus, while the financial inputs for nonbiomedical academic discoveries are fewer, they are far from insignificant.

3. *A related possibility is that academic biomedical research tends to give rise to more commercially relevant discoveries per unit of R&D funding than research in other fields.* Theoretically, this might be the case, because most government funding for university biomedical R&D has the practical mission of improving the health of the American people.[62] The only other field that might have a similar applied emphasis would be engineering, whose funding levels are only one-fourth that of life science. However, with respect to numbers of US university patents, biomedicine does not predominate to the same degree that it does with respect to R&D funding.[63] Available data do not suggest that per unit of university R&D expenditure, life science R&D is more likely to give rise to commercially relevant discoveries than R&D in other fields.

4. *Even if life science discoveries fall short of a majority of university inventions, most non-life-science inventions are licensed to established companies, as in Japan.* Unfortunately, I have not found US data that address this issue directly. However, several lines of evidence suggest that a majority of non-life science university inventions are licensed to startups or other SMEs, exclusively, rather than to large established companies.[64]

5. *Even though SMEs may have eqivalent access to university discoveries in most technology fields, the disadvantages related to size pose more severe challenges to SMEs outside the life sciences.* The success of venture companies in drug discovery indicates that the disadvantages of small size are either surmountable or are less determinative in the case of pioneering innovations—even in an industry that has high entry barriers to new companies.[65,66] However, their success depends on obtaining some degree of exclusive control over new discoveries, being able to avoid encroachment from larger companies, and having some of the large companies become their

customers and collaborators, rather than their competitors. Thus two areas that might distinguish life science from other industries are the attitudes of large companies toward collaborations with ventures and the effectiveness of patent protection.

6. *Patents enable ventures to appropriate returns from investment effectively in pharmaceuticals but less so in other industries.* Patents are important for most ventures, because they lack other means to protect their technologies from being copied by rivals. They have to outsource manufacturing and perhaps even some aspects of development, and thus they often cannot rely on being first to market (lead time) to secure a competitive advantage. Similarly, they usually are not able to offer complementary services or products to attract customers. In addition, their need to seek outside suppliers, manufacturers, and distributors compromises their ability to protect their technology as trade secrets.[67] In general, secrecy, lead time, and complementary assets protect a greater percentage of manufacturing companies' innovations from encroachment by potential competitors than do patents.[68] But being less able to employ these other means to prevent encroachment by competitors, many ventures rely primarily on patent protection and other forms of intellectual property.[69] Also the strength of patent protection may be an important factor in a large company's decision whether it will try to duplicate or engineer around a venture's technology rather than partnering with it. Thus patents are also important for ventures to attract outside financing from angels, VC funds, or corporations.[70]

Without effective patent and copyright protection most new discoveries would have to be developed in existing companies, most likely in large companies that have complementary assets and can effectively shield much of the development work from competitors. But this would foreclose the advantages mentioned at the beginning of this chapter, such as entrepreneurship, focused dedication, and assembling a workforce whose skills and motivations fit that of the venture.[71]

What do we know about the effectiveness of patent protection for ventures, particularly in nonbiomedical fields? Survey responses indicate that the costs of applying for and, if necessary, defending patents more often dissuaded small than large firms from applying for patents.[72] However, the extent to which small companies are unable to protect technologies that are really worth protecting is not clear. In the case of R&D intensive ventures, the answer probably depends on the availability of astute outside capital.[73]

In general, survey responses of corporate R&D managers indicate that patents are effective in protecting a greater proportion of a company's innovations in pharmaceuticals and medical devices than in other industries. Specifically, patents were reported to protect about half of pharmaceutical and medical device product innovations compared to, for example, about a quarter of semiconductor innovations. However, these survey responses probably underestimate the usefulness of patents for high technology ventures in nonbiomedical industries, particularly those focusing on R&D in new fields.[74] New R&D-intensive US semiconductor firms are among the most likely firms in that industry to apply for patents. Interviews with some of these companies indicated that they patent in order *to secure strong, 'bulletproof' proprietary rights to technologies in niche markets*, and thus *improve their competitive position vis-à-vis direct market rivals*—in other words, to prevent encroachment. As a corollary of preventing encroachment, strong patent protection is also important for these young firms to secure startup capital.

To an even greater degree than these new firms, large capital intensive semiconductor firms also contributed to a surge in new semiconductor patents from the mid-1980s to mid-1990s. However, in contrast to the new firms, these established firms usually patented to secure freedom to operate in an environment which had suddenly become more complex and sometimes threatening due to a general strengthening of patent rights.

This strengthening of rights was due, in part, to the establishment in 1982 of the Court of Appeals for the Federal Circuit (CAFC) as the US intellectual property appellate court. In carrying out its mandate to develop a coherent body of jurisprudence to interpret and uphold America's patent laws, the CAFC ruled more frequently on behalf of plaintiffs in infringement suits than had the nonspecialized federal appeals courts prior to the founding of the CAFC.[75] Polaroid's 1986 US$1 billion damage award in its infringement suit against Kodak[76] put companies on alert that courts were ready to mete out severe penalties against infringers.

Around this same time, several large companies began to take concerted legal action to demand substantial royalties from other companies using technologies covered by their patents. Texas Instruments lead this assertive use of its patent portfolio. Others followed suit.[77] In a complex manufacturing environment where final products may be covered by numerous patents, this combination of enhanced patent strength and the willingness of some companies to assert their patent rights aggressively led many large capital intensive semiconductor manufacturers to amass large portfolios of patents, mainly as bargaining chips. Assessed more often on the basis of quantity than their actual specifications and claims, these patent war chests were to be used defensively if their holders encountered charges of infringement and demands for high royalties. But they could also be means to access needed technologies, for example, by gaining admission to patent pools,[78] and sometimes to enhance royalty flow and block competitors. There is doubt whether such uses of patents increase incentives for R&D and innovation. Nevertheless, *even these large companies, when presented with the extreme hypothetical scenario of the US patent system being abolished, expressed concern about the negative effects on high technology ventures.* In other words, even with a clear eye to the negative effects of the patent system (increased expense and the threat of royalty demands or infringement suits holding up their operations), *large manufacturing semiconductor companies recognize the value of the patent system in enabling entry of high technology ventures into their industry.* Indeed, the implication from such responses is that *this is the primary value of the patent system*—at least in the semiconductor industry.[79]

The 1980s marked the beginning of a period of new company formation in semiconductors and related fields that is probably still continuing. A significant proportion of these companies have continued to grow, despite the 2000–3 downturn in the IT industry.[80] Several factors probably underlay this upturn in new company creation in the 1980s. However, the coincidence with the strengthening of US patents, the patenting behavior of the new firms, and the explanations offered by industry officials, all suggest that increased confidence that patents can prevent encroachment was an important factor underlying this upturn.[81]

Semiconductor products as well as other products emerging out of engineering and the physical sciences are generally complex, in that a single product usually integrates many complementary technologies. In contrast, biomedical inventions, particularly pharmaceuticals, tend to be discrete, in the sense that they are often based on a single new chemical compound or a single detection device. It has been suggested that outside of biomedicine, large companies have an advantage over ventures because the former are better able to acquire and integrate the various component technologies, and in the process they account for much of the value added of the final product.[82] However, the fact that ventures have flourished in a complex product industry such as semiconductors, thanks in part to patents, casts doubt on this assumption.

There is evidence that when new companies lack patent protection, or when they feel that patent protection is ineffective to prevent imitation, they are more likely to try to develop their technologies on their own and to compete head-to-head with established companies. But if their technologies are protected by patents, they are more likely to cooperate with other companies via licenses of their patented technologies.[83] Thus, patents have been an important factor in the vertical disaggregation of the semiconductor as well as the pharmaceutical industry. In other words, they have enabled new companies to enter the value chains in these industries with high value-added products and to negotiate partnerships with other companies in the value chain. This is also a strategy that ventures in various fields of nanotechnology are trying to pursue.[84]

Patents have also proven useful for companies that had invested heavily in research and made pioneering discoveries, and then confronted similar products developed by foreign competitors.[85]

Thus, there seems to be little clear evidence that, in the USA, patents are effective only for protecting the fruits of innovation in pharmaceutical or other biomedical ventures. Although the patent system is probably being used for purposes for which it was not intended, it is serving a useful function in enabling the development of new technologies by venture companies in a variety of industries. My interviews with Japanese nonbiomedical ventures indicated the same is true in Japan. Patents in combination with trade secrets are important to prevent encroachment and to attract financing.[86]

Also by 2005, the pendulum appeared to have begun to swing in the opposite direction, away from the tendency for large companies to amass patents mainly as bargaining chips or as weapons in a patent arms race. Some companies have begun to make some of their patented technologies freely available, especially for open-source software applications, and some companies are putting information in the public domain to prevent its private appropriation by patents.[87]

In summary, with the possible exception of access to patients and clinical investigators in university-affiliated hospitals, there seem to be no clear attributes of the pharmaceutical or other biomedical industries per se, that make them alone favorable for high technology ventures.[88]

However, in order for ventures to flourish in any industry, there must be technical and business opportunities that are not being exploited by established companies and where costs of entry are not prohibitive. In other words, they need new niches where potential demand for their output is high.[89]

Innovative university research often suggests these niches. Thus the innovativeness of university research and an effective system of technology transfer from universities are key components of a favorable environment for ventures. Ventures do not have resources to pursue basic research for its own sake. Universities do, and thus university research can open up new fields that ventures can exploit.

Established companies pursue basic research less than in the past,[90] but they nevertheless make discoveries or initiate applied R&D projects that they decide not to pursue further. Thus the ease with which employees of established companies can move to ventures to pursue projects that the established companies are not pursuing is also an important component of a favorable environment for ventures.

The importance of university connections for venture companies is clear in some industries. For example, the potential of genetic engineering and the basic underlying techniques was discovered in universities, and then the development of these new technologies was carried forward in parallel by universities and industry, but by bioventures more so than established pharmaceutical companies.[91] Universities and bioventures also pioneered the field of cartilage regeneration, and continue to cooperate closely to develop this field.[92] The development of gene sequencers described in Chapter 6 is another example. As for nonbioscience fields, the development of the Internet and Internet-related technologies such as routers, browsers, and search engines also has been largely due to university research and university startups.[93] As of 2005, a handful of ventures were trying to integrate discoveries from various universities to develop technologies to fabricate nanoscale materials.[94]

On the other hand, the founding of new companies in fields such as computers, semiconductors, hard disk drives, and IT-related software has often depended on entrepreneurial or frustrated engineers, scientists, and managers leaving large companies. Some of the companies formed in this manner have become world leaders. Intel was formed in 1968 by three employees of Fairchild Semiconductor who saw unexploited potential in some of Fairchild's technologies. It went on to weather the onslaught from Japanese memory chipmakers and become the dominant designer and manufacturer of microprocessors in the industry. Control Data Corporation was founded in 1957 by engineering managers from Sperry Rand who were frustrated by that company's failure to compete with IBM in mainframe computers. By the late 1960s, Control Data had become the leading designer and manufacturer of supercomputers worldwide as well as a major producer of computer peripheral devices. Compaq was formed by engineering managers who left Texas Instruments (TI) in frustration over the failure of TI's PC projects.[95]

IBM, which pioneered the development of computer disk drives for its large mainframe computers, did not perceive demand from its main customers for smaller disk drives, which in any case would have been slower and had less capacity. When IBM cut back R&D on the smaller drives, some of the engineers quit in frustration to form spin-offs. In a pattern which was to repeat itself over several generations of IBM progeny, these spin-offs invested heavily in R&D aiming at a niche market of small or specialty computer makers. In the process they increased the capabilities of the smaller drives until they matched those of the larger drives, thus making possible the personal computer revolution. Over forty new companies entered hard disk drive manufacturing from the 1970s to 1990s and, of those that made any money, all but four were formed by persons who had quit established firms. Faced with price competition, the new US hard disk drive companies usually outsourced manufacturing to Asia, often to Singapore. It was a strategy that their Japanese rivals did not match until much later when the technology had matured and consolidation was well underway worldwide. By 2005, about ten companies accounted for the vast majority of world hard disk production, but US companies remained among the leaders, and at least three of these companies, Seagate, Maxtor, and Western Digital, are descendents of IBM.[96] Among the R&D-oriented semiconductor ventures mentioned above, many of these are located in Silicon Valley and the majority of Silicon Valley firms are founded by employees leaving other firms.[97] As for outsourcing of manufacturing, Silicon Valley firms continue to rely on the fast and flexible manufacturing infrastructure in Asian countries, especially Taiwan, and recently mainland China.[98]

Not all ventures are based on breakthrough university discoveries or skilled personnel leaving large companies. Many university discoveries that have been successfully commercialized by startups seem to be commonsense, insightful, or assiduous applications of existing knowledge.[99] Also breakthrough university research is sometimes commercialized by large companies without involvement of ventures. For example, university research in both the USA and the UK provided the theoretical foundations for GE to develop MRI scanners. GE benefited from hiring university researchers and has had collaborative partnerships with various universities, but GE largely funded and guided this development effort on its own. The same is true with respect to the development of injection molding as a process to make plastic and other shaped composite materials.[100] Also, some successful ventures, such as Apple, Microsoft, and Qualcomm, arose on their own without close ties to universities or established companies.[101] Finally, some breakthrough university inventions have a mixed development history. One example is plasma display panel high definition television (PDP HDTV) which originated in the University of Illinois in the 1960s and 1970s. The University licensed this technology to a number of large US and Japanese companies, and then, after most of the US companies dropped out, to a startup founded by the university inventors. Short of cash, the startup was eventually absorbed by Matsushita. Although PDP HDTV sets have been on the market since the late 1990s and have been a technical and commercial success, high development costs and price competition have forced many of the manufacturers to scale back their operations.[102]

PART III: DO VENTURES UNDERMINE INNOVATION IN LARGE COMPANIES—OR AN ENTIRE NATIONAL INNOVATION SYSTEM?

The fact that many ventures in IT have been started by persons leaving other companies raises squarely the issue of competition for personnel between ventures and established companies. It also raises the possibility that the movement of personnel from established companies to ventures is undermining the former and creating a situation where all companies, new and old, are at risk.

It is ironic that one company that has raised this concern most publicly and that has initiated high profile litigation to prevent former employees from working in new ventures is Intel, a company that was itself formed by persons leaving another young high technology company, Fairchild Semiconductor. Robert Noyce, one of Intel's founders, summarized the possible ramifications of employees defecting to new ventures as follows:

If we get into a situation where Company A puts an enormous amount of money into an R&D project and Company B can simply appropriate that, then the first project will never be done again. That becomes the destruction of the American industrial base....[103]

This statement was made in the early 1980s when Intel was fighting for its life as US semiconductor manufacturers were rapidly losing market share in memory chips to Japanese competitors, and before Intel's salvation was apparent as the leading designer and manufacturer for microprocessors for personal computers.

Perhaps it is a sign of Intel's competitiveness and its commitment to innovation and protecting its core business, that it has continued, on a selective basis, to sue employees who leave to work in similar fields for other companies.[104] The persons who left Fairchild to form Intel, Sperry Rand to form Control Data, Texas Instruments to form Compaq, and IBM and its progeny to form hard disk drive companies, were apparently not prosecuted for revealing trade secrets or for violating clauses in their previous employment contracts not to engage in work that may compete with the former employer (no-compete clauses). However, outside California, high technology companies have used successfully the threat of suits alleging violations of either trade secrets or no-compete clauses to prevent employees leaving for new companies.[105] The fact that employers are willing to sue former employees, despite the risk of alienating current and potential future employees, indicates that high technology companies view the loss of talented employees to new companies as a real danger.

A well-known case concerns Gene Amdahl, one of the chief designers of the then state-of-the-art IBM 370 mainframe computer, who left IBM in 1970 to form a company to manufacture advanced mainframes less expensively than IBM. Amdahl's new central processors helped reduce the price of high performance computing power several-fold, but led to the establishment of a competitor (ultimately overseas based) that outstripped IBM's market share.[106]

Other departures that may have directly undermined high priority development efforts involved the formation of ventures to develop thin-film technologies for computer hard disks. IBM and later Burroughs and Xerox had pioneered this technology. However, IBM incorporated it in only its most expensive computers. Venture capitalists recruited key engineers from IBM and Xerox to form companies such as Komag, which became the leading manufacturer of thin-film disks, and Read-Rite, which became the leading manufacturer of thin-film disk reader heads. These ventures sold most of their output to new hard disk drive ventures such as Seagate and Maxtor, which were themselves investing heavily in R&D to develop small, high performance disk drives. In this way, the disk drive ventures and the thin film ventures surpassed IBM and other vertically integrated manufacturers in terms of technology and market share.[107]

What motivates persons to move from established companies to ventures is an important issue on which there is currently only anecdotal information. If most moves are born mainly out of frustration and a reasoned assessment that opportunities to develop technology are better in a new company, this would be an indication that such mobility benefits innovation on the whole. In the available case histories, *frustration* with the inability of established companies to pursue effectively projects in the employee's field does seem to be the most frequently mentioned motive for changing jobs.[108]

There are reasons to believe that employee mobility may benefit innovation even though companies are hurt when they lose skilled employees. Much of the success of new companies depends on assembling at the outset a core group of persons with shared goals but complementary talents. A high mobility labor market facilitates this process. From the perspective of some employees, being able to change jobs freely and perhaps to start one's own company maximizes job satisfaction. For many skilled immigrants and recent US university graduates, a free labor market has produced upwardly mobile jobs where skills and marketability rapidly increase. Information flows among companies are increased both because of actual movement of people and because of intercompany networks that are augmented by mobility. These networks, including those among immigrants from countries such as China and India, in turn aid the job matching process. In a high mobility labor market, there is an incentive to work very hard either to increase one's chances for remaining in one's current job or to increase one's reputation so that one can move to a better job.

As for the injury to companies that lose people, this is in part compensated for by the intense effort employees put into their jobs, increased access to information, opportunities for alliances with other companies, and the ability to assemble an appropriate workforce quickly and to shed labor quickly. Companies formed by defecting employees can also spread the parent company's standards and protocols more widely, create greater demand for the parent's products and even provide technology the parent itself can use.[109] Also, the patent system helps to ensure that companies retain ownership of their employees' inventions even after they have moved to other companies.[110]

Employers usually have the option to offer increased benefits to employees to induce them to stay. They also have the option to offer retention bonuses, training and promotion opportunities to longtime employees, and traditional pension plans under which benefits vest only after twenty or so years of service. The fact that they rarely take such steps, at least in regions such as Silicon Valley where labor mobility is high, suggests that they feel that the benefits of high employee mobility outweigh the costs.

The suits by Intel notwithstanding, there usually is an unwritten understanding in most employment relationships in Silicon Valley that employers will not pursue trade secret suits against employees who leave to work for a potential rival.[111] Such practices and the overall weaker effect of trade secret law in California may be one of the main reasons for the vitality of its ventures and for Silicon Valley's becoming the most important center of innovation in the USA[112] in terms of high technology employment and output of high technology companies.[113] In other words, rapid labor mobility allows resources, particularly human resources, to be deployed more quickly to projects where innovation potential and economic returns are highest.

PART IV: IS A SYSTEM OF INNOVATION BASED LARGELY ON VENTURES SUSTAINABLE?

Are the benefits of this system of rapid company creation, sometimes equally rapid collapse, and frequent job changes sustainable from even a micro perspective, that is, from the perspective of the firms in a particular industry and in a particular region, such as Silicon Valley? One criticism of the Silicon Valley model is that many of its companies lack an integrated learning base and integrated organizational capabilities. Reflecting the destruction of the American consumer electronics industry by the early 1980s and the near destruction just a few years later of the largely Silicon Valley-based American semiconductor and computer industries at the hands of large Japanese electronics companies, it has been suggested that, except in fields undergoing revolutionary change, vertically integrated, diversified companies, such as the major Japanese electronics companies, will usually prevail over companies with lesser capabilities.[114]

With the possible exception of drug discovery, as a matter of rational business practice, if a new company is to grow it usually must develop integrated competencies spanning R&D, outreach to customers[115] and usually also manufacturing—or at least the ability to outsource manufacturing.[116] In IT, many of the highly successful erstwhile ventures, such as Hewlett-Packard, Intel, Apple, Microsoft, Seagate, Sun Microsystems, Qualcomm, and LSI Logic, have such integrated R&D competencies, although as of 2005 all of LSI Logic's chip manufacturing had been outsourced. Yet compared to Japanese companies such as Hitachi, Toshiba, and NEC, even the most successful US companies have narrower business scopes.[117] There have been inevitable consolidations among US high technology ventures. But at least in hard disk drives and semiconductors, fields for which we have historical data, some of the original ventures remain and are now sales leaders in their fields while continuing to vigorously pursue R&D-focused growth. They have done so by working out sales and collaborative relationships with other companies, in most cases remaining focused on core competencies or closely related technologies, and by using IP to protect those core competencies, especially when they have to outsource final development, manufacturing, or marketing.

Perhaps because of labor mobility, there is less of an imperative for corporate expansion into different fields. Persons who have ideas for different products move to other companies. The value chain remains largely disaggregated. Large companies seem to have adjusted to this disaggregation, with more adopting an open innovation philosophy.[118] As opposed to isolated learning and organizational competencies within large, distinct companies, these competencies now reside within the system as a whole. This system-wide competency depends on free flows of information and personnel, a nonautarkic innovation strategy on the part of large companies, availability of astute financing, and IP protection that allows new companies to grow.[119] It also depends on the existence of some large, technically focused companies that do well in R&D, manufacturing, and marketing. Such companies simultaneously provide smaller, newer companies with markets, testing facilities, feedback from end users, skilled personnel, and often new ideas for products—a relationship similar to that between biotechs and pharmaceutical companies. The technical competence of the large companies is crucial for this dynamic exchange. It enables them to be creative in their own right, to absorb ideas from the small companies and to generate ideas for the small companies to develop.

An idealized characterization? Perhaps. But this seems to characterize the business relationships in Silicon Valley. After weathering severe competition from autarkic, vertically integrated overseas companies in the 1980s and then over exuberant expansion followed by the bursting of the IT bubble in 2000–1, the model still seems to be functioning relatively well, at least in the Silicon Valley of 2005. Jobs in high technology sectors have continued to decline since the bubble broke. However, salaries in these same industries are increasing. Productivity per employee is about 2.5 times the national average and continues to grow slightly faster than the national average. Net new company formation continues to increase. Venture capital investment has increased for the first time since the bubble broke, as have the number of rapidly growing companies.[120] The system no longer seems in mortal danger from large, autarkic, vertically integrated, multi-sectored companies—foreign or domestic. As of 2005, the more often discussed challenges related to regional social infrastructure,[121] the outsourcing of high technology jobs overseas, and the possibility that that the Silicon Valley model will flourish in countries such as India and China where clusters of networked new companies will soon compete even in high technology fields with Silicon Valley companies.

What about the benefits of rapid company formation and labor mobility from a more macro (e.g. national) perspective? America has benefited from Silicon Valley, which owes its success in part to the way California law facilitates job transfers between companies. Should other states adopt these same policies, and would the USA benefit if employees in all regions were as free to move as they are in California? Conversely, might not the threat of losing key employees dissuades companies from pursuing important R&D, despite suggestions from California's experience that this rarely happens? Finally, if scientists and engineers nationwide feel they are treated like commodities and have little job security, might not many middle-aged Americans leave S&E careers and fewer young Americans decide to enter such careers?[122] The answers are not clear.

More generally, to what extent does competition from venture companies undercut profits of large companies and thus prevent them from carrying out forward looking research whose benefits may not be realized for many years, and which, when they are realized, might largely accrue to other companies? AT&T's Bell Laboratories, Xerox's Palo Alto Research Center, and RCA's Sarnoff Laboratories, all produced pioneering discoveries that proved greatly beneficial but were most often commercialized by other companies. Are whatever gains from rapid employee mobility and a fertile environment for new companies canceled out by the loss of the wellsprings of such discoveries?[123] One could cite a continuing stream of innovations from new high technology companies as evidence that innovation will continue apace despite the absence of large corporate basic research laboratories. Still there is room for debate and need for investigation into whether and how universities should fulfill the role those laboratories once played.

It might be argued that high technology labor mobility is fine as long as it is within one country or there is net immigration of engineers and scientists into a country. But what if this is not the case? In recent discussions with Japanese colleagues about the benefits of labor mobility to build ventures and to diffuse technology, the most frequent cautionary response is, 'What about the Chinese scientists and engineers whom Japanese companies hire and train for important projects, who then leave to start or join a Chinese or US company that will compete with their Japanese mentor?'[124]

The USA has so far benefited from net immigration of scientists and engineers. In 2000, foreign born scientists and engineers accounted for 17 percent of the US total—37 percent if just Ph.D. holders are considered. They accounted for well over a third of scientists and engineers in Silicon Valley and nearly 30 percent of the CEOs of the Valley's high technology ventures.[125] The US biotechnology industry also depends on immigrants, with 6–10 percent of its workforce holding H-1B visas.[126] Venture companies may benefit even more from immigrants than established companies.[127]

In contrast, foreigners account for less than 1 percent of Japan's S&T labor force and the proportion is barely increasing.[128]

Even before September 2001, the flow of S&E professionals back to their homelands from the USA was increasing. However, these returnees often maintained strong ties with the USA and continued to build international networks that benefited the USA and their native countries. In other words, many of these returnees became part of a transnational movement of S&E professionals that brought about the previously cited benefits of labor mobility—better access to information, customers, manufacturing facilities, skilled personnel and capital, diffusion of standards, etc.—but now on an international scale.[129]

However, since 2001, immigration of scientists and engineers into the USA has declined sharply—largely due to stricter and more time consuming scrutiny of visa applicants.[130] These restrictions concern US high technology companies that rely on immigrants. *It is inconceivable that newly graduating US citizens and permanent residents can make up these shortfalls.* The number of those graduating with bachelors degrees in engineering has been declining since the 1980s, and the number of those graduating with doctoral degrees in any S&E field has been declining since the 1990s.[131] The recent trend to outsource increasingly sophisticated design work to countries such as India and China is driven in part by the difficulty in bringing skilled foreign workers to the USA.[132] As overseas design and manufacturing capabilities grow, fueled by outsourcing and VC investments by US companies, concern is also growing that overseas companies and regional innovation centers will soon become major competitors for US counterparts—even while the benefits and necessity of international collaboration are acknowledged.[133]

This raises the issue of the age-related demographics of high technology ventures. Are most dependent primarily on young employees who can work long hours and who, because they do not have families (or because they have a spouse with a steady job), do not mind the prospect of job changes, perhaps even occasional unemployment? There are suggestions that in the case of many Silicon Valley ventures, the answer is 'yes'.[134] Thus, as the flow of young immigrants diminishes and as the numbers of US citizen engineers and scientists remain stagnant, will the lifeblood of these companies also begin to vanish? In countries such as Japan where the population age distribution is skewed even more toward elderly persons, the demographic challenges facing ventures are even greater. These demographic challenges will only compound the difficulty ventures in Japan face in hiring skilled personnel.[135]

What about the scientists and engineers working in ventures who are over forty or who are the sole breadwinners for their families? If the ventures they work for have to shed jobs, can they find new jobs? Before the dot com bubble burst, it was said that failure was a mark of experience that would make one more eligible to obtain funding to start another company. But since the bubble burst, engineers and programmers over forty often encounter difficulty finding new jobs. Few data are available on the fate of these persons' careers,[136] or even the fate of young scientists and engineers who lose their jobs during economic downturns. If many end up taking jobs in unrelated fields, then in many cases their S&T skills will be wasted.[137] As mentioned in Part V of this chapter, this inability to retain skilled S&T personnel during downturns is one of the main problems that have plagued the US machine tool industry, an industry that is based on small- and medium-size companies. Are large, established companies more likely to retain S&T professional staff during economic downturns? Again the answers seem woefully unclear, although on such answers depends employee morale and the ability of an industry to preserve its skilled personnel for better times in the future.

Inability to find new work quickly may also entail losing health insurance. Even if they manage to find jobs sometime later, their new health insurance may not cover, for a period of twelve to eighteen months, medical costs related to medical conditions existing at the time of reemployment.[138] The high cost of health insurance is also a burden for small ventures that do not qualify for group discounts offered by health insurers to large companies. Whether the venture pays or passes the costs on to its employees, these cost can be substantial.[139] If labor mobility is to be encouraged as a means to redeploy human resources to where they can be used most effectively, and if workers of all ages are to be part of this system, these health insurance issues must be recognized as a problem that is probably unique to the USA among industrialized countries.[140]

Finally, there is the issue of inventive versus manufacturing competence. When the US semiconductor industry and Silicon Valley were in crisis in the early 1980s, one often cited flaw in the US innovation system was its lack of attention to manufacturing competence, to process rather than product innovations.[141] Twenty years on, that criticism continues to echo: Silicon Valley companies and US industry in general are great at coming up with new products, but few companies devote great care to manufacturing.[142] Of course this criticism must be balanced by observing that leaders in the IT industry, such as Intel, manufacture many of their products in the USA. Nevertheless, between 1992 and 2002, employment in US high technology manufacturing industries declined at a faster rate than for manufacturing as a whole.[143] Over this period all high technology manufacturing industries lost jobs *except semiconductor and other electronic component manufacturing* (one of three IT-related high technology manufacturing industries) and *pharmaceutical and medicine manufacturing* where employment rose 2 and 30 percent, respectively. The decline in high technology manufacturing employment is projected to continue between 2002 and 2012, although at a slower pace. Nevertheless, the three IT-related high technology manufacturing industries experienced increases in output (revenues) between 1987 and 2002 and are expected to continue to have high rates of output growth at least until 2012. Output per worker is continuing to increase for IT-related high technology manufacturing industries,[144] while employment decreases or remains steady.

But for all the other high technology manufacturing industries, growth in output per hour of work between 1987 and 2002 was close to or below the national manufacturing sector average of 3.4 percent per year. In other words, among all fields of US high technology manufacturing, there are only two growth sectors: IT, where output is increasing as is output per worker although employment is declining, and pharmaceuticals, where employment is continuing to increase but output per worker has remained steady.[145]

Because many biotech companies are classified as *pharmaceutical and medicine manufacturing companies*, it is likely that the pharmaceutical manufacturing data reflect large numbers of persons being employed in biotech companies that are pursuing mainly R&D and have little sales revenue.[146] One possible explanation of the IT trends is that, while much manufacturing continues to be outsourced overseas, much high value added, high profit margin manufacturing of relatively new products continues to be done in the USA.

If these assumptions are correct, then *the sectors where venture capital backed new company formation is most active are also the sectors where US high technology manufacturing remains strongest.*[147] More studies are necessary to show how venture companies maintain high technology manufacturing capabilities. However, the available data are consistent with such a link. In other words, the decline of high technology manufacturing in the USA may be a legitimate concern. But an environment that nourishes new company formation in high technology fields probably is more likely to create rather than undermine opportunities to develop globally competitive high technology manufacturing capabilities.

In summary, in the pharmaceutical industry there is strong evidence that innovation leadership has shifted from large pharmaceutical companies to venture companies. This shift has occurred because new companies, as a group, are probably better at drug discovery than established pharmaceutical companies, while the latter have a comparative advantage at later stage development (especially animal and human trials) and marketing—not because the new companies undermined the established companies. In other words, the ascendancy of venture companies in drug discovery reflects a mutually beneficial division of effort.

Aside from the unique advantage that university startups have with respect to access to patients and clinical research facilities provided by university hospitals, there seem to be no inherent advantages that biomedical ventures have in comparison to high technology ventures in other fields. In other words, if scientific breakthroughs with commercial potential occur in fields other than biomedicine and if institutional conditions are appropriate, venture companies should be able to assume leading roles as early stage (and sometimes also late stage) developers of these breakthroughs, just as they have done in pharmaceuticals. Their success in many fields of software and IT-related engineering suggests that innovation leadership is not limited to ventures in biomedicine.[148] Moreover, even with respect to skilled personnel leaving established companies to work for ventures, the evidence that ventures harm the long-term innovative strength of established companies, is weak. Rather than a zero-sum game where the success of one type of company undermines the other type, more often mutually beneficial relationships are worked out, not unlike the relationships between biotechs and large pharmaceutical companies.

Finally, the system of venture companies assuming lead roles in innovation while maintaining close ties with universities and other companies has endured several crises. It has contributed to overall long-term industrial progress and national economic well-being. Although care must be taken to maintain the institutional conditions for ventures to flourish, the system itself seems robust. Its greatest future challenges probably come not from large, vertically integrated, diversified overseas companies, but from the need to ensure a long-term supply of skilled, highly motivated labor despite the uncertainties inherent in a system of rapid labor mobility. This includes the need to ensure that skilled S&T labor will not be permanently lost from S&T fields during periods of downturn.[149] More broadly, it will requite the USA to maintain public support for R&D funding and R&D careers, to remain open to immigration, and to develop mechanism to help its S&T labor force deal with the stresses of job mobility. It will require all countries to recognize the mutual benefits of open science and free movements of people and capital.

The future health of American industry depends on preserving an environment where new companies continue to be created and to flourish. Is America unique or does this conclusion also apply to Japan and other countries?

PART V: TO WHAT EXTENT DOES THE DEARTH OF HIGH TECHNOLOGY VENTURES HURT JAPAN?

The above analysis focused mainly on the USA. Various factors discussed later in this chapter make the environment for ventures less supportive in Japan than in the USA. But first let us consider available evidence that Japan's global competitiveness in particular industries has eroded (or has remained weak) because it has not had vibrant R&D-focused ventures.

At least in pharmaceuticals, Japan's weakness can be attributed to its dearth of bioventures. This is clear because, as autarkic innovators, Japanese pharmaceutical companies having shown considerable innovative strength. They have discovered some of the world's most important drugs, although this is often not known because they are distributed outside Asia by US or European pharmaceutical companies that rarely reveal their Japanese origins in promotional materials.[150] They have done so relying primarily on their in-house laboratories and research teams. Relative to a particular company's number of in-house researchers, R&D expenditures, or global pharmaceutical sales, the ability of the in-house research teams of the better known Japanese pharmaceutical companies to discover innovative new drugs is no worse (and may even be better) than that of the research teams in their larger US and European counterparts.[151]

Nevertheless, Japan's recent output of new drugs, particularly innovative drugs, is low relative to the size of Japan's pharmaceutical market. In contrast, output of innovative new drugs relative to market size is high for the USA.

But remove biotech drugs, and the USA would lag farther behind than does Japan in terms of output of new drugs relative to market size. Japan, Germany and France are weak in terms of discovery of innovative new drugs. America is strong, but its advantage is accounted for entirely by discovery and early development work done in its venture companies.[152]

In the case of other fields of technology, I am struck by the frequency with which colleagues at The University of Tokyo who cooperate closely with Japanese companies identify new (or relatively new) US companies as the main competitors of their Japanese collaborators. This was the case with respect to optical switching devices and tunable lasers mentioned in Chapter 6. Although anecdotal, these are unsolicited assessments by technical experts who understand the capabilities of the companies they are collaborating with and who also are aware of what leading groups the world over are doing in their fields. The development of gene sequencers also fits this pattern. Discussions with persons in government, industry, and science and engineering research in Japan indicate deep concern that Japan's large high technology manufacturers are being squeezed between Asian competitors who can manufacture increasingly high quality products more cheaply, and new US companies (often university startups) that are discovering new technologies and developing their applications more quickly.

However, the final verdict is not in and, as described below, established companies are pursuing alternative strategies to remain competitive in newly emerging fields of technology.

PART VI: WHAT UNDERLIES JAPAN'S
STRENGTH IN INNOVATION?

If Japan now needs new companies, why did it do so well from the 1960s to the 1990s? And why is it that in some industries, for example automobiles, machine tools and some fields of consumer electronics, its companies in 2005 are still undisputed world leaders in terms of innovation and manufacturing? Succint answers are not possible. While relying on a close follower strategy[153] may have worked in the 1960s and 1970s, by the 1980s the strongest Japanese companies had become the world's leading innovators in many technologies. The fact that many have maintained this leadership suggests that autarkic innovation can be globally competitive in some industries. As suggested in Chapter 6 and Parts 1 and 2 in this chapter, such fields typically are those with high capital costs to entry, where progress is incremental in the sense that it builds closely on already commercialized discoveries rather than new technical fields, and where university discoveries or defections from large corporations are unlikely to create niches that can be commercialized before established companies do so.

Nevertheless, the success of large Japanese companies suggests that additional factors have made the Japanese environment particularly supportive of early as well as late stage innovation in large companies. The remainder of this part considers some of these factors, specifically:

- The possibility that entrepreneurial companies have in fact contributed substantially to S&T innovation in Japan behind the scenes, examining in particular the case of the machine tool industry,

- Japanese industrial policy, examining in some depth government sponsored consortium research and whether this has helped or hindered innovation, and

- The strong internal innovation capabilities of large manufacturing companies, based largely on personnel management policies predicated on lifetime employment.

This will show that some of the factors underlying Japan's strength in innovation favor established companies and are detrimental to the formation of ventures with high growth potential.

Japan's Entrepreneurial Potential: The Case of its Machine Tool Industry

One explanation is that entrepreneurial companies have indeed played a major role in Japan's economic success—that behind Japan's large companies are many small or new companies that have produced innovations that have been effectively incorporated into final products by large internationally known manufacturers.

The history of the machine tool (MT) industry shows not only Japan's potential for entrepreneurship and innovation based in small or new companies, but also, ironically, some of the weaknesses of small-company based innovation in the USA. The outlines of the story are as follows: US companies dominated the MT industry in the postwar years until the 1970s. By and large, they were small. Japanese companies tended also to be small and independent.[154] Repeated efforts by MITI to promote consolidation, in the belief that some companies would achieve economies of scale and thus become more competitive internationally, were not successful. The companies successfully resisted consolidation among themselves and also with the end manufacturers.

The Japanese companies were the first to exploit the potential of computerizing the numerical control devices in machine tools. Fanuc, originally a joint venture between Fujitsu and Makino, one of Japan's younger MT companies, built its first computerized numerical controller (CNC) in 1972. By 1975–6, these were being rapidly incorporated into Japanese machine tools. This was probably the first large-scale application of microprocessor technology—even earlier than personal computers.[155] These made MTs more versatile, as well as easier and less expensive to operate. Sales increased not only to large Japanese automakers and other large manufacturers, but also to small manufacturers. The latter could use relatively small MTs with CNCs to create high quality customized parts for a variety of end users. Throughout the 1970s and 1980s the vitality of small-scale manufacturing in Japan remained high, as did the ability of small companies to maintain independence with respect to larger companies, thanks in part to these machines and the technical skill of their craftsmen. As a result of this strong domestic demand, Japanese MT manufacturers began to benefit from economies of scale that made their products more competitive internationally.

The traditional customers of American MT manufacturers had tended to be large manufacturers who did not need the versatility of CNC machines, and military contractors that often needed precision parts but who usually did not face stiff price competition. However, by the early 1980s US MT companies found themselves competing against Japanese machines that offered better quality and versatility for price. Employment and production in the US MT industry fell sharply in the early 1980s. Since the late 1980s, US MT production has lagged behind that of Germany, Japan, and sometimes also the USSR or Russia, Italy, China, and Taiwan.[156] Consolidations and bankruptcies swept the US industry. Conglomerates acquired many firms and then, unwilling or unable to invest the effort to turn them around, sold or closed them. Workers who were laid off during downturns found jobs in other fields. When upturns came, many companies could not invest in training new workers, and even the stronger companies found that availability of skilled workers was their most severe problem.

A revival began in the mid-1980s and continued until the late 1990s as many of the remaining companies adopted state-of-the-art technologies, improved quality, produced new products in shorter development times, provided better customer service and began to pursue exports. Although import penetration was high, US manufacturers were reaping an increasing proportion of sales through exports, which helped buffer the frequent downturns in domestic demand characteristic of the industry.[157] During this revival, the larger firms tended to fare better in terms of growth in sales, employment, R&D and exports. The smallest firms were often dependent on a few nearby customers and lacked the resources to improve their product line or broaden their customer base.[158]

US MT consumption and production again declined sharply beginning in 1998. Another recovery began in 2003, but by this time the largest firms, including industry icons, had gone out of business or been acquired by foreign companies.[159] The remaining firms were even more export-oriented.[160] They were also more oriented toward the special needs of small-scale manufacturers, ironically assuming a similar strategy with respect to the global market that Japanese companies had assumed thirty years earlier. The largest of the US MT manufacturers in 2005 was Haas Automation, a relatively new firm located near Los Angeles, far from America's traditional manufacturing region near the Great Lakes. Haas eschewed the normal practice of shedding workers during economic downturns. Between 2000 and 2005 it continued to employ about 200 engineers and computer experts for design work, while its total workforce expanded from about 700 to 1,100.[161]

Similar to the US, most Japanese MT manufacturers are small, while the leading manufacturers are generally medium-size independent firms.[162] Japanese companies remain leaders in MT innovation and in manufacturing control systems.[163]

More remarkable, however, is the history of the development of this industry during the years of Japan's economic miracle from the 1960s to 1980s.[164] This period saw frequent entry by new companies and displacement of old ones. Subcontracting was common, but subcontractors usually sold to several companies. Some companies that were small subcontractors in the 1970s grew to be among the leading companies in 2005.[165] As they tried to catch up to overseas competitors and also meet the demands of a wide range of customers (whose needs were not being met by large overseas manufacturers), the larger Japanese manufacturers experimented with products and production processes. This need to experiment required them to give their subcontractors freedom and incentives to do the same, and this created an environment that encouraged innovation in small companies and formation of new companies. Fanuc's pioneering large-scale production of CNC devices for MT machines allowed small and new MT firms to respond to these opportunities. At least in the 1980s, the Japanese MT industry was among the most vertically disaggregated in the world, with the major MT firms producing considerably higher numbers of machines per employee than those of any of their major foreign competitors. They did so through a networked system of production, similar to the modular innovation and production methods that became popular later in the USA.[166] The final machine tool incorporated components from many independent suppliers. At least among smaller firms that were not direct competitors, sharing of information was common, and they occasionally shared marketing and research costs.

To some degree, this favorable environment for entrepreneurship existed throughout Japanese manufacturing. Small companies became the training grounds for skilled craftsmen and innovators. Between the 1960s and 1970s it became commonplace for young craftsmen in their late 20s to leave to establish their own companies. In the smallest firms, almost half the workforce would be expected to leave to start their own companies.[167] Their employers took such departures as a matter of course, not unlike most employers in Silicon Valley. Once they became company head, their salaries increased to above the level of craftsmen in large companies. Thus, when considered across an entire career, salary differentials between large and small companies were small. Only production employees who remained in small firms beyond their mid-30s suffered large wage gaps. Some craftsmen moved from company to company, and in some specialized fields, professional bonds among persons with the same skills were stronger than bonds to the companies in which they worked.[168] Subcontracting was common, but probably not quite so much as in the case of MT companies. In several industries the necessity for large end user manufacturers to experiment to meet customers' needs fostered experimentation among their subcontractors. In turn, the small manufacturers made the Japanese manufacturing system more flexible, compared with a dominant focus on achieving efficiency through mass production that characterized the automobile and other industries in the USA.[169]

As noted in Chapter 6, the innovative strength of Japan's automobile industry has been attributed in part to the freedom of subcontractors to develop their own designs for products and to market these to various large manufacturers.[170] Loan financing was available from a network of private and public financial institutions specifically intended to provide capital to small businesses. The Government played an important role in establishing and funding public institutions such as the Small and Medium Enterprise Corporation (1953) and the Central Commercial and Industrial Bank, as well as private mutual banks, credit associations and credit unions. Of all the government's policy measures in the postwar period, this probably resulted in the greatest benefit to Japan's MT industry and to small manufacturers in other industries.[171]

Except for the financing mechanisms, there are notable parallels with vertically disaggregated innovation in the US IT industry characterized by high labor mobility, rapid new company formation and outsourcing of innovation and R&D by large companies to independent smaller companies. Even the role of Fanuc in developing a widely used platform technology that increases innovative capabilities for new companies and provides the means for them to make additional modular innovations echoes IBM's development of the PC (or IBM's letting other companies develop compatible peripherals for its mainframes) and the development of standard operating systems such as DOS/Windows and Unix. Also, the way Japanese MT manufacturers were able to meet the latent needs of small-scale, nonmilitary US manufacturers[172] has parallels with the way small disk drive startups were able to respond to the needs of companies that IBM had ignored. These parallels aside, at the very least, the Japanese MT experience shows that, given a supportive environment, entrepreneurial innovation can flourish in Japan, large companies can be supportive of innovation in independent companies, and technically skilled people are willing to take the risks associated with careers in small companies.

In the course of investigating various topics mentioned earlier in this book,[173] I have been struck by the large number of engineering and chemical companies incorporated in the 1950s and 1960s that are independent mid-size companies today. These would have been young companies in the early decades of Japan's economic miracle. The degree to which Japan's economic growth depended on these companies is not absolutely clear, but this shows that as Japan's economy was taking off, many technology companies were being formed and they undoubtedly contributed in some measure to the miracle that became evident in the 1960s, 1970s, and 1980s.

However, there are cautionary lessons from both the US and Japanese MT experiences. The US experience shows the limitations of small size (at least in the absence of financing and good management that enable a small firm to grow quickly). The smallest firms are generally the weakest because they lack resources to improve quality and performance, develop new products, and attract new customers. As for Japan, it appears that rates of formation of new high technology companies in manufacturing in general have fallen substantially from levels in the 1970s, although it is not clear this is the case in the MT industry itself. This is due in part to the long downturn in the Japanese economy and the shifting of many manufacturing operations overseas. Also, until at least 2003, small manufacturers faced increased difficulty recruiting and retaining skilled workers as downward pressure on wages increased, fewer young people were attracted to jobs in these companies, and immigration remained tightly controlled. Also launching a new company has become more difficult. Land, buildings, and machinery are more expensive and more technical and managerial skills are needed for a viable new company. The advent of CNC-controlled machines helped small manufacturers to overcome to some extent the dwindling numbers of craftsmen with specialized skills and also to attract new young employees.[174] Perhaps, however, these positive benefits have now largely run their course, and the lack of new technical breakthroughs limits both the technical opportunities for entrepreneurial companies to exploit and makes recruitment difficult. In any case, this shows that if technical and business and demographic conditions change, vibrant entrepreneurial activity can come to an end.

It also suggests that access to new technologies is vital for ventures and probably is the prerequisite for their being competitive with respect to large companies. This in turn suggests the importance of access to university discoveries and entrepreneurial researchers for ventures. The flowering of innovative entrepreneurship in the machine tool industry, coupled with the fact that Japan's university system was and to a large extent remains closed to ventures (not to mention the barriers to forming independent high technology spin-offs), gives a hint of what innovation might have been otherwise. The loss in forgone entrepreneurial innovation probably has been great.

The Importance of Industrial Policy

It has been suggested that Japan's bureaucracy has skillfully guided the allocation of public and corporate resources so as to promote the growth of internationally competitive industries, has provided just enough shielding from foreign competition for its growth industries, and has helped to ensure that competition promotes innovation while limiting resulting destruction and waste. It is beyond the scope of this book to analyze this perspective. However, this explanation for Japan's success has been questioned. In any case, with the possible exception of consortium research projects promoted by the government, they probably do not explain the continued strength of some Japanese industries.[175]

However, government-sponsored R&D consortia deserve special attention, because they remain an important feature of the Japanese S&T landscape and, as explained in the following pages, they hinder the growth of high technology ventures by allowing large companies to preempt potential markets and the attention of university researchers.

Government Consortia, and the Preemption of New Technologies by Established Companies

Government-sponsored R&D consortia have been defined as cooperative R&D projects involving two or more companies that were initiated in part by the Japanese government.[176] The government usually plays an important role in defining the field of R&D and recruiting participants. Its support can range from a modest level of cofunding to covering all costs. Based on the recent consortia with which I am familiar, the government's contribution usually falls in the range of 50 to 70 percent of project costs, although participating companies on their own can pursue related research. The government often encourages involvement of GRIs and universities, sometimes directly and sometimes as subcontractors to the companies. Sometimes the government establishes dedicated laboratory facilities, but probably it is more common for R&D to be conducted in existing corporate, university and GRI laboratories. Government-sponsored consortia began in Japan in the late 1950s and became increasingly frequent, so that by the late 1980s and early 1990s there were over 100 consortia projects ongoing at any time.

Consortium research organized by NTT in 1973 was instrumental in enabling three cable companies, Fujikura, Furukawa, and Sumitomo Electric, to match Corning's breakthrough in low-signal attenuation fiber optic cable. However, the success of this consortium probably depended most on the strength of NTT's laboratories and the hub and spoke type collaborations between these laboratories and the three cable companies, not on collaboration between the cable companies themselves.[177] Probably the best known of the consortia has been the Very Large Scale Integration (VLSI) project, which brought together Japan's main semiconductor manufacturers in a successful effort to improve the quality and reliability of their semiconductors and manufacturing methods.[178] The project was launched by NTT in 1974 (and then joined by MITI which took on the lead role) in response to plans by IBM to develop by 1980 a new Future System computer that would use 1 megabyte DRAM chips. In retrospect, the timetable was unrealistic and IBM abandoned the project. However, in 1974–5 this seemed a mortal threat to Japanese semiconductor makers still lagging technologically behind US rivals and facing greater competition as Japan reduced tariffs on computers and semiconductors. Thus the leading manufacturers urged the government to launch a project to help them catch up, and were prepared to cooperate among themselves in some fields of R&D. However, it was probably the efforts of each participating company's researchers working independently from those of other companies—not close cooperative research among scientists and engineers from fiercely competing companies—that contributed most to success. Additional important factors were substantial METI financial support and intense research in NTT's laboratories (probably the best at the time in Japan in basic VLSI technology), the results of which NTT shared freely with three of the participants. Again this suggests the importance of public scientific resources made available to individual companies, rather than the government's ability to promote cooperation among companies that otherwise would be reluctant to cooperate.[179] In any case, by the late 1970s, the DRAMs of the Japanese participants began to surpass those of US companies in terms of capacity, quality, and price. No longer in danger of seeing its semiconductor and computer industry overwhelmed by foreign competition, it had to deal with increasing alarm from the USA which saw this fate about to befall its semiconductor industry.[180]

After VLSI, new consortium projects tended to focus more on forward looking S&T issues rather than applied research that was of immediate commercial applicability, and they also tended to be smaller scale.[181] Nevertheless, some of the consortium projects ongoing in 2005 aim at specific commercial applications, and some are quite large. For example, METI/NEDO's nanotechnology program, encompassing numerous projects organized under several R&D themes, had annual budgets of approximately US$100 million in both 2003 and 2004.[182]

METI has allocated about US$200 million from 2002–2008 to the Extreme Ultraviolet Association (EUVA) to develop technologies to use extreme ultraviolet lithography to make the next generation of chips with circuits only 45 nanometers wide. The EUVA seeks to promote collaboration among most of the Japanese companies involved in the manufacture of IC chips and the equipment for making such chips.[183] Like the VLSI consortium, METI formed EUVA in 2002 largely in response to overseas EUV lithography consortia, such as the EUV LLC initiated by Intel, and the concern that overseas rivals would obtain dominance in this technology. Like the VLSI consortium, there are indications that it is hampered by rivalry among its members.[184]

It is unclear what proportion of Japan's current S&T budget supports consortium R&D. In the 1980s and early 1990s, consortium projects accounted for approximately 5 percent of total annual government S&T expenditures.[185] However, they probably accounted for between 17 and 43 percent of *extra-mural*[186] *project-specific* expenditures, 23–84 percent if MEXT-grants-in-aid are excluded.[187] Although consortium research probably does not constitute a majority of government funding for research even in priority fields such as IT, materials, energy or environment, and biomedicine, in certain subfields where commercial applications are likely to follow closely from basic research discoveries, consortium projects do constitute a significant proportion of such funding, especially if MEXT grants-in-aid are excluded.

The following observations regarding projects initiated in 2005 under some of Japan's main S&T programs elaborate on this point:

I began by reviewing the funding announcements for new *NEDO projects*.[188] The new 2005 NEDO projects run the gamut from numerous small-scale field tests for photovoltaic solar panels, and a large number of demonstration projects for biomass energy production, to major projects in biomedicine, wireless communication, micro-electrical mechanical systems (MEMS) and nanotechnology. In between are a large number of NEDO awards for R&D projects proposed by companies and university researchers for R&D likely to have applications in various industries. Only a minority of the NEDO projects appear to qualify as consortium research. Many involve support to a single university laboratory, TLO, or company. There are many examples of support to small companies and to regional universities and their TLOs.

However, in research fields where international competition and the potential for commercial application are high, consortia-type projects linking major university laboratories to established companies is a common form of research organization. For example, the nanotechnology projects initiated by NEDO in 2005 typically involved teams consisting of a university (or GRI) and two companies, one usually a major company and the other a small company.[189] Examination of the participants in projects initiated in prior years under METI/NEDO's nanotechnology initiative suggests that large companies play an even more dominant role.[190] NEDO R&D projects in MEMS, semiconductors, digital display and networking, fuel cells, and various fields of biomedicine, show similar patterns. Typically several large companies and universities are involved in a subproject—less frequently just a single large company and one or more universities and sometimes also a small company. Projects funded by the Key Technology Center also have aimed at developing industrially relevant technologies via consortia consisting of universities and large companies, although recently in biomedicine and software, small companies have also been frequent participants.[191]

Lists of R&D projects initiated in 2005 by the *Ministry of Internal Affairs and Communications (MIC)*[192] also show a diverse range of projects and participants. About a quarter involve large companies and about half of these also involve at least one other large company, a university, or both. These latter consortium-type projects appear to be the most broad and technically advanced of all the MIC R&D projects.[193] Also, when universities are partnered with companies, generally better-known universities are partnered with large companies, while regional universities are partnered with the small or regional companies.

Most of *MHLW*'s support for extramural biomedical R&D is disbursed as grants-in-aid to university researchers.[194] However, about 25 percent is disbursed as projects funded by *OPSR*,[195] of which about 10 percent involve private companies—about half of which are large.[196] In addition, since 1988, OPSR has initiated at least fifteen incorporated consortia to pursue and commercialize R&D in fields such as drug delivery systems, gene therapy, and noninvasive diagnostic methods. The number of corporate participants in each consortium has ranged from two to twelve. All but two of the participants were founded before 1954 and have over 1,000 employees.[197] OPSR invested a considerable sum, between 7 and 40 million US$, in each consortium.

JST CREST projects usually involve 2–5 university/GRI laboratories working on a particular project. About 10 percent also involve companies. But in IT fields, about 20 percent of participants are companies, and the majority of these are large.[198]

Most other *MEXT* R&D support does not involve companies. However, its Special Coordination Funds do fund some joint research with companies, notably the *Joint Industry, University and Government Cooperative Research Program* which initiated twelve consortia in 2005. Led by university researchers, all the corporate participants are large companies.

This summary of current consortium research in several major government R&D funding programs suggests that consortia that include large companies are an important part of the S&T landscape in fields likely to have near or mid-term industrial applications—and such projects are also an important part of the university R&D landscape in such fields.[199] Compared to earlier consortia, current consortia are more likely to have small companies as participants, while participation by rival large companies is less frequent.[200] Nevertheless, in most of the projects in advanced fields of technology where commercial applications are apparent, large companies are frequent participants.

Are such consortia effective in promoting innovation? Even if they are, do they result in large companies preempting the results of publicly funded research and new market opportunities that ventures might otherwise exploit?

The first question continues to be debated the world over, and this book does not attempt a definitive answer. In the context of Japan, some studies have addressed this question by analyzing joint publications by university and company researchers, and/or patents by consortia or their member companies. The implications of these analyses are not clear. One study that compared participants with nonparticipants suggested a small benefit in terms of issued US patents in the years following the initiation of consortium research.[201] With a few exceptions, anecdotal comments prior to 2003 by academic and industry researchers about the value of consortia have ranged from noncommittal to negative, while my analysis of patents related to the OPSR consortia suggests that only a small proportion of consortia can claim success.[202] However, beginning around 2004 perceptions seemed to shift. Both university and industry researchers began to speak more favorably about *joint research*, although the industry researchers either do not distinguished between *government-initiated* consortia and *company-initiated* joint research projects, or confine positive assessments to the latter.

Studies of consortia participants indicate that consortia facilitate diversification of existing companies into high-growth industries. This diversification often is into fields occupied by either the suppliers of inputs to the consortium participants or by clients of the consortium participants. The prospect of upstream or downstream diversification is probably the primary motivation for companies to participate in consortia.[203] The consortia with which I am familiar generally support this conclusion—although government prodding often plays a role. Sometimes the diversity of participants reflects specific sub goals of a consortium (e.g. developing a particular instrument that other members can use) or the government's attempt to promote the diffusion of knowledge from stronger to weaker companies.[204] Thus *consortia may facilitate vertical integration by existing companies*—or at least they offer old companies a chance to learn new tricks.[205] While this may benefit established companies, *it tends to limit the niches available for new companies*. More specifically it reduces their potential customer base, and it limits the scope for startup formation from universities.

The following account from the CEO of a recently formed bio-startup illustrates this phenomenon. The startup specializes in a field of protein biology applicable to drug discovery and the reduction of adverse side effects. About the same time as the startup was formed, a government agency launched a consortium dealing with the same field of protein biology. The consortium members included three of Japan's largest pharmaceutical companies, six midsize pharmaceutical companies, several chemical companies, a major electronics company, three universities, one GRI, and another government organized consortium. The founders of the venture were invited to join but declined because of distance to the consortium's dedicated laboratory. If they had become involved, the startup may never have been formed. The startup has not been able to recruit clients from among the consortium participants, even though the midsize companies probably constitute a good match in terms of their needs and the services the startup can provide. The startup's CEO attributes most of this difficulty to financial and bureaucratic factors within the member companies and to loss of market opportunities caused by the consortium.

Each corporate member had to pay over 1 million USD to participate in the consortium, an amount that senior officials in the corporate hierarchy had to approve. Not being experts in the field, these officials expected that their companies would learn all they needed about this particular technology through joining the consortium. They viewed participation as an opportunity to evaluate particular compounds of interest to their companies using tests that were similar to those constituting my start up's core business. They also felt that the consortium fees had exhausted the funds they were willing to allocate to exploring the usefulness of this method of drug discovery. This in itself probably cost our company business. Without the consortium, some of the members would likely have become our customers.

But in addition, we believe our specific analytical technology is superior to the consortium's, and we offer better ongoing analysis geared to the specific long-term R&D needs of pharmaceutical companies. After all, the goal of the consortium is to pursue exploratory R&D relating to the underlying science and various analytical methods— not to be a test laboratory to meet the specific commercial needs of members. But it has been difficult to convince senior management in the member companies that we offer better analysis than the consortium provides. Even when we convince the bench level corporate scientists, it is hard for them to convince senior management that we offer more than what their company has already paid for through consortium membership.

Establishment of consortia does not always lead to preemption of a new field by the established consortia members. For example, there are several bioventures developing drug delivery technologies, even though this was the subject of one of OPSR's most successful consortia.[206] But drug delivery is a wide field. In narrower fields, the efforts of S&T agencies to promote consortia as a means to create synergies among companies and to transfer expertise to existing small companies may have negative overall effects. They thwart new companies that might otherwise have made new technologies their own niche; attracted customers, capital, and personnel; increased capabilities; and then attracted more customers in a virtuous cycle. They diffuse a certain amount of knowledge among member companies, often enough so that the members feel they can go it alone and pursue autarkic development in the particular field. But rarely does any member emerge from consortia R&D with a new commitment to advance the consortium technology, to 'run with it' or 'make it its own'. Indeed, because of the sharing of information and IP rights inherent in the consortium structure,[207] this would be difficult. In short, the government's continuing policy of promoting consortium research paradoxically promotes the persistence of autarkic innovation within existing companies, and quenches market forces that might otherwise have fostered the growth of strong new companies. It creates another tragedy of the commons.[208]

NEDO, JST, and OPSR probably seek to include ventures in their consortia whenever possible. Ventures generally welcome invitations to participate. For them the main incentive is access to government funding, not collaboration with other companies interested in the same field.[209] Usually the CEOs of ventures that have participated in consortia say they must be circumspect regarding disclosure of their technology to other members, and they must file patents on their own before collaboration.[210] For them and their investors, any requirement that the venture's inventions be jointly owned either by the consortium or by other members is usually a major disincentive to participate.[211]

As for entrepreneurial university researchers interested in forming a startup, they often practice *entrepreneurship under the gaze of large collaborators* because they are frequently involved in consortia or in joint research with large companies.[212] The more interesting and commercially applicable their research, the more likely their collaborators are to be interested in it, even though these collaborators may not follow through with a concerted development effort. This has probably led to a tendency to found startups that have a circumscribed technical focus and thus limited prospects for growth.[213] Patents and copyrights covering early-stage inventions often enable pharmaceutical and software startups to protect *final commercial products*, and thus may prevent encroachment, even at an early stage, by companies engaged in consortium or other joint research with the founder's laboratory. However, this may rarely be the case in other technical fields. Recalling the above discussion whether patents are effective protection only for pharmaceutical ventures, this perhaps qualifies my conclusion (based on the US experience) that they do protect ventures in other industries. In other words, if a startup emerges under the gaze of collaborating companies with vastly superior R&D resources, patents that do not cover final products can either be engineered around or countered by patents that cover final products. In either case, if the startup does not grow quickly and if the larger companies decide to develop the technology themselves, the startup will likely be forced into unfavorable cross-licensing situations that leave markets for high value-added products in the hands of the large collaborators.

One exception to the dearth of promising university startups outside biomedicine and software is a venture based on novel materials-chemistry technology. The founder filed the first patent applications related to this technology around 2000. Soon companies began to inquire about the technology. The founder initially refused to enter into joint research contracts with them because he knew that under the terms of such contracts, the companies could probably insist on co-ownership of related inventions. Instead, he would send samples to them and do other work for free. In 2005, shortly after the founding of the startup, approximately ten companies representing various industries were engaged in informal collaborations with the founder's laboratory. Some of these provided ideas for specific commercial applications of the startup's technology. His company has issued limited-field-of-use exclusive licenses to some of these companies. It remains to be seen whether these collaborators will become customers of the startup, or whether they will simply try to monitor R&D progress and position themselves to take over the commercialization of whatever prototype products the startup develops. After some painful lessons,[214] the founder and the startup have learned how to protect their IP position. But with a staff of less than ten, it is not clear that it can grow into a company that commands a degree of market exclusivity over high value-added products and services that it sells to many customers and improves rapidly.[215]

However, probably the greatest negative impact of consortia and other joint research on university entrepreneurship is that *they decrease incentives for faculty to start companies and for students to seriously consider jobs in ventures.* When the founder was asked if any colleagues were following his example and establishing their own companies, his reply was negative.

Although we all hope to see practical benefits to society emerge from our research, most university researchers prefer to leave development and commercialization to existing companies. It is much easier that way. Besides, cooperating with large companies under joint research contracts or consortium agreements provides a steady stream of financial support, and sometimes the collaborators dispatch researchers to the university laboratories contributing to the overall research effort. To be a university entrepreneur is still to swim against a swift current, especially when it is so hard to find skilled managers and to recruit skilled researchers. It has taken much time and effort.

At least in elite universities, the same perspective prevails among masters students, who constitute the majority of S&T graduate students. The students who are doing well on the joint research projects are often recruited into the collaborating companies. There they often can continue the same line of research. They are well paid. Their future is secure—probably. There is a reasonable chance that their efforts will result in some tangible product or social benefit. To work in a venture seems like risky small potatoes. Life in a big company is so much easier—and respectable.

It has been suggested that consortia are particularly appropriate for Japan because company creation is difficult in S&T fields. Thus the only way to commercialize new S&T opportunities is for established companies to diversify into these new fields, and consortia help facilitate this diversification. Similarly, consortia are a means of promoting interfirm communication, which would otherwise be low due to the system of lifetime employment and the tendency for autarkic innovation that it engenders.[216] However *promoting consortia helps to perpetuate the very conditions it was intended to alleviate.* How much harm consortia do to entrepreneurship and venture companies is still an open question. My sense is that the harm is significant.

If consortia provided benefits for established companies that outweighed the negative effects on ventures, promoting consortia projects could probably be justified. However, to answer the question posed at the beginning of this subsection, on the basis of available statistical and anecdotal evidence *the benefits of consortia for established companies seem modest at best.*[217]

But in the case of *privately initiated joint research*, there is considerable evidence that such research has helped large companies, and sometimes small companies as well. In the long run, the improved environment for joint research might help shift the balance in terms of access to university discoveries from established to new companies, but in the near term, it seems that it is enabling established companies to claim a significant proportion of university discoveries, at least in non life science fields.[218] Time will tell whether these closer ties with universities will enable established Japanese companies to be the engines of innovation that venture companies are in the USA.

Strong Internal Innovation Capabilities

So if consortia and other forms of industrial policy are not the reason for Japan's innovative strength, what is? Japanese writers generally mention teamwork, attention to quality, continuous innovation, and persistent long-term focus on the core business and on the laboratories, shop floors, marketing offices, retail shops, and service centers where value-adding activities occur.[219] More specifically, analyses of the computer, chemical/materials, and machine tool industries have shown closer linkages between research/design, product development, and manufacturing in Japan compared with the USA. These linkages are built largely on personnel transfers: especially from research laboratories to the more applied, product development laboratories; but also between product development and manufacturing. Such transfers can ensure continuity in the progression from research, to design, development and finally manufacturing.[220] These linkages are also built on close communication between shop-level workers and design engineers, in other words, on active feedback between R&D professionals and manufacturing/production staff.[221]

These strengths depend in large part on a long-term commitment on the part of employees to work diligently for the same company throughout most of their careers (and to be willing to forgo near term wage increases in the interest of long-term corporate profitability), and a reciprocal commitment on the part of managers to protect the interests of incumbent employees, particularly job security. The personnel management policies listed below have helped to create this commitment. Along with a high general level of education among even blue-collar workers, they have also facilitated the teamwork and the delegation of substantial decision-making authority to the team level (at least in production operations) that in turn has facilitated rapid improvements in products and manufacturing processes:[222]

1. Early, prolonged and *company-specific training and acculturation*: Most new hires are new university graduates with BS or MS degrees who enter the company at one time (usually in early April). Each year's cohort undergoes an introductory training program that lasts several months and constitutes an intense shared experience that provides the basis for horizontal communication long after they have been assigned to various different posts. The first two years of an engineer's career are spent largely in apprenticeship roles. New recruits who do not seem to fit are encouraged to leave during this initial period. Large companies still prefer to hire masters or bachelor degree graduates and to train them in house. Thus the number of corporate researchers with university Ph.D. training, and thus transferable high level skills, are limited.[223]

2. *Salary grades and job titles are uniform* throughout a corporation, so that no particular branch has more distinctive titles or higher salaries, thus facilitating the transfer of employees widely among the corporation and establishing nominal status equality among all branches of the corporation.[224]

3. Authority over recruitment and job rotations is concentrated in a *central personnel office*, unlike the USA where recruitment and initial promotions are the responsibility of individual research groups. Thus the interest of the entire corporation is paramount in these decisions. Unlike US engineers who spend most of their time in a particular laboratory developing expertise in a particular area, Japanese engineers are rotated to facilitate knowledge sharing, to instill in individual employees a broad familiarity with the corporation and a multi-development-stage perspective on problem solving, to discourage identification with a specific subunit or specialty, and to ensure continuity in the development of new products: the maxim 'to move information, move people', governs many personnel assignments. In case special skills are needed, the maxim 'make, don't buy' usually governs. Thus when Toshiba needed linguistic expertise to help develop software for a Japanese-language word processor, instead of hiring a linguist it sent one of its own engineers to Kyoto University to study linguistics for a year.[225]

4. *Promotions* are decided by relatively senior section chiefs who are usually responsible for evaluating a larger number of persons than their US counterparts. Whereas US evaluations emphasize technical achievements, sometimes over a short time on one particular project, Japanese evaluations consider performance over a longer period and emphasize social and psychological qualities—effectiveness of relationships with coworkers, communication skills, drive and initiative, dependability,

and (in the context of promotion to R&D team leader) administrative ability.[226] Not until mid-career, however, do differences begin to appear among members of the same entry cohort with respect to job rank and salary.[227] Ability and good performance are rewarded by more interesting assignments, opportunities to study abroad or to do research that might be submitted to a Japanese university for a doctoral degree, the respect of superiors and peers, and the expectation that one has a good chance to become a senior manager in a product division or (if exceptionally talented or lucky) a senior manager in an R&D laboratory. In the case of less productive workers, transfers are often arranged to subsidiary companies where salary scales are lower.

5. For employees hired before the late 1990s, it is necessary to work until at least age 50 to receive substantial *pension benefits*. In the case of large manufacturing companies, these are often paid out in lump sums on retirement, typically US$200,000 to 300,000 for a senior engineer who retires at the mandatory age of 60.[228] Although government mandated pension programs also provide benefits, when amortized over an expected post-retirement lifespan the lump sum corporate pensions usually account for about half of total post-retirement income.[229] Therefore, except in the case of recently hired employees, the pension system creates a strong incentive to keep working in the same company until retirement age. Leaving to join a venture at age 40 is a costly option, and usually there is no bridge back because large companies rarely recruit personnel from other companies in mid-career.[230] Coupled with the threat of transfer to a subsidiary, the traditional pension system creates considerable incentives to remain with large manufacturing companies until retirement.

However, thanks to changes in pension systems implemented around 1999, newly hired employees in many of the major electronics companies have fewer constraints on changing jobs.[231] Newly hired employees usually have a choice to manage all their retirement funds on their own or to accumulate benefits as they work.[232] In either case, benefits accrue approximately *linearly* and vest immediately. Thus if an employee hired after 1999 leaves at age 35 or 40 after ten or fifteen years of service, he or she will receive pension contributions from his or her employer amounting roughly to the same proportion of cumulative salary as if he or she had worked to age 55 or 60. *This represents a major change by major Japanese manufacturers.* When I asked about the rationale, the response was that this was a step to enhance employee mobility and to move away from the system of lifetime employment.[233]

Almost all young new employees select the new 'standard' plan that lets pension benefits accrue and at a steady rate, but leaves management up to the company—only mid-career hires choose to manage pension contributions on their own.[234] It is probably too early to tell if these pension reforms will affect the rate at which employees change jobs.[235]

With the exception of these recent pension reforms, the same personnel policies that have strengthened the innovation capabilities of large, established companies have also reduced growth opportunities for new high technology companies. Capable managers and researchers face financial sacrifice if they leave large companies in their 30s or 40s, unless they were hired after the late 1990s. In addition, even the best engineers in large Japanese companies are probably less likely than their US counterparts to regard themselves as *masters of a particular technology* which they could conceivably spin off into an independent venture in the way that US counterparts have.

As for partnering with ventures formed by outsiders, a strong preference probably remains, reflecting the 'make, don't buy' maxim, to have insiders take on potentially important projects rather than to leave them to outsiders. Also, large companies probably still prefer to innovate by building on an in-house knowledge base rather than pursuing breakthrough innovation that might likely require thorough reorganization of R&D teams.[236] The more finely interconnected and balanced an organization is internally, the more disruptive may be incorporation of an independent outside partner.

Finally, the flip side of being thoroughly acculturated to a particular organization is insularity. A retired Japanese electronics industry executive who became a venture business manager and consultant recently said that scientists and managers in large IT companies have difficulty conceptualizing why outside partnerships might be valuable and how partnerships might work—let alone communicating to potential partners a vision of a mutually beneficial collaboration. He also noted that obtaining permission to attend an outside conference is difficult. 'Unless you are a senior researcher, you can only go as part of a team.' This is consistent with observations that, in order to limit leakage of information to outsiders and prevent possible defections, while at the same time being able to negotiate information trades, large Japanese companies have information *gatekeepers*—senior researchers and managers with exclusive responsibility to represent the company at conferences and to negotiate transfers of information.[237]

In my discussions about the barriers to collaboration with ventures, these issues come through in a nuanced manner. Scientists and managers in large companies usually do not say that they prefer self-reliance because communication with insiders is easier than with outsiders, nor do they say they do not want to deprive their own researchers of work opportunities. More often their reasons for not seeking out collaborations with new companies tend to be that senior corporate officials do not understand or trust the technologies of the new companies, their budgets for collaborative activities are insufficient, or they simply do not have offices or personnel whose main responsibility is to initiate and manage collaborative R&D.[238] In other words, the proffered reasons for autarky are quite prosaic. However, I suspect that these budgetary and administrative reasons reflect a corporate ethos and modus operandi that are geared toward self-reliance, maintaining the integrity of the corporate body with all its employees, and engaging in outside collaborations only if they do not threaten to destabilize the corporate body.

Collaborations do occur, especially with other large companies and overseas ventures, and once an agreement has been worked out, the collaborations generally are stable and both sides are satisfied.[239] But my impression is that collaborations usually take a long time to work out and must be highly choreographed, in part to address internal personnel,[240] communications,[241] and technology-sourcing issues on the Japanese side.

Finally, large Japanese corporations seem to enter more quickly into partnerships with independent small companies when the latter's business is circumscribed. They are ready, for example, to negotiate exclusive licenses to candidate drugs from biotech companies. They are comfortable working with a university startup whose only business is to manufacture specialty IC chips in small quantities for sale to large companies for developmental work.[242] After considerable negotiations, they are willing to purchase machines that make products central to their main business, and even to invest in the company that makes the machines.[243] But as other examples in Chapter 4 indicate, they take a long time to become customers of new companies developing technologies with broad, multiple applications that relate to their core business. With a few exceptions, they do not invest in Japanese biotechs that have promising drug development programs and want to maintain their independence.[244] Partnerships with ventures to jointly develop pioneering, risky technologies that nevertheless have considerable commercial potential seem difficult to negotiate and carry forward.[245]

More information on how companies make decisions concerning collaborations would be helpful to confirm what these anecdotal cases suggest: *the autarkic innovation practices of large Japanese companies, which are to a large extent the product of their personnel management system, limit growth opportunities for new companies.*[246] For new companies, simply discovering promising niches in which to develop new technologies is usually not sufficient. Large companies also need to believe that it makes business sense to rely on new companies to develop early stage technologies and to encourage their long-term growth—including sometimes ceding to them the initiative to pursue development. Japanese companies appear to lack this ability to accept independent new companies as such a resource, even though in doing so they would be leveraging outside capital and letting others bear much of the early development risk.

But because their personnel management policies are the source of considerable strength in their core fields of business, large companies will hesitate to change these policies quickly. The following part examines other factors that inhibit change and alternative strategies for established companies to remain in the forefront of early stage innovation.

PART VII: PROSPECTS FOR CHANGE IN JAPAN

Lifetime Employment and Labor Mobility

Among Japanese manufacturing industries, employment data usually show a sharp increase in 2002 in the numbers of male employees leaving jobs in large companies, but by 2004, the rates of these mid-career job departures had decreased to 1995 levels.[247] This indicates that during the worst period of Japan's recent recession, an unusually large number of employees left large firms in mid-career, but as the economy recovered, these departure rates receded to levels when lifetime employment seemed more secure. Acquaintances in large high technology manufacturing companies generally describe the same phenomenon: after a round of shedding employees around 2002, companies are now lean and profitable and few employees (especially those in R&D) are leaving in mid-career.

The data in Appendix 7.4 cover all employees, not just those in R&D. Between roughly 2000 and 2005, all Japanese electronics manufacturers wanted to downsize their work forces by 10–20 percent.[248] Nevertheless, they tried to avoid dismissing design engineers and scientists, that is, those involved in research and new product development. One senior manager, confronted with the problem of what to do with engineers who had developed expertise in technologies his company no longer needed (and having no funds available for retraining) took on himself the task of retraining these engineers on weekends, so that they could either take on different work in his own division or move to other divisions. He described this as a 'slow, resource intensive process, which nevertheless might pay off because the engineers, although no longer young, are motivated and bright. Having them leave the company was to be avoided, if at all possible.'

Nevertheless, large electronics companies have tried various strategies to reduce even their engineering workforces. Some entered into joint ventures with other companies to spin off noncore operations. Personnel dispatched to the venture might have their salaries guaranteed for two years, after which time they would be solely under the jurisdiction of the venture. On at least one occasion, however, parents of the transferees complained so bitterly to the parent companies (they wanted their sons to continue working for a brand-name company, asserting that when their sons were hired there was mutual expectation of career-long employment), that the plan to form the venture was abandoned. Some large electronic companies have sold off noncore operations, often to foreign buyers. However, the sales contracts usually stipulate that no workers will be dismissed, at least within a defined time period. Therefore, the spun-off operations usually are not very profitable.[249] A traditional practice has been to push workers over 50 into subsidiaries. However, at the depth of Japan's recession, the subsidiaries could not absorb all the transferees. Another strategy has been to offer severance packages, in addition to accumulated pensions, to encourage early retirement. In the case of one company that offered over two years' total salary (including bonuses) most of the persons who accepted the packages were over 50 and they simply ended their working careers.[250] A final strategy is to switch employees to work under a private outside contractor.[251]

As for smaller manufacturing companies, the overall mid-career job departure rates suggest that job losses began earlier and were more prolonged than in the large companies. Generally the rates in small firms were about twice those in large firms, except in 2002 when the large firm departure rate began to approach that in small companies (see Appendix 7.4). However, some of the departures among the younger age groups (30–39 years of age) may be persons starting their own firms. Also, somewhat surprisingly, while total male manufacturing employment in large companies decreased nearly 30 percent between 1994 and 2004, it decreased only about 10 percent in companies with fewer than 1,000 employees. Perhaps this was due to job circulation among small firms and also to the smaller companies absorbing some of the persons laid off from large companies.[252] Female manufacturing employment is only 4 percent of male manufacturing employment and declined in firms of all sizes between 1995 and 2004.[253] *Thus although the overall size of the large company manufacturing sector has decreased, lifetime employment still endures despite a blip of mid-career departures around 2002.* In smaller manufacturing companies, lifetime employment (to the extent it ever was the norm) eroded more quickly and has recovered more slowly, but total sector employment has declined only slightly.

In early 2006 as a sustained recovery of the Japanese economy seemed underway, the chairman of the Japan Business Federation,[254] Hiroshi Okuda[255] summarized the perspective of many large manufacturing companies regarding R&D and employment. Mr Okuda attributed Japan's economic recovery in part to companies treating R&D as a *sanctuary* amidst the cost reductions they had to make during the recession. He also praised Keidanren companies for following *Japanese management principles* based on *respect for human beings* and *taking a long-term perspective*. As a result, unemployment did not exceed 5.5 percent during the worst period of the prolonged economic slump, while it would have reached 10 percent if companies had they followed the American model of large-scale layoffs in times of economic distress. These Japanese management principles allowed companies and society to adjust gradually to forces of globalization without large-scale disruption.[256] The truth of these assertions aside, these remarks suggest that the system of lifetime employment, particularly among R&D personnel in large companies, has not drastically changed. Thus the negative effects of this system on S&T ventures will probably persist for some time.

Responses to a questionnaire I give yearly to new graduate students in my university center suggest that very few want to work in venture companies where risks and potential rewards are high.[257] Available survey data suggest that the expectation of lifetime employment was greater in 2004 than in 1999, even among persons just starting their careers.[258]

The extent to which companies can use laws to prevent unfair competition to prevent R&D employees from changing jobs has not been clearly tested in Japanese courts. However, the principal law in this area was strengthened in 2005 primarily to prevent Japanese technology leaking to rival companies overseas (particularly in Asia). The revised law authorizes criminal penalties for the disclosure of trade secrets even when acquired in the normal course of an employee's work.[259] Judicial procedures now permit in camera disclosures of confidential business information in unfair competition cases, removing a major disincentive for companies to sue former employees for trade secret infringement.[260]

In addition, provisions in employment contracts[261] that forbid employees to engage in subsequent work that will involve disclosure of trade secrets will likely be upheld by Japanese courts if the time period is 'reasonable'. Courts would probably consider both the previous employer's investment in the technology[262] and the rate at which the knowledge will become obsolete to determine what a 'reasonable' period should be. Generally Japanese attorneys think that most courts would accept a one-year limitation. Limitations of as long as five years might be acceptable in certain circumstances.[263]

I have yet to find a court decision that deals specifically with an R&D employee leaving to work in a new company where she or he will continue the previous line of work but where the only information taken to the new job is in the employee's head. In cases that raise similar issues, however, courts seem to be reluctant to find employees who change domestic employers liable either for violating trade secret laws or noncompetition agreements. Nevertheless, the courts have implied that might find former employees liable if trade secrets are clearly defined and if no-compete clauses are clearly drafted.[264] In other words, the door seems open for companies to try to prevent their employees moving to rivals by designating broad swaths of their R&D activities as confidential to be treated as trade secrets, and then including provisions in employment/severance contracts that prohibit work in these areas with another company. Attorneys familiar with trade secret issues say that many companies are taking these steps.

Nevertheless, as indicated by the changes in corporate pension plans, companies are becoming more permissive toward employees changing jobs. But how one leaves is important, and the best assurance that one will not be sued is to reach an understanding with the current employer about the scope of one's future work. The sorts of job changes that gave birth to Intel, other Silicon Valley companies, Amdahl Computers, and some of the hard disk drive spin-offs (departures to new companies that were going to compete with the former employer) would be extremely risky in Japan.

I have heard of only one case where threats under unfair competition laws were used to prevent an employee moving to an S&T venture. Ironically but not surprisingly, this involved a potential job change between two ventures.[265]

Finally, social pressure to work in large companies, particularly from parents and wives, remains high. The factors behind this are complex but probably include the difficulty Japanese women face balancing the expectations of work and family. Families with children (or elderly parents requiring care) where both spouses have professional careers are rare.[266] Thus the family faces high risks when the main breadwinner moves from an established company to a venture.

Another underlying factor relates to the educational system, especially the monopoly on *gateways to success* held by a small number of schools and the grueling childhood-long preparation for examinations to enter those schools.[267] Early in life, largely as a result of parental pressure, learning and the notion of what constitutes meaningful life work become subsumed by the need to pass a series of examinations that stress accumulated knowledge. Children learn that the paths and gateways to success are few, and that internalization of these standards of success[268] is necessary to pass through the gateways and reach the stage where they have the social standing to chart their own destinies.[269]

However, by this time the damage has already been done in the case of many bright, energetic students. Having been forced so long to conform to these externally set criteria for success, and having learned to regard knowledge mainly as accumulation of facts in order to provide the correct response to examination questions, the natural trajectory for many bright new graduates is to continue to conform to the expectations of family and the large companies in which most S&E graduates from prestigious universities find their first jobs. To preserve or re-ignite a maverick or independent mind set after long years of accumulating facts, or simply to pursue a new line of work because one finds it enjoyable, is difficult.

Once one joins a large company, the process of in-house training, intense acculturation and frequent job rotations, does little to reignite such a maverick spirit. Of course, even those who are bent to the system do develop their own interests up to a point. Among those serious about learning and who desire professional careers, a few are willing to pursue their own interests even to the point of having nonconforming, unconventional careers. I have taught some such students. I have interviewed others who are now running their own companies and some who still remain in large companies. But my sense is that the numbers are small. In other words, the educational system, family pressures, and the system of personnel management in large companies, make the likelihood that a bright, energetic scientist, engineer, or R&D manager will want to leave a large company (or a secure university position) to work in a venture company considerably less than in the USA.

This is not to argue that the US educational system is superior, or that Japan's educational system will tend to evolve naturally toward that of the USA. (Indeed, in the latter regard, the opposite is more likely.[270]) Japan's primary and secondary schools have produced graduates with higher proficiencies in mathematics and science than those in the USA.[271] The flip side of not having to cram for one exam after another is that US students tend to enter university with less proficiency in core areas than their Japanese counterparts, and that valuable time must be spent on remedial courses. The confidence that helps Americans to form ventures may be a confidence born out of a seemingly naive lack of awareness of their own limitations, which their Japanese counterparts do not share because they have gone through grueling knowledge accumulation, testing and ranking against their peers.

There are some fundamental changes that favor ventures. Unlike the situation described twenty years ago,[272] my conversations with S&T professors and students indicate that graduating students now have primary responsibility for finding jobs. They are the ones who initiate contact with potential employers.

Their professors are willing to recommend company contacts, if asked, and a letter of recommendation from the professor can clinch a job. But at least in the case of graduates from top tier universities, students are now responsible for charting their entry into the working world. This is important, not only because it indicates they are assuming responsibility for their own careers, but also, if they should later want to leave their initial job, they no longer have to feel they are sullying their professor's reputation with their first employer.[273]

Also, it no longer seems the case that leaving a company to join a new company is automatically regarded as *betrayal*. The former chairman of the entertainment subsidiary of a large S&T company recently said,

Things are different now. It is possible for an important employee to leave to start his or her own company and even continue to have business relations with the former employer. What matters is employee's reputation before departure. Was s/he reliable and of good character? Was the decision to leave made with serious consideration?

Others have remarked, 'How one leaves is important'. I believe this means that if one has the reputation of being capable and sincere, if one can convince the company that the decision to leave makes sense from the standpoint of the employee's inner motivations and his or her business plans, and if one will not act against the company's interests, then departure can often occur amicably. Nevertheless, changing jobs probably still raises more sensitivities on both sides than in the USA.[274]

As noted earlier, some major manufacturing companies have changed their pension systems so that pension funds accrue and vest linearly. Therefore an employee who changes jobs can collect accrued pension funds on departure and does not forgo pension benefits. This ought to decrease disincentives to changing jobs for employees who want to leave to join ventures, as well as those whom the company wants to ease out. However, any positive effect for ventures probably will not become apparent until about 2010 or later.

Many companies have adopted merit pay systems. However, whether these systems will undermine lifetime employment and promote labor mobility is not clear.[275]

The rate of cross-company and long-term shareholding has decreased in Japan.[276] Thus in theory, Japanese companies are exposed to more pressure from shareholders to show profitability—and to make potentially painful workforce reductions in order to do so. However, it appears that so far such pressures have been muted. Reciprocal obligations between employer and employee are still widely felt, and widely regarded as a positive social norm.

Of course, only a few persons need become entrepreneurs in order for entrepreneurial companies to thrive. The cases in Chapter 4 indicate that a small stream of experienced engineers and managers are leaving large companies to form the backbones of a small but steadily growing number of IT hardware ventures. But for the reasons already mentioned, it is unlikely that the small stream will grow wider soon.

Lack of skilled researchers and managers moving from pharmaceutical companies to bioventures has been one of the latter's greatest problems, although since 2005 such transfers have been more frequent. Whether this is a one-time infusion due to mergers and reduction of R&D in some pharmaceutical companies, or is a sign of increasing interest among pharmaceutical employees in changing jobs is an open question. However, considering the advantages of bioventures relative to other ventures,[277] if the flow of personnel from pharmaceutical companies will continue to grow, then their overall long-term outlook is probably fairly bright.

Universities: Wellsprings of New Companies or Contract Researchers for Old?

University and GRI research is becoming increasingly important for both established and new companies. If few persons will leave large companies to form independent spin-offs, then universities have to be the origin of most new discoveries that form the basis of new companies. As corporate basic research declines,[278] universities also become the source of most fundamental discoveries that will be the basis of many of the next generation products by established companies. In the USA, venture companies now are the main intermediary between university discoveries and pharmaceuticals. This probably is also the case in other industries.[279]

The environment for university startups in Japan has improved greatly thanks to reforms of the system of IP and technology management in universities and also to increased access to venture and government financing. Nevertheless, the overall system of university–industry cooperation still favors transfer of university discoveries to large companies. As shown in Chapter 3 and the discussion of consortia in this chapter, the reasons have little to do with laws, but much to do with:

- the weakness of university administrations and of most TLOs,
- hesitancy on the part of even the strongest universities to confront companies and demand higher license fees and stronger control over sponsored research,

- long established patterns of university–industry cooperation, in particular the preference of professors to work directly with companies without university interference and to favor large companies as partners,

- the active presence of researchers from large companies in many leading university laboratories,

- career preferences of graduates and their families that favor large companies,

- the tendency of Ph.D. programs and postdoctoral programs to be filled primarily with persons desiring academic careers—students interested in careers in industry having left after a bachelors or masters degree for company-specific training and acculturation in established companies, and

- issues related to promotions and research funding in universities, particularly peer review and university recruitment systems that are not geared to detect, and promote novel scientific research; emphasis on applied research that meets the needs of existing companies; and the continued prominence of consortium research.

In other words, the reasons that large companies are still the favored recipients of university discoveries have to do with institutional and social factors that are not easy to change and to the very success of policies to encourage closer collaboration with industry.

As for large Japanese manufacturers, since 2004 their senior officials frequently stress the need to engage in more cooperative research with Japanese universities in order to acquire new 'seeds' to develop into successful products. However, cooperation with Japanese ventures or other SMEs is hardly ever mentioned. The upsurge in joint research projects with universities, the prevalence of industry researchers on campus, and the high proportion of large company–university joint inventions, show this strategy is being pursued in earnest. Large manufacturing companies are now speaking favorably about collaborations with university laboratories. Researchers from Continental European countries at the University of Tokyo remark with admiration and a twinge of envy on the close cooperation between large companies and University of Tokyo laboratories.

This close collaboration, however, is resulting in the preemption of a large proportion of university discoveries, as well as the time and energy of university researchers. This is not to argue that such collaborations decrease the overall quality of university science, although this issue deserves further investigation. My impression is that scientists respected for their academic work and fundamental insights in fields relevant to industry are also those that have many cooperative projects with industry. Ventures are not excluded from university–industry partnerships, but if they are included at all in cutting edge projects, it is most often as junior, limited-role partners with large established companies.

Thus we have in the making a great natural experiment comparing two different innovation systems. Will Japan's large companies rejuvenate themselves and become pioneer innovators in newly emerging fields of technology by partnering with universities? In new fields of IT, materials, and energy-related technologies will they take back much of the discovery initiative that now seems to be held by new US companies? Will they compete effectively with large integrated companies in other countries, such as Samsung, that itself has dense, close ties with universities in its home country? We should have answers within a few years.

But one cautionary observation from Japan's own history is that its large companies have always had close ties with universities. Yet with a few exceptions, such as NTT laboratories assisting in the development of fiber optic cable and the VLSI project, or universities assisting in the development of amorphous silicon-based solar cells,[280] universities and GRIs probably were not critical contributors to Japan's economic miracle of the 1960s to 1980s. So if these close ties have always existed but were not crucial in the past, what has changed to expect that they will be crucial in the future? Perhaps now, with more of their employees on campus and faculty more attuned to commercial interests, the interaction will prove to be more productive. Also many of the joint research projects as well as many of the government funded consortia projects could be classified as translational research.[281] Assuming that there are more translational research projects now than ten years ago,[282] maybe this increase in publicly subsidized translational research will be sufficient to encourage established companies to pursue the development of new technologies arising in universities. Nevertheless established companies could have done all these activities under either joint research agreements or donations twenty years ago and received essentially the same degree of IP protection they obtain today. For established companies, the degree of access to university research has not changed, only the degree of public cofunding and perhaps the level of commercial interest among faculty.

The coming years will also provide insight into the most effective form of university–industry cooperation. Is the optimal system to commercialize university discoveries based on entrepreneurial faculty and strong entrepreneurial university administrations that control technology transfer via formal licensing and close oversight over contract research—realizing that this type of technology transfer system can lend itself well to creation of startups but sometimes results in adversarial relations with existing companies? Or is the Japanese system based upon direct cooperation between professors and established companies, allowing the latter a free hand in technology management and downplaying entrepreneurialism, a better way to promote the development of new technologies? The answer depends on whether new or established companies are better at early-stage innovation.

Japan's system is not necessarily inimical to startups. However, its recent experience suggests that, in a weak entrepreneurial environment, a formal, US style technology transfer system promotes the formation of startups by providing an alternative to direct pass through of university discoveries to established companies and by raising faculty consciousness about the commercial potential of their discoveries. On the other hand, in an entrepreneurial environment where venture capital and other resources for startups are readily available, a system that leaves discretion about technology management largely in the hands of university inventors may be equally likely to promote startup formation (or even more likely, if TLOs are not competent).

PART VIII: CONCLUSION

Japan: Shumpeterian End Game?

For the foreseeable future, Japan will continue to be a country that relies on large established companies for innovation.

With a few exceptions, the Japanese government has already enacted most of the feasible legal reforms necessary to create a more hospitable environment for ventures. But the differences in innovation nevertheless persist. Whether large Japanese companies with their enhanced ties to universities can become innovation leaders in new fields of technology will become clear in the next few years.

However, it may be difficult for Japan to remain at the forefront of countries in terms of early stage innovation relying solely on its established companies, even if they have close links with universities. In other words, without vibrant new companies, Japan probably will not develop and commercialize early stage discoveries, especially in new fields of technology, as quickly as countries with such companies. The reasons include the following:

First, the handoff of new technologies from universities to industry is problematic. If an established company is already pursuing a particular line of R&D, the young researchers it sends to universities may learn useful things about this field which they will take back to their companies and use. However, they will be less able to focus on new university discoveries with which they are relatively unfamiliar and even less able to convince their seniors in their companies to devote corporate resources to developing these new areas. Even if their seniors are present in the university laboratories, it will still be hard for them to convince the upper echelons of their companies to pursue promising university discoveries, unless the discoveries are so clearly winners that they would be developed in any innovation system. In contrast, university startups and independent spin-offs have a built in growth dynamic focused on new technologies. They are founded by university researchers or former employees of existing companies and their financial backers who believe there is commercial potential (often not immediately obvious) in particular university discoveries and who are committed to developing those discoveries.

Second, internal corporate decision-making and funding processes decrease the likelihood that new discoveries will be developed rapidly. Large established corporations tend, perhaps with good reason, to focus on existing lines of business and the products or services they are already producing well. For them to devote resources to develop new technologies in-house is a decision they have to undertake with some caution. Thus the approval and funding process tends to be bureaucratic and slower than in the case of a new company, which focuses just on one or two new technologies and (provided it operates in an open, rather than autarkic, innovation environment) aims to develop them for a wide variety of uses and markets.

Third, and perhaps most speculatively yet importantly, both small size and newness tend to enhance the motivation of employees and cooperation among them. Researchers and managers identify closely with the goals of their company and cooperate readily with each other to achieve these goals. At least in the case of managers and main researchers, working in a small company provides a sense of control over their destinies that most cherish and, for the preservation of which, they will work concertedly.[283]

The desire for independence and control over one's destiny is probably just as strong among Japanese as among westerners. As counterpoint to the oft-heard refrain that Japan's is fundamentally a village society—closed, hierarchical with strong communal obligations—one need only look out over a typical Japanese urban landscape to notice that most people live, not in apartment buildings, but in densely packed individual houses. In this respect, the contrast to the ubiquitous large apartment buildings of Singapore, Hong Kong, any modern mainland Chinese city, and even New York, Paris and Geneva is striking.[284] Families and individuals want their own space and privacy, limited as they may be. It may be that crowding in Japan's small habitable land area has engendered a degree of sensitivity to others and a tendency to self-censorship that are unusual in other cultures. Nevertheless, desire for independence is not far below the surface.[285]

Perhaps one of the tragedies of Japan's modern systems of industrial organization and education is that they have not accommodated this desire and motivating force—at least not in realms that offer the greatest prestige and access to resources. It is as if the structures of government, education, and industrial organization took on too readily the closed, communal, and hierarchical characteristics of feudal/village Japanese society, relegating the individualistic aspects to the periphery.[286] As the section on machine tools indicated, small businesses have long been a refuge for 'individualists' in Japanese society.[287] Yet, except for a few cases of entrepreneurial companies rising to industry leadership,[288] such companies (and individual initiatives in general) seem to have a second caste status in Japan.

It has been said that the style of in-house innovation in Japanese corporations, where most employees expect to work until retirement and horizontal coordination is encouraged,[289] is superior to the innovation system in 'typical' large Western manufacturing companies so long as technology and the business environment is changing at a *moderate* rate.[290] However, as university R&D and globalization increase, in many fields the pace of technical and business change will increase, and it is not clear that even this more flexible type of in-house innovation will be quick enough to compete with new companies that can network quickly and effectively with other companies–and then, when their technologies have been eclipsed, let their employees join other companies.

Japanese companies will remain strong in manufacturing at least for some time to come. They may be able to shield themselves from lower cost overseas competitors in some high technology fields that require delicate, black-box know-how that they can keep in house.[291] Yet such a tacit-knowledge-based innovation strategy seems to offer limited opportunities for synergy with universities or other companies. It also probably has nearly zero tolerance for mid-career departures of the key persons who understand the black-box aspects of the technology. If the key black-box knowledge resides in a relatively small number of craftsmen or engineers, the same strategy can probably be replicated by small companies in China, India, or the USA.

The fundamental problem facing Japanese industry is that it is not well suited to take the initial steps to develop new technologies, and this is largely because it lacks new high technology companies. Some companies have been good at developing a steady stream of new high quality products. But companies such as Toyota, Canon, Toray, and Chugai cannot possibly develop all the promising early stage technologies in their fields. Japan cannot rely on them for all its innovation needs. It needs more independent high technology ventures to increase the speed and flexibility of innovation by its industry. Without them, it risks being squeezed between countries that can rely on new companies to bring new technologies to proof of concept stage quickly, and countries where manufacturing can be done at lower cost with almost the same level of quality.[292]

American Entrepreneurship: Born Out of Adversity

The reasons that ventures have thrived in the USA are of equal interest as the reasons they have not thrived in Japan. Maybe it is not Japan that is unique, but rather America. Moreover, while the past three or four decades have shown that the US system of innovation based largely on new companies is robust, its continued vitality cannot be taken for granted. This book has noted multiple factors that are necessary for vibrant S&T ventures: financing, managerial talent, pools of skilled researchers willing to work in ventures, customers that include large companies, strong IP protection, generous yet astute public support for basic university research, competent university technology management that is supportive of startup formation, effective management of conflicts of interests that preserves academic integrity and core academic values, and the ability of R&D researchers and managers in established companies to leave to set up independent spin-offs. Take away any of these elements, and the system is undermined. There are many ways to fail, but only a few ways to succeed.

Regions of the US where ventures are scarce offer some insights. Recently a manager of a VC fund returned from meetings with local venture businesses and officials of a particular state's main university and medical school. Accompanied by a business adviser to that state, his goal was to assess prospects for bioventures and investment opportunities for his firm. He remarked that some of the discoveries emerging from the medical school seemed quite interesting. However, financing was problematic, an observation confirmed by some of the entrepreneurial scientists whom he met, who had to work very hard to obtain funding for their companies. This particular state is geographically distant from major centers of VC financing. Although a few VC funds have representatives near the university, attracting their interest or pulling together investment syndicates has been difficult. Also managers are scarce. One of the most successful ventures is still run by its scientist-founder. Most Ph.D. students are not interested in careers as entrepreneurs. However, what struck both the VC financier and the adviser most was the lack of focused efforts by university officials to improve the environment for ventures. Inventories of university research that might have commercial potential were not available. Neither were lists of university scientists that had expressed interest in forming companies. Plans to provide incubator space were still on hold. The TLO was preoccupied with normal licensing and material transfer agreements and did not have experience or resources to focus on inventions that might be the basis of a successful startup. Such proactive steps may not have been necessary in a region imbued with venture culture and infrastructure, such as the San Francisco Bay area, but their absence in this state confirmed in the mind of the VC investor that this was not an attractive investment environment.

A colleague at the University of Tokyo who is from the bioscience department of a well-known university in another US state remarked that, although success rates on NIH grant applications are fairly high, few scientists think about forming companies. Awareness of the commercial potential of their discoveries is low. 'It is much more entrepreneurial here at the University of Tokyo', he remarked.

Even the continued vitality of entrepreneurial regions should not be taken for granted, nor the importance of chance and serendipity. George Rathmann, Amgen's first CEO, noted recently how hard it was to obtain investments from large pharmaceutical companies, as well as the danger that investing companies would want control, as they usually do in Japan.

... In the early days, there was a lot of cynicism on the part of Big Pharma. They called it 'hype' and 'flash in the pan.' We went out on tour and could barely get interviews with pharmaceutical companies, despite that Genentech was already a success. ... Then we developed bona fide products, and suddenly we got respect....

I had a lot of fear eight or nine years ago that Big Pharma would own the biotech industry.... It turned out a lot better than we had thought.... But a lot of the money [now] is being spent in fields where the payback is instant. [Big Pharma] is looking around for companies that are in late-stage clinical development. That doesn't help build good R&D. They need to take a gamble and respect R&D.[293]

There are also longer-term challenges, especially due to declining numbers US graduates in S&T fields and declining immigration of foreign scientists and engineers. The age demographics of ventures still are not clear. Can they still be innovation leaders if the pool of young scientists and engineers begins to decrease? Could an older workforce cope with the prospects of layoffs or frequent job changes that are supposedly part of the venture scene? Do ventures depend on immigrants more than established companies? Can venture companies or their employees pay the high premiums for individual or small group health insurance, and will the high cost of health insurance and the prospects of being without insurance for long periods following job changes dissuade persons from working in ventures?

The rebound of venture funding after sharp declines from 2001 through 2003 and the rise in funding even for new ventures shows the system is resilient.[294] The proportion of total industry R&D that is accounted for by small companies continues to grow and is now over 20 percent.[295] High technology venture investment constitutes a larger proportion of GDP in the USA than any other country except Israel. In contrast, in Japan this proportion is the lowest among major OECD countries, although it is also low in most countries of Continental Europe.[296]

For R&D personnel, working in large companies is probably more attractive relative to ventures in Japan and Continental Europe compared to the USA. The prestige of working in a large Japanese company has been discussed. Salaries tend to be somewhat higher than in ventures.[297] This is probably also the case in Continental Europe.[298] But probably the most telling difference between the USA on the one hand, and Continental Europe and Japan, on the other, is job security. The preceding discussion shows that lifetime employment is still the standard in large Japanese manufacturers, especially for R&D employees. On the other hand, job stability fell in the USA due to layoffs during economic downturns.[299] In the USA, legal protections against layoffs are the lowest among OECD countries, while in the countries of Continental Europe they are the highest. In Japan they are lower than in most European countries but higher than in the USA. Indeed, there is a rough inverse correlation between the level of such protection and VC investment in high technology industries as a proportion of GDP— although Japan and Sweden do not fit this correlation relationship well.[300] In other words, the easier it is for companies to dismiss employees, the more suitable the environment for entrepreneurial high technology companies. The most likely explanation for this phenomenon is that weak protection against dismissal is an incentive for employees to be more mobile, and this mobility is vital for high technology ventures to meet their personnel needs.

This is not to say that risk of dismissal from a venture is much higher than from a large company in Japan or Continental Europe. There have been few outright failures among Japanese ventures in biomedicine, the field with which I am most familiar. There have been some consolidations, and there has been movement among senior managers. However, as noted in Chapter 4, probably just one of these companies, MBV, has over 100 employees and only about 5 others had over 50 employees in 2005. As with all Japanese companies, they are reluctant to dismiss employees—perhaps more so than established companies because they know their ability to recruit in the future will be damaged by a reputation for involuntary dismissals. Also to some extent, government contracts provide some cushion when income from private sources is scarce, and to some extent the overlap with university R&D allows scientists and technicians to rely on university salaries for some of their company-related work. As for German biomedical ventures, a large proportion are focusing on research tools and research services where they have a steady, modest income stream, as opposed to drug discovery and development where the needs for R&D personnel are high but the risk of failure (and attendant large-scale layoffs) is also high.[301] Among Japanese bioventures focusing on drug discovery and development, a common strategy is the *no-wet-lab* venture which in-licenses candidate compounds that other companies have decided not to develop and then contracts out all the optimization, animal, human, and other laboratory work. In this way the number of employees dependent on the venture for their livelihoods is kept low.[302] In other words, ventures in countries with low labor mobility avoid employing persons who may later have to be laid off, but this diminishes their chances for success in risky but potentially lucrative fields such as drug development where rapid ramp-up of employment is often necessary.

Job security is also less for US academics compared with their Japanese and perhaps also European counterparts. US junior faculty (assistant professors) must begin to obtain grants to fund their salaries and research expenses soon after they start work. Japanese faculty rely instead on the resources of the professor's laboratory (kouza). Even tenured US university faculty must keep obtaining competitive grants in order to receive their full salaries, have time for research, and to pay for laboratory facilities, equipment, graduate students, and postdocs. Although more emphasis is being placed on obtaining competitive research grants in Japan, salaries are still independent of outside funding and, to a larger extent than in the USA, so are laboratory space and students. Thus in comparison to an academic career, working in a venture may seem more attractive in the USA than Japan.

However, high job security may be just one prong of policies that enable large companies to hold on to their employees for most of their careers. The other prong is that large Japanese and European companies may be more attractive places to work—not only in comparison to ventures in their own countries but also in comparison to typical large US companies. I have already mentioned the reciprocal obligations felt by Japanese managers and their subordinates, and the renewed commitment within high technology manufacturing companies, to lifetime employment. In Germany, nationwide collective labor contracts have continued to give workers in large companies generous wages and benefits that many small companies have not been able to match. Large employers agree to these generous benefit packages *in order to guarantee peace, predictability, and labor cooperation, which have become dear to firms in an era of just-in-time production.*[303] In contrast, anecdotal reports suggest that employees in US corporations, including skilled engineers, are more often viewed as interchangeable or replaceable commodities.[304]

The following comments by a senior software engineer may reflect the frustrations and uncertainties US high technology workers feel working in some large companies and how these frustrations encourage movement to ventures. The company in question is a leading developer of software for a particular industrial application. It established its leadership position when it was still a young venture, but it was then taken over by a large diversified company. The engineer joined it during its independent growth years and has witnessed its transition to a fully controlled subsidiary of the large corporation:

When worker morale is high, when communication is good and there is a sense that 'We are all in this together,' workers deliver not just 100 percent but 180 or 200 percent. They cooperate well and they put in 16 hours of work a day to get the job done. But when they feel management is not paying attention to their ideas and makes decisions that compromise the basic goals of the company and its dedication to its customers, then performance falls not just to 80 percent of what is expected, but much lower. And performance is even lower if layoffs are threatening. I have seen this happen in my company after its incorporation into XXX Corporation. It's not that my company's management is especially incompetent or self-serving. But within the enlarged corporate structure, capable dedicated employees no longer feel that their efforts on behalf of the company are rewarded, recognized or even utilized for the benefit of the company.

What are layoff rates among engineers and scientists in ventures compared to large companies? How does job satisfaction in ventures compare with that in large companies or academia? Is job satisfaction among scientists and engineers higher in large Japanese or European companies compared with their US counterparts? These are questions that cry out for answers. Despite accounts of exploitation and summary dismissals in ventures, there are indications that many engineers and scientists prefer working in small companies.[305] The software engineer quoted above added:

Most of the software companies related to my industry were formed as spin-offs. The pay is good in my current company. But especially now when everyone is concerned about receiving pink slips, there is a good chance I will soon join one of these new companies or form one myself. There is a process of renewal in both a personal and professional sense that occurs when one joins a new company.

Thus, the strength of US ventures may depend on factors that undermine innovation in large companies. Specifically, the availability of skilled people willing to work in ventures depends to some extent on pressures large companies face to shed employees during economic downturns, which reduces their ability to sustain R&D initiatives through such downturns.[306] The departure of R&D personnel in turn undermines R&D and manufacturing projects, interferes with quality control, and reduces in-house tacit knowledge important for ongoing R&D and product improvement. If human resources really ought to be shifted to other fields to take advantage of new technical and business opportunities, this sometimes involuntary mobility may be good for the economy as a whole. However, it may sometimes result in well-integrated, productive R&D teams being broken up sooner than necessary, and the premature departure of skilled scientists and engineers from S&T occupations.

Conversely, the attractiveness of large companies as places to work in Japan and Continental Europe may be one of the main factors holding back the flowering of S&T ventures in those countries. In other words, the adversity of employment conditions in established companies in the US has contributed to its uniquely strong entrepreneurship.

Innovation Systems

The discussion above may provide further insight into why the environments in different countries seem to be conducive to either *radical* or *incremental* innovations, a distinction put forward in the *varieties of capitalism* literature.[307] In particular, US firms are said to excel in radical innovations while German firms excel in incremental innovations.[308] This is attributed to US companies being able to shift resources quickly to pursue new technologies even if it means disrupting ongoing projects. German companies cannot shift resources so easily but, because they are assured of a stable supply of capital and skilled labor, they can devote more effort to long-term improvement of existing products.

Deeper insights might, however, be gained from examining which innovations arose from new-or-small versus large-and-established companies. I suspect this would show that the differences between the innovation propensities of the USA and Germany in various industries (and also, between the USA and Japan) are largely explained by whether new companies are major innovators in these industries. In other words, the industries Hall and Soskice designate as characterized by radical innovation are probably those where a significant proportion of patents are emerging from new (often US) companies, while those labeled as incremental innovation industries are those where the contributions of new companies to innovation are minor and most patents are being obtained by established companies.[309] If this is indeed the case, it suggests that the differential impact of the two main varieties of capitalism is mediated largely through their effects on new companies. In other words, from the standpoint of the vitality of an innovation system as a whole, the greatest benefit associated with a liberal market form of capitalism may be that it creates a more favorable environment for innovation by new companies.[310]

Furthermore, this analysis might shed light on the question that began this chapter, 'are new-or-small companies more innovative than old?' Distinguishing between radical and incremental innovations must be done with caution.[311] However, if reasonable criteria are used to distinguish radical and incremental innovations,[312] I suspect that this would show that industries with a high proportion of new companies are those characterized by a high rate of technological change. It might even show that within these industries new companies are the source of a high proportion of the most innovative discoveries.[313]

To Alter Destiny

The preceding discussion suggests that the Japanese system cannot change quickly, and that America will have to continue to rely on new companies in order to remain in the forefront of innovation. The source of each country's innovation strength with respect to either established or new companies is also the source of its weakness with respect to the other type of company. Japan has done most of what it can in terms of government policies to try to create a more favorable environment for ventures. While it is still too early to judge with certainty their effects, they appear modest. For the foreseeable future Japan will have to rely on its large companies for innovation.

I suspect that the situation in Continental Europe is similar, in terms of long-standing factors limiting labor mobility and the limited improvements that recent reforms are likely to bring in the short term. Both Germany and France rely on large companies for innovation. In both countries, legal barriers to dismissing employees are higher than in Japan.[314] Japan has no system of national unions that negotiate employment terms applicable across the country as does Germany. The role of Japanese corporate unions is less formalized and less influential than the legally mandated German works councils ensconced in companies to represent workers' interests.[315] Yet, in Germany, as in Japan, there are pressures on corporations to pay more attention to profitability. A 2003 German law reducing taxes on sales of corporate cross-holdings led to a large unbundling of cross-held corporate shares.[316] Equity financing by German corporations has increased.[317] But, although it may be premature to judge, these reforms do not seem to be translating into increased labor mobility.[318] How much more it will take in Japan, Germany, or France to create the degree of labor mobility needed by ventures to thrive is not clear. The US case suggests this may require substantial dismantling of legal and social barriers against dismissals. In Germany it appears that this would only come about after a difficult political process, and in Japan only after a sea change in social expectations with respect to large companies. The fact that both the Japanese and German economies seem to be in sustained recoveries means that the pressure to change is now low, although this may create an opportunity for political change in Germany.

The situation may be different in China, India, and perhaps also Eastern Europe where new companies offer some of the most attractive employment opportunities for bright energetic scientists, engineers, and S&T managers.

In contrast to Japan and countries such as France and Germany, the USA will be a country that relies to a large extent on new companies for innovation, particularly in areas where university discoveries hold the key to new commercially relevant products. Nevertheless, it is not certain whether ventures can be engines of innovation in emerging fields of technology outside life science, software, and some IT-related fields. These are the areas in which they have already proved themselves and also, coincidentally or not, where US high technology industry is strongest in terms of manufacturing and employment growth. If they cannot, it is not clear that established US companies, innovating on their own, can be internationally competitive either.

Despite these barriers to change, Japan should try to create a better environment for high technology ventures in ways that do not require drastic reshaping of its innovation system. Too much is at stake, and there are too many warning signs concerning the innovative capacity of established companies, to proceed further into the new millennium relying only on large, established companies for early stage innovation. The most appropriate place to start probably is universities, where there is room for reforms. The following suggestions are not far-reaching (except for perhaps the first and last), but they should equalize the playing field between new and established companies with respect to access to university discoveries, encourage more entrepreneurship among university researchers, and give startups more breathing space vis-à-vis large companies. They should also increase the quality of university science.

The primary recommendation is for the government to stop cobbling together research consortia and earmarking large chunks of research funding for consortia. The evidence that consortium projects result in meaningful cooperation or technical progress that would not otherwise have occurred is scant. On the other hand, there are real grounds for concern that government organized consortia are preempting new areas of technology that would be fertile fields in which ventures could develop. This is a variation on *trying to pick winners* that is unnecessary in view of the advanced technical level at which Japanese companies are competing internationally, beyond the capacity of the government to do effectively, and inherently anticompetitive. Of course researchers from various universities and companies should be free to submit joint research proposals to government funding agencies, but these proposals and the claimed synergies from collaboration should be evaluated case by case on their merits.

Second, funding agencies should carefully scrutinize university funding programs that tend to support applied research. In general, only if there are clear reasons to believe that private industry on its own cannot fund such research, or that government funding is necessary to ensure benefits are widely available, should such university projects be funded. Otherwise there is substantial risk that public resources will be wasted or end up subsidizing established companies, and opportunities will be lost to make progress in new areas of science—areas that might be fertile grounds for ventures to develop.

Third, improve the quality of peer review for all competitive funding programs open to university researchers. At stake are taxpayers' funds; the quality of scientific research; and the careers of university researchers. Although the peer review system has improved over the past ten years, much room for improvement remains. The need for expert, objective peer review is especially high in the case of applied research projects, because of the risk of misdirecting resources mentioned above, and also because the results of applied research projects may be less subject than basic research findings to critical evaluation in academic publications and other open fora.

Fourth, with respect to technology transfer and joint research, universities should make sure that the scope of joint research projects are well defined and commensurate with the funding from the sponsors, that sponsors obtain IP rights only to inventions that fall within the scope of projects, and that claims of company coinventorship are scrutinized. If sponsors want exclusive rights to inventions, they should negotiate for them case by case.[319] To these ends, universities must develop their own professional competence in contract management and hire persons whose careers are devoted to this field and who are paid salaries competitive with those in private business. Government and universities have to be clear that universities' primary goals are education and research, and that, while joint research that serves these goals is welcome, universities are not to be contract research laboratories for companies seeking to leverage taxpayer financed resources to conduct research of primary value to them alone.

Fifth, the remaining barriers to universities taking equity in lieu of cash for licenses to startups should be removed.

Sixth, there should be real, substantial moves away from the kouza system so that young researchers have the means and the independence to pursue their own research ideas. New young faculty must have their own laboratories and graduate students and reasonable prospects for tenure. This in turn requires reforms with respect to funding, peer review, and how faculty are recruited. The present patronage system will not do. There must be some system to incorporate objective external evaluations into the appointment and promotion process.

Beyond the realm of universities, tax deductible loss carryovers for angel investors should be easier. Also, the JPO and judicial agencies should keep track of suits initiated under the Law to Prevent Unfair Competition (LPUC), to determine the extent to which that law prevents employees of established companies from starting their own spin-offs. If the threat of suits under that law are indeed preventing persons from changing jobs to new domestic R&D companies, then consideration should be given to interpreting the law narrowly in such cases.[320] Also the JPO and Japanese courts should be cautious about broadly interpreting Japan's already generous prior use rights with respect to patents, lest the ability of ventures to prevent rivals from encroaching on their technologies be undermined. The current guidelines of concessionary stock exchanges that strongly recommend that bioventures have alliances with large companies prior to public listing of their stocks should be interpreted flexibly. Finding alliance partners is harder for independent ventures in Japan than in the USA. Public investors should be justifiably on guard if a bioventure does not have alliance partners. Nevertheless, requiring such partnerships tilts the bargaining tables even more in favor of established companies. In view of the autarkic innovation modus operandi of large Japanese companies and the way this increases the financial challenges facing Japanese ventures, Japan must be receptive to foreign investment in high technology companies. Some of Japan's most successful ventures owe their business breakthroughs to foreign companies.

America, for its part, should realize the extent to which it relies on new companies for innovation—to develop new ideas (wether originating in universities, corporate laboratories or the minds of independent inventors) to the point where larger companies are willing to commercialize them, or the new companies themselves are able to commercialize them. It should also consider the possibility that the primary social benefit of a user-friendly patent system that provides strong protection for inventions; a system of university IP ownership that encourages university, faculty and student entrepreneurship; and even its system of liberal market capitalism, is that these facilitate the creation and growth of new companies. These systems have their faults, but reforms should not undermine this basic benefit and function.[321] It is fortunate that, at least in the case of patent system reforms, advocates often acknowledge the need to ensure that the system supports high technology ventures.[322]

In addition, America needs strong public support for basic science research in nonbiomedical as well as biomedical fields. Its science agencies need to remain strong and to keep in touch with the other branches of government, the scientific community and the general public, in order to set reasonable scientific priorities and plan how to achieve them. They also need to maintain effective systems of peer review so that research funds are allocated fairly to researchers most likely to make progress. Despite legitimate concerns about terrorism, espionage, S&E wage increases being held down, brain drains and reverse brain drains, America must keep its doors open to immigration of scientists and engineers. Available evidence suggests that America's high technology ventures depend disproportionately on such immigrants, and that their international mobility benefits both America and their countries of origin in the long run. Either at the federal or state levels, mechanisms should be worked out to allow small venture companies to pool the health care costs of their employees so that these costs are manageable. Finally, the USA must also provide a healthy environment for its *large* high technology companies, because new companies cannot be engines of innovation without strong large companies as customers and as sources of discoveries that sometimes end up being left to new companies to develop.

Developing countries should plan national development strategies that encourage the growth of new technology oriented companies. While a short-term development strategy might justifiably emphasize low-cost manufacturing that might best be done by large companies, the infrastructure to support entrepreneurship should also be nurtured. This involves establishing effective university technology management systems, caution with respect to reliance on industry to fund a substantial portion of university R&D, strengthening IP laws and facilitating their use by small companies to protect their discoveries, facilitating establishment of VC funds and other financial institutions to support new companies, strengthening equity markets for small high technology companies, keeping open the door to foreign investment in high technology companies, establishing a system of portable individual pension accounts to allow employees to accumulate retirement funds as they work and to transfer these to new accounts if they change jobs, and caution in adopting laws that limit the ability of employers to dismiss and hire workers—while doing what they can to provide a safety net for persons who lose jobs (and their families) and facilitating retraining and information networks to help them find new jobs.

APPENDIX 7.1

Table 7A1.1. Intellectual property holdings according to size and age of Utah and New York bioscience companies

	years since incorporation	≤50 employees	>50 employees	per firm means		per employee means (unweighted)	
				≤50 employees	>50 employees	≤50 employees	>50 employees
# firms in sample	≤10	74	12				
	>10	52	45				
# patents	≤10			6.9	51.2	0.9	0.4
	>10			4.5	34.2	0.4	0.2
# trade secrets	≤10			8.2	37.1	1.3	0.3
	>10			20.7	46.0	2.0	0.3
# total IP items*	≤10			20.5	154.6	3.0	1.5
	>10			36.1	120.3	3.7	0.9

* Total IP items means the sum of patents, trade secrets, copyrights, trademarks, and licenses & options.

Source: Willoughby (1997, 1998).

APPENDIX 7.2: ORIGINS OF NEW DRUGS APPROVED BY THE
US FDA 1998–2003

Table 7A2.1. Numbers of new FDA approved drugs by type and type of organization where inventors worked

	Pharma	Biotech	University licensed 1st to pharma	University licensed 1st to biotech	Total	Total nonpharma origin
Priority NMEs	35.05	11.25	4.65	12.05	63	27.95 (44%)
Nonpriority NMEs	61.40	7.40	5.70	6.50	81	19.60 (24%)
NTBs	3.00	13.17	1.10	7.73	25	22.00 (88%)
Total	99.45	31.82	11.45	26.28	169	69.55 (41%)

Table 7A2.2. Number of new FDA approved drugs by type of drug, type of employer institution, and location of laboratory where the inventors worked (with share of global pharmaceutical market as a benchmark reference)[1]

Country	Share of 2003 world pharma mkt.	All new drugs			Nonpriority NMEs			Priority NMEs			NTBs		
		Total	Biotech	% biotech	Total	Biotech	% biotech	Total	Biotech	% biotech	Total	Biotech	% biotech
USA	44.4%	75.7	46.7	62	22.5	10.0	44	32.0	17.4	54	21.2	19.3	91
UK	3.8%	17.6	2.4	14	10.3	1.0	10	5.8	0	0	1.5	1.4	93
Japan	12.3%	16.9	1.0	6	12.4	0	0	4.5	1.0	22	0	0	
Germany	6.0%	16.6	1.0	6	10.9	0	0	5.7	1.0	18	0	0	
Switz	0.6%	10.4	1.0	10	5.1	1.0	20	5.2	0	0	0.1	0	0
France	5.7%	6.0	0	0	3.0	0	0	1.8	0	0	1.2	0	0
Canada	2.1%	4.0	2.0	50	2.0	1.0	50	2.0	1.0	50	0	0	0
Other[2]	25.1%	21.8	4.0	18	14.8	0.9	6	6.0	2.9	48	1.0	0.2	20
Total	100%	169.0	58.1	34	81.0	13.9	17	63.0	23.3	37	25.0	20.9	84

[1] See notes 25–8 and accompanying text for data sources and methodology. 'Biotech' includes drugs invented in universities that are licensed/transferred to biotechs.

[2] The other countries of origin are Italy, Finland, Sweden, Denmark, Israel, Czech Republic, Hungary, Belgium, Australia, Spain, Korea, Austria, and Egypt (listed in approximate order).

Table 7A2.3. Share of new FDA approved drugs from regions with few biotechs compared with regions with many biotechs (with share of 2003 global pharmaceutical market as a benchmark reference)

	Fr, Ger, Jpn, Switz	Fr, Ger, Jpn only	Canada, UK, USA	USA only
No. new drugs (all types)	50.0 (30%)	39.6 (23%)	97.2 (58%)	75.7 (45%)
No. NTBs + priority NMEs	18.5 (21%)	13.2 (15%)	62.5 (71%)	53.2 (60%)
% world pharma market	25%	24%	50%	44%

APPENDIX 7.3: THE JAPANESE MACHINE TOOL (MT) INDUSTRY

Table 7A3.1. Leading Japanese machine tool companies ranked by 2004 sales

World rank	Name	Inc. Year	Employees 2005	MT revenue FY 2004 (M USD)	Total revenue	Comments
2	Yamazaki Mazak	1919	5,000	1,576	1,576	
5	Fanuc	1972	4,549	1,280	3,079	Builds mainly CNC controllers. Fujitsu affiliate.
6	Amada	1946	1,599	1,256	1,674	Has many affiliates for which it acts as designer & distributor
7	Mori Seki	1948	3,012	1,137	1.137	Small subcontractor until 1970s
8	Okuma	1918	1,897	1,073	1,073	
10	Toyoda Machine Works	1941	3,878	772	2,272	Affiliate of Toyota Motors and its weaving machine parent.
11	Makino	1937	907	734	879	
27	Aida	1917	1,087	395	407	

Source for revenue data: 2005 Machine Tool Scoreboard.

In comparison, the largest US companies in 2004 were Unova (a mini-conglomerate which sold off its MT operations in 2004–2005) ranked eighteenth with MT sales of US$470 M and Haas, ranked nineteenth with sales of US$464 M. The precise rankings should not be emphasized, however, because they fluctuate.

APPENDIX 7.4: EMPLOYMENT TRENDS IN JAPANESE MANUFACTURING BY INDUSTRY, SIZE OF COMPANY, AND AGE RANGE

The following are representative graphs of my analysis. In total I analyzed age-specific employment trends for men in companies with at least 1,000 total employees in all manufacturing industries combined, and then in each of nine specific industry categories (electrical machinery & instruments, precision machinery & instruments, general machinery, chemicals, plastics, nonferrous metals, metal products, transport machinery, and steel). I did the same analysis for all manufacturing industries combined for each of the following size categories of companies (total number of firm employees): 300–999, 100–299, 30–99, and 5–29.

I am grateful to the Ministry of Health, Labor and Welfare (MHLW) for providing raw data on numbers of employees and numbers of departures by industry, gender, 5-year age groups, and year.

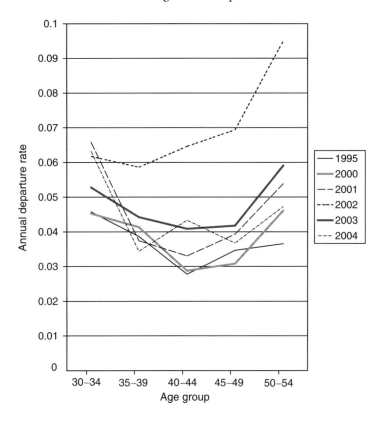

Figure 7A4.1. Age-specific mid-career departure rates for men in all manufacturing firms with ≥1,000 employees

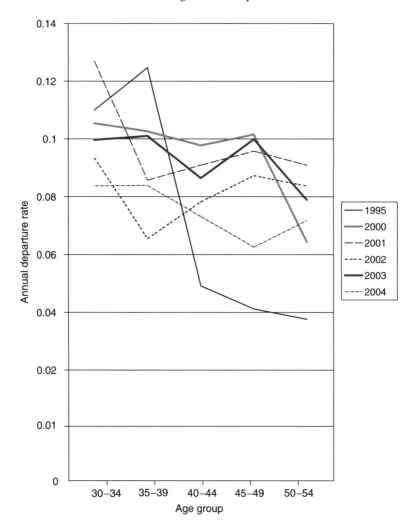

Figure 7A4.2. Age-specific mid career departure rates for men in all manufacturing firms with 30–99 employees

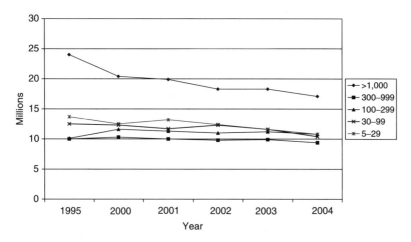

Figure 7A4.3. Total male employees in manufacturing industries by firm size (no. of employees)

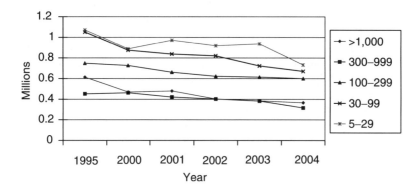

Figure 7A4.4. Total female employees in manufacturing industries by firm size (no. of employees)

NOTES

1. At times I use the terms *new* and *small* interchangeably, because it is not clear whether newness or smallness is most closely tied to innovation potential. Acs and Audretsch based their analyses on size. My analysis of patents in Chapter 1 is based on age, mainly because I wanted to include new firms that had grown to considerable size along with new firms that are still small. Also where to divide small from large or new from old varies. Acs and Audretsch use 500 employees as their dividing line, consistent with the US Small Business Administration's definition of 'small businesses'. Japan's official definition of a small- or medium-size enterprise is a firm with no more than 300 employees.

   In the Chapter 1 patent analysis and the analysis later in this chapter of the origin of new drugs, I make 30 years of age the dividing line between old and new firms. This is partly because I feel that the organization and culture of a 30-year-old firm probably has more similarities with a 10- than a 60-year-old firm, at least in pharmaceuticals. Also, I wanted to be able to include any relatively new Japanese firm, there being so few in the entire 0–30 age range that are patenting in high technology areas (at least in the areas that I sampled) or discovering new drugs.

   One of the few datasets with innovation data for very small firms has been compiled by Willoughby. It covers all manufacturing or R&D-oriented bio-science firms (including those focusing primarily on medical devices, bioremediation, bioprocessing, etc., as well as pharmaceuticals) in Utah and New York (see Table 7A1.1 in Appendix 1). As expected, these data show that patents and other IP increase with firm size, although as firms employ more people, the number of patents and other IP per employee declines. However, for firms of approximately the same size, newer firms hold more patents than older firms, and this trend remains even on a per-employee basis. (There is a slight trend for trade secrets to increase with age, even on a per-employee basis.) Thus these data are some indication that, when measured by patents, younger firms may be more innovative than older firms, and thus age rather than size probably is a more crucial determinant of innovativeness.

2. See generally Christensen (1993), Cohen and Klepper (1991), Hyde (2003), Rothwell and Dodgson (1994), Scherer (1991), Wilson (1985*a*), as well as earlier writings of Richard Nelson and Michael Tushman.

3. Branchflower and Oswald (1992, 1998) and Branchflower, Oswald, and Stutzer (2001). These studies are based on UK and US survey data. In interviews with entrepreneurs Wilson (1985*a*) found that 'freedom of action' and 'having your hobby be your job' were the main motivations and rewards of forming new businesses (p. 212). Friedman (1988) and Whittaker (1997) make clear that the desire to be one's own boss is probably equivalently strong in Japan. My interviews with the CEOs of Japanese ventures indicate similar motivations are strong.

4. Two examples of such constraints are offered later in this chapter. One is IBM's failure to develop small disk drives because its main customers did not need computers with small drives. The other is the development by large diversified vertically integrated Japanese electronics companies *for the Japanese market* of IT systems, architectures, and customer support systems—which later proved inappropriate for overseas markets.

   Another example is a small Japanese company developing and distributing software for oil and gas exploration. Its main clients are Japanese energy companies. This company has declined alliance overtures from overseas programmers that might allow it to offer substantially improved software (e.g. to improve offshore exploration capabilities) and expand its client base beyond Japan. Its main clients are satisfied with the service it currently provides and they have their own in-house programmers that can adapt the small company's products to fit their needs.

5. See Karen Pallarito, *Best Places to Work 2006: Industry*, The Scientist (April 2006 online). This latest annual survey based on responses from readers in the biomedical industry shows that all the top ten ranked companies are biotechs formed no earlier than 1976, except for Monsanto (ranked 10th). Moreover, 6 of the top 10 had fewer than 1,000 employees, while the top 2 (Tec Laboratories of Albany, OR and Transform Pharmaceuticals of Lexington, MA) had fewer than 100. The highest ranked pharmaceutical company as well as the top ranked non-US company was Astra Zeneca (twelfth).

   Chesbrough (1999) emphasizes the ability of venture companies to give key employees generous financial incentive packages. He quotes from Block and Ornati (1987) the following remarks of a venture capital manager, 'The only reason for our existence is the inability of corporations to provide the financial incentives which can be achieved in an independent start-up.' This comment implies that the main motivator for entrepreneurship is financial. It may also imply that the motivating factor that is most important for venture success is the incentive package for the venture's top officers, who may be seconded from a lead investor VC fund. However, the studies mentioned in note 3 above cast doubt on the assumption that entrepreneurship in either the USA or Japan is motivated mainly by quest for financial gain.

   Of course, the assumption that success of ventures hinges primarily on the compensation packages for top officers raises the question about compensation for all employees, which in turn raises questions about the distribution of stock options. It is clear that abuses of stock options have occurred, and managers have taken advantage of employees. (See Hyde (2003: 194) for an account of how Steve Jobs used his control of Pixar to deprive key engineers of their stock options once the success of the company became evident.) However, Hyde (2003), Barley and Kunda (2004), and my own conversations with persons who have worked in ventures suggest that, even though compensation packages may differ widely between the CEO and 'average' employees,

employees of ventures are sensitive to issues of fairness and the possibility of exploitation. If a company has the reputation of treating its employees in an underhanded manner, the repercussions in terms of defections and diminished performance can be severe.

6. e.g. see the account in Chapter 6 of the race to develop gene sequencers.

7. See generally Acs and Audretsch (1991), Branchflower and Oswald (1998), Chandler (1990, 2001), Chesbrough (1999), Merges (1995), Scherer (1991), and Shane (2004).

8. Although sometimes, as in Chapter 1, the only measure available for comparative analysis is number of issued patents.

   An alternative approach has been to examine the proportion of total sales accounted for by products newly marketed (or substantially improved) within the past five years. Using this approach for a sample of fifty-one German metal working firms in 1977–9, FitzRoy and Kraft (1991) found that this ratio tended to be higher for smaller firms, indicating that smaller firms have newer product lines.

9. These consist primarily of (*a*) data on 635 US inventions between 1970 and 1979 compiled by Gellman Research Associates (on contract to NSF) from 14 trade journals (but primarily the magazine *Industrial Research and Development* which annually announces awards for the 100 most innovative newly marketed inventions), (*b*) data compiled by Scherer from *Industrial Research and Development*'s annual top 100 inventions probably from 1963 until the late 1980s, and (*c*) data on 8,074 new inventions introduced in the US in 1982 compiled by the Futures Group (on contract to the US Small Business Administration) from over 100 technology, engineering, and trade journals covering nearly all the manufacturing industries listed under the 247 four-digit standard industry classifications (SICs) (Scherer 1991; Acs and Audretsch 1988, 1991).

10. The Science Policy Research Unit (SPRU) of the University of Sussex compiled a database of 4,400 newly marketed inventions by UK companies between 1945 and 1983 on the basis of queries to industry experts asking them to identify significant technical UK innovations (Rothwell and Dodgson 1994; Pavitt, Robson, and Townsend 1987).

11. The share of small firms rose from 24.3 to 38.4%. That of mid-size firms fell from 33.6 to 19.2% (Rothwell and Dodgson 1994).

12. According to the UK SPRU data, in 1975 firms with 100–499 employees accounted for 2% of total manufacturing R&D expenditure in the UK but produced 21% of total inventions for an R&D efficiency ratio of 10. Firms that employed more than 10,000 persons accounted for 80% of national R&D but accounted for 43% of inventions, for an R&D efficiency ratio of 0.5, one-twentieth that of the small firms.

   The US Gellman-NSF data show a similar declining trend of inventions per unit of R&D expenditure in relation to firm size. Firms with less than US$100 million in sales in 1985 had on average 0.09 new products per

US$1 million of R&D while firms with over US$4 billion in annual sales had less than 0.002 new products per US$1 million in R&D, approximately one-fiftieth the rate of firms in the smallest sales category (Rothwell and Dodgson 1994). As for innovations per employee, the US Small Business Administration's dataset shows that the ratio is about 2.4 times greater for small firms than for large (Acs and Audretsch 1988, n. 6).

13. Large firms are defined as those with at least 500 employees, small firms have less than 500 employees.

14. Electronic computing equipment (large:small firm innovation ratio 0.7) was first with 395 of the 8,074 inventions in 1982, while process control instruments (innovation ratio 0.7) was second with 165 inventions (Acs and Audretsch 1988).

15. Acs and Audretsch (1987).

16. Rothwell and Dodgson (1994).

17. (Scherer 1991). A somewhat different line of analysis suggests that while R&D tends to increase in rough proportion to firm size, innovations per unit of R&D expenditure tends to decrease (Scherer 1991; Acs and Audretsch 1988, 1991).

18. i.e. innovation tends to be correlated with proportion of the industry workforce employed in large companies.

19. Acs and Audretsch (1988).

20. Rothwell and Dodgson (1994).

21. Scherer (1991).

22. Of some interest, pharmaceuticals was not one of these industries detected by these early surveys. This is another area where new companies have emerged to overtake large companies in innovation leadership.

23. Echoing one of the main themes of this book, Rothwell and Dodgson (1994) also note the importance of partnerships for small companies to realize their innovation potential.

    One can only speculate about the insights that would be gained if these datasets were also analyzed according to age of the innovating companies.

24. This total excludes ten drugs approved during this period that were either imaging agents, dialysis solutions, chemical warfare skin protection agents, or contraceptive rings or patches.

25. Fortunately, in the case of 144 new molecular entities (NMEs, i.e. small molecule drugs) the FDA approval process has a built in filter to designate the most important patents. Under US law, companies submitting NMEs to FDA for marketing approval must designate active US patents covering the NMEs. These patents are listed in the FDA Administrative Correspondence (AC) related to the NME. If I had any doubt about whether the patents listed in the AC were the key patents, I cross-checked with O'Neil (Merck Index) (2006). When the AC listed multiple patents for an NME, I gave greatest weight to those covering the therapeutically active chemical compound upon which the drug is based. Among these, I gave greatest weight to the earlier patents

so long as they indicated proof of concept in a mammalian system. If the only patents listed in the AC or Merck Index pertained to manufacturing methods, methods of clinical use, delivery systems, etc., I weighted these according to their importance with respect to discovery of a workable therapy. Three NMEs did not have any patent record, but I determined their likely origins with the help of a colleague in the orphan drug industry and the Merck Index. As for the 25 new therapeutic biologics, Rader (2006) provides information on their development histories and key patents. More information about methodology is in Kneller (2005*a* & *b*) and a forthcoming article.

26. I was able to do so for over 90% of all the listed inventors. The most frequent method was, for each inventor, to obtain at least one scientific article co-authored by the inventor (preferably on the same topic as the invention), written close to the time of the initial patent application, and that identified the inventor's affiliation. My main sources to find such articles were Google Scholar and NIH's PubMed database. Very often, one or two key articles would be co-authored by many of the inventors on the same patent, thus reducing the number of articles I had to obtain. In the case of the 18 NMEs that had Japanese inventors, I also used various online Japanese sources to determine the inventors' employment.

27. For the purpose of this analysis, I use the terms biotechnology company (biotech) and bioventure synonymously to mean companies either (*a*) with fewer than 500 employees at the time the drug was discovered or (*b*) formed after 1975 that concentrate on science-based discovery of drugs and other bio-medical products/processes. Thus this term includes (*a*) small pharmaceutical companies and (*b*) relatively new, independent, companies that, at least until later in their development, did not devote substantial resources to later stage clinical trials and, in particular, to marketing. I did this with the express purpose of including some companies that have grown quite large, such as Genentech (founded 1976) and Amgen (founded 1980), in the set of companies that I contrast with traditional pharmaceutical companies. My reasons include an intuitive sense that age is a more important differentiator of innovative potential than size (see also note 1), and there are residual influences related to founding within the past thirty years that result in a different innovation environment than exists in a large company that has been in existence since before World War II. Moreover, if I excluded relatively young companies that have become large from the set of greatest interest, I am deleting precisely the types of companies that I postulate this group is inherently likely to give rise to. In other words, I would be penalizing this group for the very success I want to measure. In addition, the drugs that I classify as biotech origin that were approved by FDA between 1998 and 2003 were drugs that have been in development by biotechs for approximately ten years. This makes their origin the 1980s or early 1990s. Even the oldest and largest biotechs would be considered young at the time they began discovery/development of these drugs.

   With respect to US-origin drugs, the inclusion of small pharmaceutical companies within my definition of *biotechs* is immaterial, since all drugs classified as *biotech origin* were from companies formed after 1975. However, a few of the European drugs classified as *biotech origin* were from small pharmaceutical companies formed before 1975.

28. My allocation among the various categories was usually according to tenths (one decimal place). In the case of drugs with more than one key patent, I combined the allocations of the individual key patents according to the weighting system described in note 25 to calculate a weighted overall allocation for each drug, where the total value assigned to each drug was "1." This "1" was allocated separately according to type and location (country) of inventing workplace.

29. See Appendix 2, Table 7A2.1. In the case of US university-origin drugs, the number of transfers from universities to biotechs is ninefold greater than the number of transfers to established pharmaceutical companies.

30. Pharmaceutical companies obtained FDA marketing approval for 121 drugs, and biotechs obtained approval for the remaining 48. Of these, at least one quarter (13) were transferred from at least one biotech to another prior to approval.

31. Appendix 2, Table 7A2.2. Only 11.4 drugs attributable to university discoveries were passed directly to pharmaceutical companies. The proportion of such drugs varies by country. The UK is highest, with such drugs constituting 15 percent of its total. The USA and Japan are among the lowest, although for different reasons, as discussed in the text. See also note 57.

32. 63 (44%) of the 144 NMEs approved between 1998 and 2003. Appendix 2, Table 7A2.1.

33. 37% compared with 17%. Appendix 2, Table 7A2.2.

34. The USA accounted for 32 of 63 priority NMEs (51%), but only 28% of nonpriority NMEs. As a benchmark, in 2003 it accounted for 44% of world pharmaceutical purchases. Appendix 2, Table 7A2.2.

35. NTBs usually replicate or modify naturally occurring proteins such as hormones, or are antibodies designed to affect naturally occurring enzyme systems or neutralize disease causing substances. Most NTBs are much larger in size than typical NMEs. The US FDA approved twenty-five NTBs between 1998 and 2003 (Appendix 2).

36. McKelvey (1996) and Zucker and Darby (1997).

37. See Crabtree (2006), citing studies by the Tufts University Center for the Study of Drug Development released in 2005 and 2006, respectively, on the costs of developing NMEs and NTBs.

38. Adalimumab (Humira®, approved 2002 for rheumatoid arthritis) was co-discovered by researchers at Cambridge Antibody Technology (a UK biotech) and Knoll Pharmaceutical's Massachusetts (USA) laboratory using antibody technology from the Australian biotech, Peptech. At the time of Humira's discovery in the mid-1990s, Knoll was a subsidiary of BASF, which sold the Knoll research facilities to Abbott in 2001 along with rights to Humira.

39. Appendix 2, Table 7A2.2.

40. Data published by the Pharmaceutical Research and Manufacturers of America (PhRMA) under PhRMA's *New Medicines in Development* series shows that of approximately 346 new drugs in clinical trials for cancer in 1999, 62% were sponsored by biotechs and only 20% by pharmaceutical companies, with drugs sponsored by NIH, or other government or nonprofit organizations accounting for the remainder. Of 123 drugs in clinical trials for cardiovascular diseases in 2001, 55% were sponsored by biotechs and 45% by pharmaceutical companies. Of 251 drugs in clinical trials for infectious diseases in 2002, 77% were sponsored by biotechs and 22% by pharmaceutical companies. However, not all of these drugs are undergoing trials for first time approval, some are in trials for additional medical indications. See Kneller (2003) for a more complete discussion of these data.

41. As for non-US origin NMEs, biotechs outside the USA (and non-US universities licensing to biotechs), discovered 19 percent of priority NMEs but only 6.7 percent of non priority NMEs, a nearly three-fold difference. Appendix 2, Table 2.

42. Rescula® (unoprostone isopropyl ophthalmic solution), approved in 2000 for glaucoma, was discovered by R-Tech Ueno, a small independent Osaka company founded in 1989, which licensed the drug to Ciba Vision. Sacrosidase (Sucraid® approved 1998 for congenital sucrase-isomaltase deficiency) was discovered by clinical researchers in a Munich hospital in the 1980s. There is no history of any patents. Orphan Medical, Inc. (Minnesota) sponsored the drug for FDA approval.

43. Perhaps this suggests the large companies in these countries are *crowding out* ventures by preempting human resources, capital, distribution networks, etc.

44. However, in the case of NTBs (which were pioneered by biotechs rather than pharmaceutical companies) from the beginning it has been common for the biotechs to complete development and even undertake marketing.

45. This was cisplatin for cancer. See the discussion in Cockburn and Henderson (1997) of the basic discoveries underlying twenty-one of the most therapeutically important drugs initially marketed during this period.

46. Since the latest date of synthesis of any of the active compounds was 1988 (sumatriptan for migraine) with the second latest date being 1985 (erythropoietin for anemia), it would not be expected than many of these drugs would have been developed by biotechs. Among the twenty-one drugs, only erythropoietin appears to have been developed by a biotech, Amgen.

47. See, e.g. Florida and Kenney (1988), especially n. 29; and Wilson (1985*a*: Chapter 13). These writers raised this issue primarily in the context of IT and other nonbiomedical industries, and I return to this argument in Part III of this chapter.

48. Rathmann (1991).

49. McKelvey (1996). As another example, see Ganguli's history of Alexion (2006), developer of eculizumab to treat paroxysmal nocturnal hemoglobinuria and other diseases caused by poor regulation of the complement immune system.

50. A well-known university researcher may hesitate abandoning an academic career to work for a small company that may not exist in five years, but would be more likely to do so if he or she could head a large research organization and have many years of committed research support. For example, in 2002, Novartis recruited Dr Mark Fishman, to head its new Institute for Biomedical Research in Cambridge, MA, to pursue drug discovery research. Dr Fishman was previously chief of cardiology and director of cardiovascular research at Massachusetts General Hospital and professor of medicine at Harvard Medical School. Academics who retain biomedical university positions consult frequently for large pharmaceutical companies, just as they do for biotechs.

51. In some nonbiomedical fields, there is evidence of real annoyance and harm to established companies caused by employees leaving for new ventures—as discussed in Part III.

52. Shane (2004).

53. Probably more bothersome than royalty obligations is universities' frequent insistence that they have the right to terminate an exclusive license if the pharmaceutical company is not developing a candidate drug and to relicense it to another company. The director of the IP department of a major Japanese pharmaceutical company remarked, 'Sometimes after investing millions of dollars investigating a candidate drug, rational business reasons lead us to halt development. It is not fair if the university can then terminate our rights and re-license to one of our competitors.' However, universities also have a legitimate interest in preventing licensees, that turn out not to be serious about developing candidate drugs, from using their license rights to prevent other companies from developing the drugs. Some pharmaceutical officials have said that if the university license gives them three to five years of exclusive control over the compound, this is sufficient, and beyond this time it would be reasonable for a university to request return of IP rights if the pharmaceutical licensee was not developing it.

54. Confirming this phenomenon, licensing officials at the US NIH report that most licenses for NIH-discovered candidate drugs or drug targets are now licensed to biotechs. Big pharma is simply not interested in early stage licenses they report. In contrast, approximately ten years ago when I was working in technology development at NIH, a substantial proportion of such licenses did go to big pharma.

    In addition senior managers in one of the largest medical device companies told me in 2006 that in the field of devices, their company also has come to rely mainly on ventures for new technologies, many of these ventures being university startups.

55. This entrepreneurialism has been characterized as greed resulting in excessive payments to universities akin to nonproductive rents, especially when universities occasionally reap high royalty windfalls from licenses for compounds that later turn out to be the basis for successful drugs (see e.g., Leaf 2005).

In Germany at least up to 2002 and in Japan even to 2006 (and perhaps also in France and Switzerland), universities are less demanding than US and Canadian universities about ownership of IP rights and obligations to pay royalties. Thus, pharmaceutical companies could have chosen to fund substantially more research in Japanese and German universities and, following minimal negotiations with university authorities, they could have obtained complete control over resulting lead compounds and yet not have had to pay royalties on the scale they would have had to pay to US universities. Nevertheless, despite the more demanding and entrepreneurial nature of US, Canadian, and perhaps also UK universities, pharmaceutical companies probably direct more of their research monies to them than to docile Japanese and Continental European universities (based on conversations with pharmaceutical and university officials).

Jensen and Thursby (2001) and Thursby and Thursby (2003) indicate that financial incentives for university researchers and TLOs are important for technology transfer from universities to industry to be effective.

56. Appendix 2, Table 7A2.3. See Casper's (2000) description of the tendency of German biotechs to focus on research tools and other platform technologies rather than drug discovery.

57. Among the 11.4 drugs discovered in universities and licensed directly to pharmaceutical companies, half are non-priority NMEs, while among the 26.3 university-discovered drugs that are licensed to biotechs, three-quarters are NTBs or priority NMEs (Appendix 2, Table 7A2.1). With only a few exceptions, all the NTBs and priority NMEs originating in universities were licensed to biotechs. The principal exceptions are the NME, sirolimus (Rapamune®), approved in 1999 to prevent transplantation rejection, which was invented by Roy Calne of Cambridge University and licensed to Wyeth; the NME temozolimide (Temodal®), approved as an anticancer drug in 1999, which was co-invented by researchers in Nottingham and Aston Universities, Charing Cross Hospital, and the pharmaceutical company May and Baker (all in the UK), and later licensed to Schering Plough; the inhaled gas NME, nitric oxide (INO max®) approved in 1999 to treat pulmonary hypertension in new borns, invented by Swedish and US academic researchers and transferred first to a large manufacturer of anesthesia equipment, Ohmeda, and then to a Swedish industrial gas company, AGA AB; and finally the NTB, Rebif®, the most recently FDA-approved version of recombinant interferon beta-1a for multiple sclerosis, which was based upon discoveries at the Weizmann Institute (some made in collaboration with a researcher at the Pasteur Institute), which were licensed from the Weizmann Institute to the Israeli subsidiary of the Swiss pharmaceutical company, Serono.

58. Goodman and Myers (2005). Note 313 summarizes this study's conclusions.

59. Murray (2002) has documented the close links between bioventures and universities that benefit from access to patients, as well as the clinical research of some of the academic advisers and founders of the ventures.

60. Confirmed, e.g. by discussions in July 2004 with the former director of corporate licensing for GE.

61. These percentages are for 2001 when total US university S&E R&D expenditures were US$32.7 billion. The federal government's share of this support ranged from 57% for engineering and 58% for life sciences to 70% for the physical sciences. Most university life science research is funded by NIH which, with an annual budget of over US$30 billion, has by far the largest nonmilitary R&D budget of any US government agency. However, other US agencies, notably NSF, the Department of Energy, and the Department of Agriculture also fund significant university life science R&D (National Science Board, Science and Engineering Indicators 2004, A5-8, 9).

62. While in other fields, economic theory and public policy generally holds that government R&D support should be confined to basic research, leaving private companies to fund R&D with commercial applications. Otherwise government funding for commercially relevant R&D would tend to favor certain companies or industries and distort market forces. On the other hand, because the commercial payoffs from basic research are usually uncertain and far in the future, private companies are unlikely to fund basic research. Moreover, basic research is akin to a public good in that the findings and newly trained scientists can benefit many companies. Therefore, according to economic and public policy theory, government support of basic research is appropriate and necessary, but this is not so in the case of R&D with direct commercial applications, barring some other public purpose such as promotion of health, defense, or alternative energy sources.

63. Of 31,570 patents identified as issued to US universities between 1980 and 2001, 12,648 (40%) were classified by the US PTO into one of the following categories: pharmaceuticals, biotechnology, agriculture, medical equipment, and medical electronics. Pharmaceutical and biotechnology patents accounted for 28% of the total. Even considering only the 3,094 patents issued to universities in 2001 (the most recent year for this dataset), life science inventions accounted for 42%, with pharmaceuticals and biotechnology accounting for percent 34%. (Data courtesy of Professor Diana Hicks, chair, School of Public Policy, Georgia Institute of Technology.) In contrast, 59% of university R&D funding was life science related. (National Science Board, Science and Engineering Indicators 2004, A5-11.)

64. The vast majority of university patents (90%) are licensed exclusively, either in their entirety or (more commonly) for a specific application or geographic region (Shane 2001 & 2004 and follow-up conversation with Shane in 2004), and startups or other SMEs are the recipients of most university exclusive licenses (76%, 2004 AUTM Survey Summary, Table US-14). Thus, even assuming that approximately half of US university startups are life science focused (as suggested in Shane, 2004), the fact that about 60% of US university patents are non-life-science related(previous note) indicates that a large proportion of these are probably being licensed exclusively to startups or other SMEs. If this

were not the case, universities would have to be licensing the vast majority of their life science patents but only a small proportion of their non-life-science patents. This would not make sense from a technology management perspective, and there is little evidence this is occurring.

65. Pharmaceuticals have traditionally been a vertically integrated industry where the time, expense, and risk associated with development are high. Returns to scale in terms of R&D are presumed to be high, at least this is one of the oft-cited rationales for the recent series of pharmaceutical mergers. It requires strong complementary assets, including facilities for high throughput screening, personnel and outside networks to conduct clinical trials, networks with regulatory agencies, and large sales forces. Also, some therapeutic fields have become quite crowded.

66. An opposite reason for small size to handicap ventures less in biomedicine is that many nonbiomedical discoveries have short product development cycles and commercial life spans. Thus it would not make sense to invest the effort to build a company to develop such discoveries, when they could be licensed to companies already in existence with resources to handle development, manufacturing, and distribution. However, licenses could be made to existing SMEs that already have done similar development work. Also, the sorts of pioneering inventions with which this book is mainly concerned probably are not inventions whose market will be short lived.

67. i.e. in-house know-how that is not to be shared with outsiders.

68. Cohen, Nelson, and Walsh (2000).

69. See Merges (1995), Hall and Ziedonis (2003), Cohen, Nelson, and Walsh (2000) and the following text. In the case of software companies, copyright can also be a means of protection. US copyright law protects computer source and object codes and US courts have generally adopted a 'substantial similarity' test for infringement, even in the absence of actual duplication of source code (Gorman and Ginsburg 1993: 719).

70. Cohen, Nelson, and Walsh (2000). Willoughby (2005*b*) found that bioscience firms that receive VC financing have more patents per employee than do firms that do not receive such financing.

71. Merges (1995).

72. Cohen, Nelson, and Walsh (2000) found such responses across a wide range of industries.

73. The cost of obtaining a US patent is roughly US$10,000. The cost of obtaining protection in the US, Japan, and major European countries is approximately US$100,000. The legal costs of infringement litigation can be much more, and an adverse final judgment can be devastating for a small company, although it can also bring a great windfall. On the other hand, there were probably over 50,000 investments totaling over US$22 billion by American angels in 2004. A median range angel investment has been estimated to be on the order of US$300,000. VC investments, although fewer (2,876 in 2004, most of which were mid or late stage) were larger (median per financing round about

US$7 million). Total VC investment in the USA was about US$21 billion in 2004 (Angel Capital Association, Aug. 2005, using data from Dow Jones VentureOne/Ernst & Young). Thus, while expensive, the cost of obtaining patents is probably not prohibitive in relation to the funding that promising ventures can usually expect to raise from outside sources.

The data in Appendix Table 7A1.1 show that even small, young New York and Utah ventures in biotechnology, medical devices, and other life science fields have on average about seven issued patents.

74. The survey findings of Cohen, Nelson, and Walsh (2000) are based on responses to a 1994 questionnaire from R&D managers in over 1,100 corporate laboratories in a cross section of US industries, asking them to report the percentage of their product and process innovations for which each of a variety of mechanisms had been effective in protecting the firm's competitive advantage during the prior 3 years. The specific mechanisms surveyed were: secrecy, patents, other legal mechanisms, lead time, complementary sales and services, and complementary manufacturing capabilities. The survey included firms with as few as twenty-five employees and with annual sales as low as US$1 million. However, firms without either US$5 million in annual sales or a business unit of at least twenty people were excluded. Also, Fortune 500 firms were over sampled. Since the initial sample frame was R&D laboratories or business units, some laboratories included in the final analysis are within the same large firm. Thus the overall results generally reflect the responses of large companies.

Within individual industries, the number of total respondents is not large. For example, with respect to product innovations, there were only forty-nine responses for pharmaceuticals, sixty-seven for medical equipment, twenty-two for computers, eighteen for semiconductors and related equipment, and thirty-five for precision instruments. Thus in specific industries where patents are important to innovative small or new firms, their responses would be obscured by responses from large firms.

The authors note when responses from small companies differ from the overall responses. However, if among small firms in a particular industry relatively few are engaged in the discovery or development of new products, but those that are include innovation leaders for the entire industry, responses by these few firms to the effect that patents are important will be obscured by responses of the majority of small firms across all industries that are not pursuing early stage innovation and do not rely on patents.

This likelihood is increased because the survey sampled firms engaged in manufacturing as well as those concentrating primarily on R&D. Manufacturing-related innovations (e.g. manufacturing processes, products that are intermediate in the manufacturing process, and refinements or combinations of existing products) probably are less reliant on patents than are R&D related innovations, especially R&D innovations in new fields of technology. Thus the proportion of a company's innovations that rely on patents to ensure appropriability may diminish as more of a company's value-added shifts toward manufacturing as opposed to R&D (see Hall and Ziedonis 2003). Therefore, industries where manufacturing accounts for a high proportion of the value added will report that patents are effective in protecting a lower proportion of their innovations, compared to industries where a relatively high proportion of firms are engaged mainly in R&D. Drug development is probably in this latter category, because the value added by many bioventures consists mainly of R&D. However, in industries such as precision instruments, where a large majority of firms (large and small) are probably engaged in manufacturing, responses applicable to manufacturing-related innovations probably obscure responses applicable to R&D-related innovations. The authors acknowledge that by asking for the percentages of total product or process innovations that were protected by various mechanisms, a firm may indicate that patents were effective in protecting a relatively low percentage of innovations, even though patents were effective in protecting its most import innovations.

75. The following table is based on data from Kortum and Lerner (1999):

District court decisions	Percent upheld by	
	Federal appeals courts 1953–1978 (%)	CAFC 1982–1990 (%)
Holding patent to be valid and infringed	62	90
Holding patent to be invalid or not infringed	88	72

76. The patents at issue covered Polaroid's instant cameras and related technologies.

77. For example, Motorola, ATT, and IBM began to inventory their patents and collect royalties more deliberately (Hall and Ziedonis 2003).

78. Patent pools are formed by organizations that each have rights over various technologies that need to be combined to develop a particular line of products. For example, the MPEG system for digital compression and transmission of video data involves technologies covered by patents held by several companies and universities. Typically, members of a patent pool have preferential rights compared to outsiders to use the pooled patents.

79. The four preceding paragraphs are based largely on Hall and Ziedonis (2003), although I am also indebted to Cohen, Nelson, and Walsh (2000), Barnett (2003), Merges (1995), and Fransman (1995). Except for the last two sentences, which are my own, the sections in *italics* are near verbatim quotes from Hall and Ziedonis's work. Any misinterpretation of their work is, however, solely my responsibility.

   As for the ability of the Japanese patent system to protect the interests of ventures, see Chapter 4, notes 52–61 and accompanying text.

80. Hall and Ziedonis (2003) describe the growth of R&D intensive, design focused, and mostly recently formed semiconductor firms up to about 1990.

Hyde (2003) discusses in more general terms the continuing birth of Silicon Valley IT firms up to the bursting of the IT bubble in 2001. He notes that the newer firms are often fabless operations (i.e. they focus mainly on design and do not do commercial manufacturing), or else their manufacturing is done overseas (i.e. a large proportion of the newer firms are R&D intensive design firms).

A partial list of the firms included in Hall and Ziedonis's survey is available in an earlier version of their paper at http://jonescenter.wharton. upenn.edu/papers/2000.htm. This list includes forty semiconductor firms whose first issued US patent was filed after 1980. The following table shows 1994 and 2004 sales data for the subset of these firms whose 1994 revenues were over US$300 million:

Name of firm	Year of first successful patent application	1994 sales ($M)	Approx. 2004 revenue ($M)* or fate of firm
LSI Logic	1982	902	1,700
VLSI Technology	1982	587	bought by Philips in 1997 for $\sim$ $1 B
Integrated Device Technology	1982	422	390, to merge in late 2005 with Integrated Circuit Systems
Cypress Semiconductor	1984	406	950
Atmel	1987	375	1,650
Xilinx	1984	355	1,600

* In some cases, revenue is for fiscal rather than calendar year. Also if the 2005 fiscal year ends before July 2005, I report FY 2005 sales. SEC reports and online investor information are the sources for sales data.

However, I have follow-up information on all forty companies. Among these, seventeen were still independent firms in 2005 and had 2004 revenues at least double their 1994 sales. (These seventeen do not include LSI Logic which has the highest revenue among all these firms.) Only six of the forty had ceased operations or had 2004 sales less than 80% of their 1994 sales. Twelve of the remaining companies had been bought by other companies—a few when in a state of distress, but probably a majority when revenues were at least equal to 1994 revenues. In summary, the forty new companies Hall and Ziedonis surveyed in the 1990s do not appear to have done poorly over the subsequent ten years.

81. Other factors include greater availability of venture capital made possible by a 1979 ruling by the US Department of Labor clearing the way for pension fund managers to invest in VC funds for the purposes of *portfolio diversification*, reductions in capital gains tax rates, the culture and infrastructure of regions such as Silicon Valley, which provided a fertile environment for the formation of S&T companies, and profit squeezes that forced manufacturing and financial corporations to seek high returns from their financial resources.

In addition, the restructuring of many large corporations in response to profit squeezes and conglomerate type mergers may have caused skilled employees to leave for ventures.

The 1980 Bayh-Dole amendments to US tax law probably facilitated the formation of university startups, because it encouraged universities to apply for patents on their discoveries and facilitated the transfer of exclusive IP rights to companies, particularly small businesses.

Most of the case studies in Chapter 4 suggest that patent rights have also been important for the formation of Japanese ventures.

82. See, e.g. Shane (2004) and Cohen, Nelson, and Walsh (2000).

83. Gans, Hsu, and Stern (2000). See also Arora and Merges (2004).

84. Examples include Nanosys in the field of fabrication technologies and Nanophase Technologies which is developing powders for sunscreens and other uses, NVE, which is developing a system to store and transmit information based on electron spin rather than electron charge and several new companies working in nanoscale drug delivery systems.

85. Fransman (1995: 173, and 203–54) provides two examples: Between 1971 and 1977 NEC, which had been following Intel closely, produced microprocessors that were modeled on Intel's and were Intel-compatible. However, shortly after this period, NEC switched to its own proprietary microprocessor design, which was incompatible with Intel's, largely out of concern that the Intel-compatible microprocessors would infringe Intel's intellectual property. Despite the switch, Intel did sue NEC for infringing its microprocessor code. Other Japanese electronic companies also developed their own proprietary microprocessors and microcontrollers. This fragmented the domestic Japanese market not only for hardware but also for software, and made these microprocessors unattractive overseas where there were few incentives to develop software for the various proprietary Japanese microprocessors and their operating systems. NEC's microprocessor became neither the industry standard that Intel's became for PC applications, nor could it benefit from the standards that Intel set.

The second example is the development of high efficiency (low signal attenuation) fiber optic cable. The breakthrough method to limit signal transmission loss was worked out by Corning Glass in the late 1960s. In response, NTT (then a subsidiary of the Ministry of Posts and Telecommunications) launched a cooperative R&D project with Japan's three major cable companies, Sumitomo Electric, Furukawa Electric, and Fujikura, Ltd. to produce a Japanese alternative to Corning's method and invent around Corning's patents. Building on prior competence in optical research and information from AT&T/Bell Laboratories (which was also doing research in fiber optic signal transmission and which could use Corning's patents under a preexisting cross licensing agreement), NTT researchers developed a similar method which they shared with the three Japanese cable companies, which also made important improvements. By the late 1970s, world records in low signal attenuation were being set in NTT's laboratories. In the early 1980s Sumitomo Electric began exporting fiber optic cable to Canada and setting up a production plant in North Carolina. Corning sued, alleging patent infringement. In 1987, the New York District Court ruled that Corning's patents were valid, that Sumitomo's technology was *equivalent* to Corning's, and that Sumitomo had to cease production at its North Carolina plant. Sumitomo entered into a joint venture with AT&T under which AT&T received a majority stake in the North Carolina plant and could sell its cable output.

In the late 1980s with the entry of MCI into the newly deregulated US telecommunications market and MCI's decision to install fiber optic cable connections extensively in the USA, Corning's persistence began to pay off as sales of fiber optic cable finally took off. In 1987, Corning held 40% of the North American market for fiber optic cable, AT&T (using Japanese technology) 43%. Corning's Japanese applications for the same patents that were upheld in New York were not approved by the JPO. The 1987 Japanese market was split largely between Sumitomo (35%), Fujikura (22%), and Furukawa (21%).

Although the 1987 ruling of the New York District Court was a winner-take-all verdict based on inherently slippery judicial concepts (i.e. the *equivalence* of inventions), it nevertheless resulted in a deserved reward for Corning, justifying its risky decision around 1970 to invest heavily in a new technology for which market demand was uncertain. At the same time, the Japanese companies were also rewarded (and their scientific backer and main customer, NTT, vindicated from an industrial policy perspective) for their rapid improvement of the basic method, which pushed Corning to devote more resources to its own improvement efforts, which resulted in this new technology becoming more rapidly available.

In 2005, high technology Japanese companies were involved in several high stakes suits alleging IP infringement by lower cost competitors in Korea and China.

86. The companies summarized in Chapter 4 generally said that patents in combination with trade secrets are important to prevent encroachment. The same probably applies to US nonbiomedical ventures (Cohen, Nelson, and Walsh 2000). See also Appendix 1, Table 7A1.1 based on data from Willoughby.

Approximately one-third of the companies in Willoughby's dataset are working primarily in medical devices in contrast to pharmaceuticals or biotechnology. Preliminary analysis of these data indicates that this phenomenon of mixed trade secret and patent protection is a characteristic of new small medical device firms as well as those focused on pharmaceuticals and biotechnology (see Willoughby 2005*a*, 2005*b*).

87. IBM has been a leader in making large numbers of patents freely available for use, especially by developers of open source software. Other companies such as Nokia, Red Hat, and Sun Microsystems have also decided not to assert some of their patent rights (Cukier 2005, 'An open secret'). For an analysis of

private parties putting patentable discoveries in the public domain, see Merges (2004).

88. One additional possible advantage (suggested by McKelvey 1996: 271) is that meeting pharmaceutical regulatory requirements requires drug companies to devote a great deal of their resources to this downstream activity, leading perhaps, to pharmaceutical companies being more locked in to particular lines of research than large companies in other industries. This leaves new ventures with more niches to exploit with respect to new drugs, than is the case in other industries. However, the extent to which this phenomenon actually exists in pharmaceutical companies is not clear. Clinical trials and procedures for manufacturing and quality control (activities that may have to be tailored to a particular type of drug) do indeed account for a large proportion of pharmaceutical companies' expenses [about US$9.3 billion (44%) of total US pharmaceutical R&D expenses of US$21 billion in 1998]. But marketing, which presumably could adjust easily to different types of medicines (at least for the same disease category), accounts for even more, US$12.3 billion in 1998 (PhRMA, Pharmaceutical Industry Profile 2000: 26 and 2002: 18). My impression is that pharmaceutical companies' networks of CROs, physicians, etc. can adapt relatively quickly to carry out clinical trials for either small molecule drugs or antibodies to treat rheumatic disease, just as one example. These networks may even be able to adapt to drugs to treat different types of diseases. In any case, it would be surprising if lock-in due to downstream development infrastructure is significantly greater for pharmaceutical companies than for large companies in other industries.

89. This hypothesis is supported by Lerner's analysis of the patenting behavior of biotechnology companies (1999). It is also supported by McKelvey's account of the development of recombinant DNA technology and how favorable perceptions of ultimate technical feasibility and the medical/market demand for genetically engineered drugs guided both academic and corporate scientists (many of the latter in Genentech) to pursue R&D in this area.

90. The relative scale of corporate basic research is less in 2005 than it was in the 1960s. Gone are the days when large groups of scientists and engineers in AT&T's Bell Labs, RCA's Sarnoff Center, and Xerox's Palo Alto Research Center pursued largely curiosity driven, often fundamental, research—that nevertheless led to many of the products we enjoy today. As of 2005, IBM still did some fundamental research, mainly at its Thomas J. Watson Research Headquarters in Yorktown Heights, New York, as did General Electric at its Global Research Center near Schenectady, New York.

   As for the central research laboratories of large Japanese electronic companies, some industry observers contend they never have done fundamental exploratory research—with the possible exception of NTT's Basic Research Laboratory and Communications Science Laboratory. In any case, their central research laboratories now depend mainly on contract research from their design and production divisions, and engineers and scientists are being transferred to these divisions. Of course all large high technology corporations do research, but usually applied research with specific commercial potential in mind. Intel, for example, is known as a company whose future depends on fast paced R&D. But Intel's policy is that all research should have a definite business objective in mind.

91. The university research that gave rise to genetic engineering technology culminated around 1973 when Professor Stanley Cohen of Stanford and Professor Herbert Boyer of UCSF jointly developing a series of techniques to cut and paste genes from humans or other organisms and to insert these into bacteria so that the bacteria would produce the proteins (e.g. human insulin or growth hormone) coded for by those genes. Stanford and UCSF jointly applied for a patent on this invention in 1974, which was granted by the US PTO in 1980. Cohen's and Boyer's discovery was preceded by other university research, primarily in the USA and UK. Boyer went on to found Genentech in 1976, thus providing a bridge between university research and industry R&D efforts in this area, most of which were conducted by bioventures. Later some large pharmaceutical companies tried to master this technology, and Lilly notably made good progress (McKelvey 1996: especially Chapters. 4, 5; Zucker and Darby 1997). However, the above analysis of NTBs approved by the FDA indicates that venture companies are still the unquestioned leaders in this field.

92. Murray (2002).

93. For an account of government funded university research that prepared the way for the Internet see Roessner et al. (1997). See also the histories of companies such as Cisco Systems and Sun Microsystems, both of which are based on Stanford technologies.

94. For example, Nanosys has obtained exclusive licenses from MIT, UC Berkeley, Lawrence Berkeley National Laboratory and other universities and GRIs, of patents covering nanomaterials and nanofabrication methods. As of 2005 Nanosys's staff of approximately thirty-five was developing customized applications of these technologies for companies such as Sharp, Intel, DuPont, and Matsushita, as well as US government agencies.

95. These examples and others are from Hyde (2003), Chandler (2001), and Christensen (1993).

96. The historical account up to 2000 is based on Christensen (1993) and Hyde (2003). Seagate, Maxtor, and Western Digital had 2004 revenues of approximately US$8, US$4, and US$4 billion, respectively. Hewlett Packard also manufactures disk drives, as does Adtron, founded in the 1970s and based in Arizona. Non-US manufacturers include Fujitsu, Hitachi, Samsung, and Toshiba. In 2005, Seagate claimed to sell more units than any other company.

97. Hyde (2003: 31). My review of the forty ventures in Hall and Ziedonis's 1990s analysis (see note 80) also indicates that most were formed by persons with experience in other R&D or device manufacturing companies.

98. Saxenian (1999).

99. See Association of University Technology Managers (AUTM) Product Stories at http://www.autm.net/aboutTT/aboutTT_prodStory.cfm

100. Roessner et al. (1997: Chapters 2, 3).

101. Steven Wozniak and Steven Jobs founded Apple in 1977, sensing both the commercial potential for a small computer and the feasibility of building one by combining many new component technologies. Microsoft was founded in 1975 by Bill Gates and Paul Allen to provide control programs for radio controlled model airplanes (Chandler 2001).

   Irwin Jacobs founded Qualcomm in 1986 to develop digital wireless communication. He soon decided to use code division multiple access (CDMA), developed for military communications, as the mechanism to transmit digitalized mobile telephone messages. Qualcomm was neither a university startup, nor a spin-off from a large company, yet it went on to commercialize the mobile wireless applications of CDMA. By 2005 it held more of the essential patents covering third generation wireless communication technology than any other company (Goodman and Myers 2005). However, Dr Jacobs's previous experience as head of another venture, Linkabit, which was involved in scrambled communication over satellite terminals, may have provided him and his early Qualcomm colleagues with the same sort of hands-on industry knowledge that was valuable to the engineers who left IBM and Intel to start their hard disk drive and semiconductor spin-offs.

102. The university startup, Plasmaco, was short of funds from the time of its founding. Its founders managed to show proof of concept at a 1994 industry convention. Plasmaco then entered into a joint development program with Matsushita, which bought Plasmaco two years later. Other Japanese companies also cooperated with University of Illinois scientists. Fujitsu, in particular, also made important improvements to the basic technology. However, by 2004 Fujitsu had sold most of its stake in PDP televisions to Hitachi and then in 2005 Hitachi partnered with Matsushita, essentially leaving Matsushita as the largest manufacturer (Hutchinson 2002; 'Flat-panel TV makers doing deals', *Nikkei Weekly*, Feb. 14, 2005, 14; Takato Daisuke. 'Price Competition Drives Pioneer to a Loss', *International Herald Tribune*, Nov. 1, 2005, B2).

103. Quoted in Wilson (1985a: 191).

104. See examples in Hyde (2003: 27–8, 38–40).

105. California law is unique in that its courts will not enforce no-compete clauses. Whether employers in California can use trade secret law to the same effect by arguing that for former employees to work for a competitor inevitably will disclose the former employer's trade secrets is still an open question. However in most other US jurisdictions, both these weapons are likely available for employers, particularly the use of no-compete clauses, provided they are of *reasonable duration and geographic scope* (Gilson 1999). See Hyde (2003) for an analysis of these laws and their effects on ventures and the people who work in them.

106. Wilson (1985*a*: 215).

   Unable to raise sufficient venture capital, Amdahl turned to Fujitsu for financing. Coincidentally, MITI had just designated Fujitsu and Hitachi to build a computer equivalent to the 370, as the center piece of MITI's New Series Project to maintain Japan's competitiveness in computer technology. Fujitsu invested over US$50 million in Amdahl's company. By 1976 it was marketing a mainframe similar to the 370. Amdahl relinquished control in 1979 when Fujitsu raised its equity stake in the company to 47%. Fujitsu's acquisition of IBM mainframe technology via Amdahl assured the success of the New Series Project. After this Japanese computer manufacturers came to dominate the European mainframe market (Siemens-Nixdorf also came to rely on Fujitsu for technical and product development expertise for mainframes sold under its brand name), and by 1996 they held nearly two-thirds of the world market. That year Fujitsu lead the world in mainframe computer sales, its sales of mainframes exceeding those of IBM's by 15%. In 1999 Siemens-Nixdorf was incorporated into a joint venture with Fujitsu (Chandler 2001, quoting from computer historians Marie Anchordoguy and Kenneth Flamm).

107. This paragraph summarizes and includes quotations from Christensen (1993: 579).

108. See the histories of the formation of Intel, Control Data, Compaq, Amdahl, and the hard disk drive companies in Chandler (2001) and Christensen (1993). In Amdahl's case, the source of frustration was his conviction that IBM could have produced a cheaper mainframe, not that it was failing to develop the technology.

109. Daisy Systems, a pioneer of engineering workstations in the early 1980s, recruited at least 50 engineers from Intel. But by helping companies speed up designs of complex integrated circuits, it probably helped Intel's R&D (Wilson 1985*a*). The defections from Intel to other chipmakers, from Cisco to other communications device firms, and from IBM to hard drive and other computer component makers, helped create products that were compatible with the parents' products and that used the parents' protocols and thus, at least in some cases, increased demand for the parents' products.

110. Most employers in most industrialized countries require, as a condition of employment, that employees assign to them rights in any patentable inventions or copyrightable works make in the course of their employment. Usually the employer applies for patents and maintains ownership of issued patents even if the employee leaves. These protections do not work, however, if an invention is not patentable, if employers use trade secrets rather than patents to protect their technologies, or if employees fail to disclose inventions to their employers. Hyde (2003) describes cases of Silicon Valley employees failing to disclose inventions and other information when the end of their employment was in sight.

111. Hyde (2003: 68).

112. More so than the Route 128 corridor outside Boston.

113. See Saxenian (1994), Gilson (1999), Hyde (2003), and note 105.

114. This is the main thesis of Chandler (2001: 236). It is also echoed in Aoki (1990: 9) and Fransman (1995: 313–16 and concluding chapter). Both Aoki and Fransman note the importance of lifetime employment and internal job rotations to build long-term learning and organizational competence. However, the failure of US consumer electronics companies, such as RCA, Ford Philco, and GTE Sylvania, to withstand Japanese competition was probably due largely to incompetent management that became distracted from the companies' core business.

115. Including marketing, customer support, and feedback.

116. However, despite their success in drug discovery, only a few biotechs have developed integrated R&D competencies. In other words, many do not conduct clinical trials, most do not have sales forces and, except in the cases of NTBs, manufacturing is usually outsourced.

117. This point is made graphically in Fransman (1995: ch. 9, Fig. 9.1).

118. See Chesbrough (2003) and Cukier (2005, especially 'An open secret').

119. It has been suggested that, as companies engage in patent arms races, established companies might use their large patent arsenals to harass or drive out of business new companies they perceive might become competitors. So far evidence of such behavior targeted particularly against new companies seems scant. More often one hears new companies say that patent protection is essential for their early survival. See text above as well as Cukier (2005).

120. The high technology sectors of Silicon Valley are in order of number of employees: software, semiconductors and related equipment, computer and communications hardware, innovation services, biomedical, electronic components, corporate offices, and creative services. Overall Silicon Valley employment in these sectors declined 3.2% over the twelve months ending in June 2004, while average salaries in these sectors rose 8.2% over the twelve months ending in June 2003. The balance between firm creations and deaths and entries and departures from Silicon Valley, was a net gain of 12,600 companies in 2002, the largest net gain since 1992. VC investment in Silicon Valley was US$7.1 billion in 2004, higher than for any year except 1999, 2000, and 2001. In 2004, thirteen publicly traded companies reported revenue increases of at least 20% per year over each of the previous four years, a higher number than any year since 2001. Corporate R&D spending as a percentage of sales was 12% in 2003 compared with 14% in 2001 and 2002, but this was still a higher rate of R&D investment than any year prior to 2000 (when it was also 12%), and four times the national average of 3% (Joint Venture: Silicon Valley Network 2005).

121. High housing costs, problems assimilating immigrants, mediocre performance by school children, falling local government revenues, and (perhaps surprisingly) lower rates of residential broadband access than other high technology US regions (Joint Venture: Silicon Valley Network 2005).

122. Begley (2002) summarizes the apparently widespread frustration among US engineers, who feel they are treated as commodities, in part because of large

numbers of young foreign-born engineers (who presumably can change jobs more easily) and who hold down salary increases for senior employees. Many engineers interviewed for the Begley article said they discourage their children from entering the profession. Whatever the merits of their other claims, there is little evidence that foreign scientists and engineers bid down wages, or that they cause significant unemployment. Foreign-born scientists and engineers tend to have higher incomes than US-born scientists and engineers. See Hyde (2003: 345), Dahms and Trow (2005), and Anderson (1996).

123. Of course, employee mobility and competition from ventures may be less important reasons for the demise of large corporate basic research laboratories than competition from other companies and antitrust actions, that reduce profits and force stricter cost accounting, pressure from shareholders, etc.

124. One might reply that the hurdles such persons face obtaining senior positions in Japanese companies or permanent jobs in Japan, and Japan's reluctance to accept skilled professionals and their families as immigrants, are the root causes of Japan's inability to attract and retain more skilled immigrants.

125. Saxenian (1999) and Hyde (2003) for 1990 data. Wulf (2005) for 2000 US-wide data. For comparison, twenty years earlier in 1980, 24% of Ph.D. scientists and engineers were foreign-born. Because Silicon Valley is known to attract a large number of foreign scientists and engineers, the percentage of foreign-born scientists and engineers in the Valley in 2005 is probably higher than the US national average.

126. These allow US companies to employ foreign workers in specialty fields such as science, engineering, medicine, nursing, law, and accounting for up to six years. The annual limit of new visas was 65,000 until 1999 when it was raised to 115,000 and then to 195,000 in 2001. However, under sunset provisions, it reverted to 65,000 in 2004. (Dahms and Trow, 2005).

127. This is probably true nationally, because Chinese and Indian S&E professional immigrants are more concentrated in California than any other state (Saxenian 1999: 12) and a high proportion of California's S&T employment is in venture companies. However, I know of no data on whether, within any particular high technology region, the ratio of immigrants among S&E professionals in new S&T companies is greater than the ratio in established S&T companies. (Generally, on S&E immigrants in Silicon Valley, see Saxenian 1999; Hyde 2003.)

128. Data on immigrants in Japan is classified according to type of job, not level of education. The relevant categories are 'researchers' and 'technical personnel'. The former applies to foreign scientists, engineers, and technicians in universities and GRIs, the latter to those in other organizations. (Students, trainees, physicians, nurses and other medical personnel are excluded, as are permanent residents.) In 2004, the total number of foreigners in these two categories combined was 25,758 (2,548 researchers and 23,210 technical personnel). Compared to the total number in 2000 (19,465) this suggests a net annual increase of about 1,600 persons (Ministry of Justice and Ministry of Foreign

Affairs). Assuming that the Japan's total S&E labor force as a percentage of its total population (127 million in 2004–6) is approximately the same as that of the USA (3.8%), suggests that Japan's total S&E labor force is around five million. Thus foreign nationals probably account for about 0.5% of Japan's S&E labor force.

129. See Saxenian and Hsu (2001). However, Japanese-born engineers account for only 4% of the foreign-born Silicon Valley engineers, which probably is one reason Japanese regional innovation centers have not grown in parallel with Silicon Valley, the way centers such as Bangalore, India, and Hsinchu, Taiwan, have.

130. Rather than being triggered by concern that the visa applicant will engage in terrorist activities, this stricter scrutiny usually is aimed at preventing transfers abroad of technologies with military applications. Strict scrutiny is triggered if consular officials think a visa applicant's background or likely study/work in the USA will involve access to one of a long list of *sensitive* technologies listed on the Government's Technology Alert List (TAL). However, following the Sept. 11 terrorist attacks the TAL was substantially expanded to include, for example, almost all fields related to infectious diseases, environmental planning, and landscape architecture. Also consular offices were instructed shortly after Sept. 11 to be more vigilant concerning sensitive areas of study. In 2003, it often took over 60 or even 120 days for consular offices simply to receive a recommendation whether to issue the visas. F-1 visas issued for full-time students fell 20% (from 293,357) in 2001 to 2002 and 8% from 2002 to 2003.

    As for professional foreign nationals recruited by US companies to work in special technical fields, between 2001 and 2002, H-1B visa recipients (a large proportion of whom are foreign scientists and engineers) fell 27% from 161,643 to 118,352. Between 2002 and 2003, they fell an additional 9% to 107,196. Reductions in H-1B visas issued to Indians and Chinese, who make up approximately half of foreign-born engineers in Silicon Valley (Saxenian 1999: 11–12), are even higher than these overall averages. In October 2004, with the reversion of the H-1B visa ceiling to 65,000, H-1B immigrants fell sharply again (Paral and Johnson 2004; Dahms and Trow 2005).

131. In 2000, 55,000 US citizens and permanent residents received bachelors degrees in engineering, off from a high of 71,000 in 1985 and lower than any number since 1977. In 2000, 18,000 US citizens and permanent residents received doctoral degrees in science and engineering, off from a high of 21,000 in 1995 and lower than any number since 1993 (National Science Board 2004).

132. See Flanigan (2005) reporting on the decision of Conexant (the 1999 spin-off of Rockwell International's semiconductor operations, that makes microchips for modems connecting home computers and other devices to the internet) to outsource two-thirds of its semiconductor design and other high technology engineering work to its subsidiary in India, because too few American engineers are graduating each year and Indian graduates are staying home and finding work in India due to post-9/11 visa restrictions.

133. e.g. see Markoff (2005) and Rai (2005).
134. Hyde (2003).
135. See Chapter 4 and Chapter 3 page 60.
136. Hyde (2003: 228).
137. Although their knowledge may be valuable, for various reasons (lack of information during the job search process, salary expectations that are out of line with what potential new employers are willing to pay, preconceptions by potential new employers about what constitutes a desirable mix of ages, personalities, etc.) companies may not hire them, or by the time they are able to hire them (following an economic upturn) they have found other work unrelated to their former field.
138. Health Insurance Portability and Accountability Act of 1996 (Public Law 104-191) §701.
139. In 2005, the out of pocket cost for health insurance coverage for a healthy family of four living in the Washington, DC, area (principal breadwinner about age 50 with no significant risk factors) was US$5,000–6,000 annually. Even at this rate, deductibles and co-payments (the costs of various procedures that the family had to pay for out of pocket) were high, almost US$2000, mostly for routine visits–not for any major ailments. This insurance would not have covered medical conditions (e.g. diabetes, circulatory diseases, and so on) existing at the time insurance was purchased for a period of twelve to eighteen months following the purchase (enrollment) date.
140. Most other industrialized nations have a nationwide health insurance system that provides, if not for public funding of self-employed persons and employees of small businesses and their families, then for pooling of health insurance payments from such persons so that risk can be spread among a large number of persons. Except for persons over 60 who are covered by Medicare or those who are quite poor and eligible for Medicaid, the USA lacks such a nationwide insurance system, relying on employers to provide health insurance, but not requiring (at least on a national level) that they do so.
141. Wilson (1985*b*).
142. Hyde (2003: 63) quoting the vice-president of one of Silicon Valley's best known companies.
143. Employment in high technology manufacturing industries declined by 15% compared with 9% for manufacturing industries as a whole (Hecker 2005).
144. Hecker (2005). The three IT-related high technology manufacturing industries are *computer and peripheral equipment manufacturing, communications equipment manufacturing,* and *semiconductor and other electronic component manufacturing,* and these experienced average annual increases in output per hour of work of 25%, 10.5% and 21%, respectively, between 1987 and 2002.
145. Between 1987 and 2002, average annual output per hour in pharmaceutical manufacturing increased only 1%, while over the decade up to 2002, employment increased by 30%.
146. In Hecker's analysis, most biotechs were classified either to *pharmaceutical and medicine manufacturing* or to *scientific R&D services.*

147. 70% of US venture capital disbursements in 2002 were in IT-related companies. 24% were in biotech and other fields of health and medicine (National Science Board (2004, appendix table 6-15)). The distribution of angel investments shows a similar pattern, with perhaps a slightly higher percentage for companies in non-IT engineering fields (energy, instruments, etc.). (Online reports from the Center for Venture Research and the Angel Capital Association.)

148. e.g. Qualcom developed the strongest portfolio of essential patents covering 3G wireless technology (Goodman and Myers, 2005). Of course it would not be feasible for a venture company to finance an extremely expensive project, e.g. a space station as a platform to develop microgravity manufacturing processes. However, if some other organization could finance a space station, a venture company might be able to quickly develop such processes and pay a reasonable use fee to the space station operators.

149. The system will have to be flexible enough to ensure that people return to a particular field when downturns are over, or, shift to other S&T fields (or related management) if downturns in a particular field will be prolonged. The Japanese system has been to keep persons within the same company even when the company no longer needs their expertise. The US system has been to rely on these persons to find new employment on their own. So the Japanese system is good at preserving its S&T labor force but not at redeploying it to areas of greater need. The US system is better at redeployment to areas of greater need, but it runs a greater risk of losing persons permanently from its S&T labor force in times of prolonged downturn.

150. See Chapter 2, note 16 and accompanying text.

151. Kneller (2003) summarized in Chapter 2.

152. Of course, if biotechs did not exist, US pharmaceutical companies would carry out some of the drug discovery work currently done by biotechs. However, the discussion in Part I of this chapter suggests that the output of innovative drugs would probably be less.

153. i.e. observing closely innovations made in overseas companies and then trying to improve on those products while manufacturing at competitive cost .

154. In 1977, there were over 1,300 companies manufacturing machine tools in the USA. Their average employment was 62. Only 10 plants had more than 1,000 employees. The average size of a Japanese machine tool company was even smaller, 21, and only 6 had more than 1,000 employees (Carlson 1990).

155. Apple did not market its first microcomputer until 1976 or 1977, Commodore and Tandy not until 1977. IBM did not begin marketing the PC until about 1982 (Chandler 2001).

156. In 2004, the leading producers of MTs were, in order, Japan, Germany, Italy, mainland China, Taiwan, and the USA (2005 World Machine Tool Output and Consumption Survey).

157. Around 2000, over 60% of the value of US MT purchases were imports, but about 30% of US production was exported, a rare phenomenon for a 'mature' industry (Kalafsky and McPherson 2002).

158. As the median number of employees for all US firms in 2000 was around sixty, the smaller firms all had fewer than this number of employees. These small firms tended to be clustered in the states bordering the Great Lakes and to depend on customers who were themselves old rust belt manufacturers— at risk of moving to other regions and thus further depleting the small MT companies' customer base (Kalafsky and McPherson 2002).

159. UNOVA, a data automation and communications company descended from the conglomerate, Litton, had acquired Cincinnati Machine (formerly Cincinnati Milacron, one of the largest independent US MT companies), Lamb Technicon and Landis Grinding as well as a host of other MT companies. Together, as UNOVA's Industrial Automation Systems (IAS) unit, they made UNOVA the largest US MT producer from the late 1990s until about 2004. However between 2004 and 2005 UNOVA divested its entire IAS unit, the Cincinnati and Lamb components going to an investment firm and the Landis operations to a French industrial engineering consortium, Groupe Fives-Lille. Ingersoll Milling Machine Co. was America's second largest MT maker in the late 1990s. It went bankrupt in 2003. The larger part of its assets were acquired by the Camozzi Group of Italy. A smaller part was resuscitated as Ingersoll Production Systems by the Chinese MT company, Dalien. In a similar vein, Thyssen, Germany's largest MT maker bought Giddings & Lewis in 1997, saving G&L from dismemberment at the hands of the American company Harnischfeger that was interested only in G&L's service parts business (2005 Machine Tool Scoreboard and various media reports).

160. 40% of US MT production was exported in 2004 (2005 World Machine Tool Output and Consumption Survey). See also Kalafsky and McPherson (2002) and Uchitelle (2005).

161. Uchitelle (2005). Haas was founded in 1983 and is among the youngest of the major US MT companies. It eschewed layoffs because they go against its corporate culture, and because it wants to have skilled employees on hand when orders are high. While increasing exports, it is also increasing automation in order to keep labor costs in check.

162. See Appendix 3, Table 7A3.1 for a ranking of the largest Japanese and US companies. Of the 90 members of the Japan Machine Tool Builders Association, one of the most important industry trade groups, only 9 had more than 500 employees. The median number of employees was about 100, equivalent to the US median of 60 in 2000. Germany has vied with Japan for leadership in this industry, and its MT firms also tend to be independent SMEs, an exception being Siemens which was the sales leader in 2004, mainly, like Fanuc, on the strength of sales of numerical controls and measuring systems.

163. See Tom Beard (2005), 'World Machine Tool Review' (Special Report, Jan. 9). *Modern Machine Shop*, www.mmsonline.com

164. Unless noted, this account is based on Friedman (1988).
165. According to Friedman (1988), this was the case with Mori Seki (no. 7 in the world in 2005) and Matsuura (2004 sales about US$100 million and 301 employees).
166. Carlson (1990). Modular innovation in the US arose in part from various companies making IBM compatible computer peripherals and the advent of standard computer operating systems such as DOS/Windows and Unix (Christensen and Raynor 2003: 137–41; and Chesborough 2003).
167. Friedman (1988) citing studies by Kiyonari and Koike.
168. Friedman (1988) and Whittaker (1997) are the source for most of this paragraph. This particular example is from Whittaker.
169. This is the main thesis of Friedman (1988).
170. See Asanuma (1992), Dyer (1996), and Odagiri (1992) under Chapter 6, References.
171. Friedman (1988) would probably argue it is the only policy that offered significant benefit. Loans by the much larger Japan Development Bank (JDB) and government research subsidies did not account for a large proportion of capital investment (pp. 86–91).

    Calomiris and Himmelberg (1995) have analyzed lending to MT firms by not only the JDB, but also the Export–Import Bank and the private, development-oriented Industrial Bank of Japan and the Long-Term Credit Bank. They conclude that JDB and Ex-Im Bank loans may have helped to promote private investment, but in general their conclusions agree with Friedman's—these large-scale government loan programs provided at best marginal benefit to MT firms.

    Tariffs on MT imports also cannot account for the flowering of the Japanese MT industry in the 1970s and 1980s. These were instituted in the late 1950s. Many were rolled back in the 1960s. By the 1970s, rates were equivalent to those in the USA. In 1983, all tariffs were removed.
172. Needs not being met by the main US MT manufacturers.
173. For example, the nonbiomedical, nonsoftware companies that had IPOs 2000–4 listed in Chapter 5, Appendix 1, Table 5A1.1; also the companies engaged in patent infringement or unfair competition suits (see notes 264 in this chapter and 58 in Chapter 4).
174. See Whittaker (1997, Chapters 7 and 8). See JSBRI (2003, Chapter 4 references) for rates of SME entry and exit in manufacturing.
175. The importance of a capable, interventionist bureaucracy was stressed by Johnson (1983). Fransman (1995) described the virtues of managed competition, often mediated by a government monopoly-corporation, such as NTT, that has advanced research resources to share with Japanese companies and is a guaranteed purchaser of their products. For counterarguments to the competent industrial policy thesis, see, e.g. Callon (1995), who questions the effectiveness of MITI-inspired R&D consortia once Japanese companies were beyond the catch-up phase. See also Friedman (1988), who describes how

the machine tool industry prospered despite repeated government efforts to exert greater control over that industry—efforts that probably would have been counterproductive had they succeeded.

As an example of recent Japanese research on this issue, Takahashi (2007) concludes, on the basis of interviews with former officials of MITI and the Ministry of Post and Telecommunications, that while ministry guidance was of some value in the catch up phase, it failed to anticipate key changes in the IT industry and it hindered Japanese telecommunication companies once they had attained globally competitive capabilities in the 1980s.

176. This appears to be the definition adopted by Mariko Sakakibara (1997: 449), who has published analyses of data from 1959 to 1992 covering most government consortia. Sakakibara defines her sample as follows: 'government-sponsored R&D consortia in the sample include all significant company-to-company cooperative R&D projects formed with a degree of government involvement'.

However, to my knowledge, projects are not classified by funding agencies as to whether they are *consortium* projects. In other words, with the exception of some large-scale projects such as VLSI, New Sunshine, and PERI that are clearly consortium projects, decisions as to which projects to classify as consortium projects probably have to be made on a case-by-case basis. Thus, how to define *consortium projects* becomes important. A review of all NEDO projects funded in 2005 shows a wide range of projects, that may or may not be considered to be consortium research depending on the precise definition. There are subprojects within projects. The subprojects may involve a single university or company, but the larger project involves several companies and universities. There are projects involving local government-affiliated companies and private companies. There are (sub)projects that involve only one company but a group of universities or GRIs. There are projects involving an industry association (often METI sponsored) such as the Japan Bioindustry Association or the Japan Fine Ceramics Center whose members include a large number of private companies. Finally, there are projects or subprojects involving a single company which subcontracts to universities and/or other companies. As indicated below, subprojects that involve two or more companies are relatively rare even for NEDO (rarer still for JST-CREST and OPSR projects). However, for high priority frontier fields of R&D where successful research is likely to find commercial applications (e.g. IT, nanotechnology, and some fields of biomedicine), (sub)projects involving two or more companies (usually large companies) are fairly common.

177. See Fransman's account (1995), and note 85, above.

178. This following account is from Callon (1995) and is corroborated by discussions I have had with persons in the semiconductor industry. The five corporate participants in the project were Hitachi, Fujitsu, NEC, Toshiba, and Mitsubishi. The former three, as long-term main suppliers to NTT, also collaborated closely with NTT's laboratories during the project.

179. Callon emphasizes that the amount of information shared among the participants was limited because of concerns about leakage of each company's confidential information, although toward the end of the project, sharing of information increased as new technologies evolved, including technologies developed by the project itself.

I have heard the same basic story with respect to the amorphous silicon solar cell consortium. Advances were achieved mainly by cooperation between individual corporate participants and university scientists, not by cooperation between scientists in different corporations.

180. Ironically, in 1980 at a time of high trade tension involving automobiles as well as electronics, and amidst charges by the US government and US semiconductor manufacturers of secrecy and unfair government subsidization of R&D, MITI and NTT decided to open all VLSI consortium patents to any company to license nonexclusively. Since IBM and Texas Instruments had cross licensing agreements with the consortium members, they were able to use almost all the VLSI patents for free (Callon 1995: 168).

181. Callon (1995) and Sakakibara (1997).

182. See www.nedo.go.jp/nanoshitsu/project/index.html and note 190, below.

183. These include most of Japan's major chipmakers (Fujitsu, NEC, Toshiba, and Renesas (a joint venture between Hitachi and Mitsubishi)), three main producers of EUV light sources (Komatsu, Ushio, and Gigaphoton (a joint venture between Komatsu and Ushio)), and two of the main manufacturers of optical equipment for lithography, Nikon and Canon. Seven universities are also participants along with AIST.

184. The EUV LLC was initiated in 1997. In addition to Intel, it includes Advanced Micro Devices (AMD), Micron, Motorola, IBM, Infineon (the 1999 spin-off of Siemen's semiconductor operations), and ASML (a Dutch company whose main rivals are Nikon and Canon), and three DOE laboratories, Lawrence Berkeley, Lawrence Livermore, and Sandia. Various European consortia have been initiated, some privately and some under the EU's Eureka/Medea+ Program. In 2005, IBM organized the International Venture for Nanolithography (INVENT) consortium to pursue further development of EUV (and also to develop 193-nanometer immersion technology to enhance the capability of normal or deep ultraviolet lithography). Besides IBM, the INVENT consortium includes the University of Albany and the university's Albany NanoTech R&D center, AMD, Infineon, Micron (all of which use ASML scanners), and ASML itself, as well as Applied Materials, VEECO, and dozens of other metrology, photoresist, and other equipment and materials suppliers.

Apparently no Japanese or Korean companies are included in the EUV LLC or INVENT Consortia (although Tokyo Micron has a separate collaboration with the Albany NanoTech Center). A Japanese industry scientist familiar with these consortia attributed the absence of Japanese companies to METI's concern that Japanese technology would leak to foreign rivals, the inability of Japanese companies to deduct consortium expenses from income subject to

Japanese tax (whereas they could deduct expenses for Japanese consortia), and linguistic and cultural hurdles.

This scientist also observed that cooperation among the member companies of the Japanese EUVA has been more difficult than in the VLSI project, and also more difficult than among the corporate members of EUV LLC and INVENT. He attributed this to the EUV LLC and INVENT consortia having been formed voluntarily by private initiative without the aim of including (and sharing technology among) all relevant companies. In contrast, as it the case of VLSI, METI sought from the outset to include all major Japanese companies and thus had to force collaboration between companies that were rivals in many of the fields in which they were to be cooperating. Research is constrained by the requirement that it not be likely to help one particular company more than another. At the same time, the scientific and technical uncertainties still facing all EUV lithography development efforts worldwide are greater than those that faced the VLSI consortium members in 1976. The foreign threat in 2005 seems less immediate than in 1975, and the Japanese EUVA companies are reluctant to commit large human and financial resources to projects where the payoffs are uncertain and where the government's top-down push for collaboration is not reinforced by grassroots agreement on important areas on which to cooperate.

Some of the EUVA members are working out cooperative arrangements outside the consortium framework. But, as in the case of the VLSI project, cooperation seems easiest among companies that do not have competing technologies and business goals. Toshiba, which unlike the other members is concentrating on NAND flash memories, seems to be a 'natural' partner and it has agreed with Hitachi and Renasas to prepare to begin joint production of 65 nanometer system chips in 2007. NEC and Matsushita were included in the original negotiations but have backed off from the project. However, NEC and Toshiba are discussing joint R&D on 45 nanometer chips. ('Chipmakers study Joint Production'. *Nikkei Weekly*, Jan. 9, 2006: 14).

185. To my knowledge, the only full accounting of government consortia spending is by Sakakibara (1997). The last year for which she has published these levels is 1992, when consortia spending amounted to approximately 85 billion yen, about 4.2% of the total government S&T budget for that year of 2,023 billion yen (NSF Tokyo Office, annual Japanese S&T budget summaries). 85 billion yen is high compared to consortia spending in the 1960s and 1970s but it is off from a peak of about 100 billion yen in 1988.

186. By limiting my analysis to extramural expenditures, I am trying to exclude funds that are budgeted for ongoing research in GRIs. In other words, I am most interested in consortium research as a percentage of discretionary government R&D funding devoted to specific projects. Thus I also want to exclude from the denominator ongoing large budget projects related to research and exploration in space, deep oceans, and the Antarctic.

187. In order to make these estimates, I assumed the validity of Sakakibara's finding that 4.2% of the total 1992 government S&T budget was devoted to

consortium research (see two notes previous), and also that this percentage is approximately the same circa 2005.

To calculate the denominator for the lower limit percentages, I started with the *allocations to promote science and technology* in the *regular budget* and subtracted out the expenditures for space, deep ocean, and Antarctic research. I had precise amounts for these expenditures for 1994, because I have a detailed official analysis in Japanese of the entire S&T budget for that year. (As background, the S&T budget consists of three components: (*a*) *regular budget allocations to promote science and technology*, (*b*) *other regular budget allocations* related to S&T (mainly administrative and infrastructure expenditures, including salaries for full-time employees), and (*c*) allocations from *special account budgets* of which four are relevant to the S&T budget: one to support national universities, two related to energy and one for industrial investment.) Next, I added 1994 expenditures from the *other regular budget allocations* and the *special accounts* that I thought might be used to fund R&D projects that might have commercial applications, even though indirect and sometime in the future. The three largest add-back items, all from special accounts budgets, were funding for the New Sunshine Project (alternative energy), nuclear power technology, and the Japan Key Technology Center. This adjusted 1994 S&T promotion budget was 569 billion yen, or 24% of the total 1994 S&T budget. $(4.2 \times 100)/24 = 17.5\%$, before excluding MEXT-grants-in-aid. However, this percentage is an underestimate, because the denominator overestimates project-specific R&D expenditures, because it contains expenditures for training, infrastructure development, transfer payments to local governments, and the basic, non-project-specific R&D-related expenses of universities and GRIs.

The upper-limit estimates assume that all consortia research projects are a subset of competitive research projects, and that competitive research funding is approximately equivalent to extramural project-specific research expenditures. These assumptions allow using total funding for competitive research as the basis (denominator) against which to compare consortium research expenses. The major competitive research programs in 2002 are listed in Chapter 3 Appendix 1, Table 3A. All competitive programs together accounted for 9.8% of the 2002 total S&T budget, a proportion that has been fairly steady since 1990 according to annual Cabinet Office reports. $(4.2 \times 100)/9.8 = 43\%$, before excluding MEXT-grants-in-aid. Are the assumptions underlying this upper-estimate valid? Most of the consortium research projects that I found (see following text) are funded under competitive programs such as JST's CREST or NEDO's Industrial Technology Research Projects. However, I am not sure if former Key Technology Program projects such as New Sunshine and Rite are considered competitive projects. Also it is possible that some other major consortium projects (e.g. EUVA) are not included within competitive funding programs.

188. Available at http://www.nedo.go.jp/informations/press/index.html. In 2005, there were over forty such announcements.

189. The projects newly initiated in 2005 all related to the subtheme, 'R&D for the practical use of advanced nano-technology components [Nano sentan buzai jitsuyouka kenkyuu kaihatsu]'. Sixteen teams were funded in 2005.

190. I randomly selected several projects within METI–NEDO's overall Nanotechnology Program launched in 2000, which had about US$100 million in funding in both 2003 and 2004. (A list of the main themes and subthemes under this program is available at www.nedo.go.jp/nanoshitsu/project/index.html. However, finding the names of the projects and participants required additional searching under each subtheme.) Under the Nano-material/Processs subtheme, the main participants in the 'Fine Macromolecule Technology' project are twenty-three companies, mostly large, fourteen university laboratories and AIST. Also under the same subtheme, the main participants in the 'Development of Nano-carbon Products' project are eight companies (again mostly large), the Japan Fine Ceramics Center (a METI organized industry group), four universities, and AIST. Under the Nano Manufacturing and Measurement subtheme, the main participants in a project to develop conductive ceramic materials amenable to low temperature molding are five companies (all large), four universities, AIST, and a METI organized industry association originally focused on robotic manufacturing. Under the same subtheme, a project to use non-destructive electron beams for nanoscale manufacturing involves six large companies, and three universities. Of the thirty-two NEDO nanotechnology projects highlighted in an English language publication (available at http://www.nedo.go.jp/kankobutsu/pamphlets/nano/nano_eng.pdf), most are headed by professors at elite universities (or by AIST researchers) and most involve mainly large companies as collaborators.

191. The Key Technology Center was formed in 1985 to invest proceeds from the partial privatization of NTT into socially and industrially relevant fields of S&T and R&D. It has funded some of Japan's largest consortium projects; including the Protein Engineering Research Institute (PERI), the Advanced Telecommunications Research Institute (ATR), the Research Institute for Innovative Technology for the Earth (RITE), and Helix (genome research); some of which continue today as major research centers. In 2003, the Center was disbanded and its operations assumed by the Key Technology Promotion Group within NEDO. Information about recent Group projects are at www.nedo.go.jp/kibanbu.

192. MIC is the successor agency to the Ministry of Posts and Telecommunications, and exercises substantial control over the partially privatized NTT and NTT's affiliated laboratories.

193. These consortium-like projects involving at least one large company and one university, or two large companies, account for fifteen of the 135 MIC projects initiated in 2005. One of these was commissioned by MIC to develop home and mobile applications for Internet Protocol version 6. Six others funded R&D proposals solicited under the competitive Program for Strategic Promotion of Information and Communication Technologies. Eight

others were commissioned by MIC in a variety of telecommunications fields. In addition to these fifteen consortium-type projects, sixteen other projects in advanced fields of telecommunications involve a single large company (or in two cases a small company and a large company). An additional forty-seven projects support university/GRI R&D without corporate collaborators (ten involved disaster preparedness and nine were for international collaborations). Finally, 57 projects support R&D by small or regional companies.

194. By *extramural R&D*, I mean research support made available by government agencies to outside organizations (companies, universities, and in some cases GRIs) for specific research projects, usually on a competitive basis.

195. OPSR is roughly MHLW's equivalent of NEDO or JST with respect to R&D support. In 2005, OPSR's R&D support functions were transferred to MHLW's National Institute of Biomedical Innovation. (See Glossary)

196. The remainder were either new bioventures or incorporated consortia created by OPSR (see next note). About half the OPSR projects with corporate participants had more than one corporate participant. Lists of projects and participants are available at www.pmda.go.jp/kenkyuu/kisoken. I reviewed the final reports of projects initiated in 1996–9, as well as those of a special series of projects in genomics and proteomics initiated in 2001.

197. The youngest participant was a government-affiliated organization formed in 1989 to promote industry–university–government cooperation in the Tohoku region. The youngest private sector participant, Nihon Medi Physics, was formed in 1973 as a joint venture between Sumitomo Chemical and a US company. The others, all formed before 1954, included pharmaceutical, chemical, medical device/instrument, foodstuffs, electronics and optical companies. See Okada and Kushi (2004) for more information about these consortia.

198. The CREST program funds fairly large-scale projects, usually by teams from several institutions (see Chapter 3). Lists of CREST projects and participants are available at http://www.jst.go.jp/kisoken/crest/report/heisei16/html/kenkyuu_houkoku.htm.

    Among the other JST programs, ERATO projects (fewer in number but larger in scale) usually include industry researchers, while PRESTO projects (smaller in scale, usually for less well-known researchers) usually are to single university laboratories.

199. Sakakibara (2001: 1004) also found that most of the participants in the consortia initiated between 1959 and 1992 were large companies, although her published data do not elaborate on this issue, and she provides no data on university participation.

200. Or erstwhile rivals might be assigned separate fields in a project so that their rivalry is limited, as in the case in the broadband optical communication project mentioned in Chapter 6.

201. See Hayashi (2003), Okada and Kushi (2004) and Branstetter and Sakakibara (2002). One problem is lack of an appropriate group with which to compare companies that participate in consortia. The latter study tried to address this by comparing consortia participants with a 'broadly comparable' set of non-participants. This study found a small (approximately 5%) greater number of US patents for participants compared with nonparticipants. Other problems related to the validity of jointly authored publications and patents (especially mere patent applications) as indices of innovation are discussed in Chapter 1. These problems are particularly acute when the consortia participants and the agencies funding the consortia know that their activities will be evaluated on the basis of such indices.

202. I have discussed this issue with about ten persons who are personally familiar with one or more consortia, in addition to the persons whose interviews are reflected in the Chapter 4 case studies and in Kneller (2003). Although most were skeptical about the value of consortia in terms of either the discovery/development of useful products/processes, or the improvement of participants' ability to conduct R&D, they were more likely to say that the consortia were beneficial from the perspective of producing advances of open scientific knowledge.

Not all projects/programs are assessed negatively. VLSI and some of the ERATO projects are widely regarded as successful consortia. As for lesser known consortia, former participants have cited the silicon solar cell project involving several companies and universities (mentioned in Chapter 2) as a success. Another is the Drug Delivery System (DDS) Institute, the first of the fifteen OPSR consortia mentioned above. The DDS Institute consortium operated 1988–95. The participants were Asahi Kasei, Ajinomoto, Eisai, Daiichi, Meiji Seika, Shionogi, and Tanabe. One of the participating companies volunteered that this consortium had advanced its R&D program and resulted in technologies that it aimed to commercialize.

Analysis of patents related to the DDS consortium support these favorable comments. More US patents were issued to this consortium (thirteen) than to any of the 14 other OPSR consortia. At least seven of these thirteen patents had inventors from two or more of the consortia participants, suggesting a fairly high degree of intercompany collaboration. Five of the consortia patents were cited by later patents issued to four member companies doing research on their own. This suggests that at least these four companies (Ajinomoto, Asahi Kasei, Daiichi, and Tanabe) found the consortium research relevant to their continuing in-house drug delivery R&D. A fifth consortia member, Shionogi, does not have patents that cite consortium patents—but one of its researchers is an inventor on consortium patents, and also on ten patents belonging only to Shionogi (many drug delivery related) that span the time before, during, and after the consortium, suggesting that this experienced researcher contributed his expertise to the consortium. Some of the DDS consortia patents have been cited by non-Japanese inventors. *However, none of the other OPSR consortia* approach the DDS Institute in terms of favorable comments from participants or a patent record suggesting close cooperation among members and continuing benefits to members, and citations by inventors worldwide. (However, I have complete (or nearly complete) lists of the patents of only the first eight of the fifteen OPSR consortia, because the latter seven did not end until 1999 or later. Thus some US patents may still be pending while others have had little time to be cited. A partial analysis of registered Japanese patents does not suggest results different from the analysis of US patents.)

203. Sakakibara (1997, 2001). Responses to a 1993 survey of R&D managers in sixty-seven companies participating in eighty-six consortia between 1959 and 1992 indicated that companies generally participate in government consortia to obtain complementary knowledge from other participants and to enter a new field of technology/business. They considered the main benefits to participation to be researcher training, breakthroughs in or accelerated development of a technology and increased awareness of the importance of R&D related to the technology, increased internal funding of R&D, and increased awareness of the subject. Sakakibara, who conducted this survey, believes these responses indicate consortia facilitate diversification. Her findings with respect to movement from low to high growth industries and vertical integration arise from a comparison of the self-identified primary three-digit SIC (standard industrial classification) of 627 manufacturing companies participating in 186 consortia over the 1959–92 period with the three-digit SIC classification that most closely matches the activity of each of the 186 consortia.

204. In the case of ventures and other SME participants, the most commonly mentioned motivation they cite for participation is access to government cofunding to pursue their own core R&D.

205. The history of the Protein Engineering Research Institute (PERI) provides insight into the actual reasons companies participate in consortia and their interactions with each other. PERI was a Key Technology Center consortium established in 1986 to develop technologies related to protein structure, synthesis, and function. With its own building, an annual budget of over US$10 million, fourteen participating companies and approximately forty-two scientists, it was one of Japan's larger and longer enduring consortia. It probably had a somewhat more basic science orientation than most. In 1992, seventeen of its scientists were from universities (generally entry-level faculty or postdoctoral-level researchers) and twenty-five were from its fourteen member companies: Ajinomoto, Fujitsu, Kaneka, Kirin, *Kyowa Hakko*, *Mitsubishi Kasei*, Nihon Digital Equipment, Nippon Roche, Showa Denko, Suntory, *Takeda*, *Tonen*, *Toray*, and Toyobo (*italics* indicate core/founding members, companies that generally had the strongest protein engineering capabilities at the start of the consortium). All of these are established companies with at least 1,500 employees. Originally scheduled to operate until 1996, its operations were essentially extended for ten more years, while its focus was somewhat broadened, under the name Biomolecular Engineering Research Institute (BERI).

In order to maximize cooperation among the scientists and to provide a foundation for wide improvement in Japan's protein science capabilities, almost all the projects focused on basic research issues. Evaluation of the consortium was to be based primarily on publications in international scientific journals, and company scientists were enjoined from reporting individually the specific contents of their research to their home companies. (How this was enforced in practice is not clear.) By the early 1990s, PERI's output of international journal articles per researcher was fourth among all of Japan's bioscience institutes, and in its first ten years of existence, it contributed half of Japan's input to international databases on three-dimensional protein structures. In addition, Japanese scientists and research managers outside PERI generally felt it was beneficial in raising awareness, particularly within the pharmaceutical industry, of protein engineering as a way to solve various problems.

However, there was widespread concern among the corporate members about losing researchers to a basic research project for two to three years. Some of the companies without much expertise in protein biology/chemistry felt this was outweighed by the training their scientists received. Some of the larger companies, however, felt participation offered little practical benefit aside from building up the R&D capabilities of weaker companies (Protein Engineering in Japan 1992; Coleman 1999).

Three of the core members, Kyowa Hakko, Mitsubishi Kasei, and Takeda, as well as Nippon Roche (via its Swiss parent), already had considerable expertise in the application of protein biology and chemistry to pharmaceuticals. For junior members Ajinomoto, Kirin, and Suntory who had recently begun pharmaceutical R&D, participation in PERI probably was a way to acquire expertise to help them diversify into a new field. Kanebo may have been interested in applications of protein chemistry to its main cosmetics business. Kaneka, Showa Denko, Tonen, and Toray are all primarily chemical companies and were probably interested in applications of protein science to industrial processes—i.e. not so much diversifying into new disciplines as adapting new disciplines to their core businesses. Fujitsu and Nihon Digital Equipment were responsible for computational and database aspects of PERI research. Fujitsu was also an important participant in some of the instrumentation projects—such as development of an electron microscope that could operate at near absolute zero thus enabling imaging of protein crystal structures—one of the most highly regarded projects at PERI around 1992.

206. OPSR's DDS consortium included only established companies. To my knowledge, none of the ventures developing DDS benefited from the research of the OPSR consortium, although some have received other forms of government support.

207. Typically, the consortium is supposed to own or co-own patents arising from consortium research. Members typically have a license to use the patented

inventions for their own purposes. Licenses to outside companies usually require consortium approval.

208. My contention that rarely does any company emerge that pushes forward consortium research on its own does not apply to established companies, such as those involved in the VLSI, EUVA, and photovoltaic cells consortia, that already have strong business reasons to pursue R&D in these fields. Beyond this qualification, I base my assertion on anecdotal accounts, such as those mentioned in the text above, and the apparent dearth of new products or vigorous new companies that trace back to consortium R&D. One exception that proves the general rule may be a gene therapy vector that constitutes the core technology of a venture company that was originally formed as an OPSR consortium to develop this and other gene therapy technologies. Another may be the DDS discoveries referred to above. However, to my knowledge, most of the DDS consortium-linked discoveries are being pursued by large pharmaceutical companies that have long been interested in improved drug delivery systems. Although the DDS consortium (in particular the government funding) may have advanced the pace or range of discoveries in this area, this is a field the large companies could have pursued (and indeed are pursuing) either on their own or with other companies in privately initiated collaborations.

209. This is the consistent response from the ventures I have spoken to. See also the Chapter 4 case studies of Big Crystal, Fine Molded Plastics and Internal Search Engine. A secondary incentive mentioned by some ventures is access to university research. Cooperation with other companies is usually not mentioned as a benefit of participation, except in so far as they contribute instrumentation or support services that are useful to the venture.

Of course this raises the question whether government funding for ventures creates overdependence and diverts them from goals that offer the best chance for long-term commercial strength. I know of examples where this is probably the case, yet I also know of examples where government funding seems to have provided a necessary bridge to attracting more substantial private funding.

210. For example, see the Chapter 4 case study of Molecular Visualizer which decided not to participate in consortia out of concern that large companies would absorb its know-how and technology. Note also the importance most of the companies featured in the Chapter 4 case studies attach to patent protection.

211. But note that Big Crystal (Chapter 4) participates in consortia even though IP must be shared.

212. At least this seems to be the case in elite universities with strong S&T departments. In lesser known universities, a relatively small percentage of S&T faculty may participate in consortia or joint research with large companies.

213. See the observation in Chapter 4, that outside biomedicine and software, the technology and business potential of most startups from the University of Tokyo, Keio University, and AIST seem to have limited technical scope and business potential.

214. One of the early informal collaborators, a large well-known electronics company, applied for a patent on one of the founder's inventions, without even naming him or any of his university colleagues as an inventor. The application was later challenged but the opportunity for the startup to receive IP rights to this technology was lost.

215. At the end of 2005, about a year after its founding, the startup had seven employees. It hopes to increase the number to twelve by mid-2006. But attracting qualified Ph.D. level researchers and capable managers is one of its biggest challenges.

216. See Sakakibara (1997) and remainder of this chapter.

217. Here I am referring to the benefits that would likely be achieved beyond simply making government funds available to individual companies or leaving companies to work out collaborative R&D on their own.

   This raises the question of the benefits of government organized consortia in the USA and other countries. I am not aware of many studies that try to assess concrete, practical benefits of government organized consortia. SEMATECH, one of the best known US consortia, was formed in 1987 with the mission to rebuild a strong infrastructure for domestic semiconductor manufacturing equipment. Improving communication and institutional linkages between semiconductor manufacturers (which tend to be large companies such as Intel and Motorola) and the more numerous manufacturers of the various types of semiconductor manufacturing equipment (which tend to be smaller companies) appears to have been SEMATECH's main achievement. One vehicle for coordination was the development and periodic updating of a National Technology Roadmap for Semiconductors, which defines agreed-on industry goals and the technical developments necessary to achieve those goals. SEMATECH also helped improve buyer–seller transaction processes. It established a process to certify whether semiconductor manufacturing equipment met industry-agreed-on standards, and thus eliminated expensive duplicative testing. It also established a standard auditing process for equipment manufacturers (Corey 1997: 1). Thus it seems SEMATECH's main achievements lie more in improving the institutional framework for cooperation than in specific technical advances. Japanese consortia generally are more applied, in the sense of aiming for specific technical advances, and deal with topics closer to commercial applications.

218. As described in Chapter 3.

219. The Japanese term for these value adding corporate workplaces is *gemba*.

220. The former pattern, transfers from central/basic research to development laboratories, is described by Westney and Sakakibara (1986*a*, 1986*b*) in the case of three major computer companies. Each company had a central (primarily basic) research laboratory that was the initial posting for many new engineering graduates. After 6–7 years, they were transferred to laboratories of the various manufacturing divisions where they worked on scale-up and commercialization of projects they began in the central research laboratory, providing important continuity for each project.

In the latter pattern, described by Kuzunoki and Numagami (1998) in the case of a large chemical/materials company, product development is the job transfer nexus linking research, process development, and manufacturing.

In contrast, US engineers tend to have much more control over their careers, but to spend all their time in a research laboratory, to have less contact with production staff, and often to view production staff with disdain (Westney and Sakakibara 1986*a* & *b*; Lynn, Piehler, and Kieler 1993).

Absent from these accounts of typical job rotations for engineers are posts that involve customer contacts. An executive from a major IT company remarked that while Japanese engineers come to understand their companies well, they rarely have contacts with customers. Most have no experience trying to understand needs outside their companies or how things are done in other companies.

221. Chuma (2001) documents differences in the degree of active feedback between R&D and production staff in US, Japanese, and German machine tool companies (with the German practices generally being closer to the Japanese).

222. Based on Aoki (1990), Westney and Sakakibara (1986*a*, 1986*b*) and Kuzunoki and Numagani (1998). This summary is specific for engineers and R&D managers, but similar management policies apply to other scientific and technical personnel in large corporations. In the case of blue-collar workers, rotations are usually decided by shop floor or factory supervisors rather than the central personnel office.

   Aoki stresses the importance of frequent and easy communication among teams (or among individuals in different teams) to solve problems, what he calls Japanese style *horizontal* coordination as opposed to Western style *hierarchical* coordination. Others (e.g. Doeringer, Evans-Klock, and Terkla 1998) confirm these contrasting styles in the context of Japanese manufacturing transplants in the USA compared with new plants of indigenous US companies in the same industries.

223. Kneller (2003) documents the low proportion of Ph.D. new hires by pharmaceutical companies. Discussions with large companies in other industries as well as graduate students and professors indicate that most leading manufacturing companies still hire MS or BS graduates for their R&D laboratories. One university professor who has collaborations with many of Japan's leading electronic companies observed that today, as opposed to ten or twenty years ago, large companies do not regard university Ph.D. training negatively—at least in laboratories oriented toward fundamental research. However, neither do they regard a Ph.D. favorably. Students who obtain Ph.Ds. and then take jobs in industry usually are those who were undecided between a career in academia or industry and ended up choosing industry, often because of the scarcity of academic positions.

224. This factor is stressed by Aoki (1990). With the introduction of merit pay systems in many companies, this uniformity may be breaking down.

225. This example is from Westney and Sakakibara (1986*b*). The quoted maxims are from Westney and Sakakibara (1986*a*, 1986*b*). Recent communication with one of the authors suggests that these policies are probably still in effect in most companies, although follow-up studies are needed to confirm this.

226. Young Japanese engineers are generally free of administrative personnel matters and thus have more time to devote to research than their US counterparts who have to submit personnel evaluations even though they may have only two or three subordinates. The fact that Japanese engineers in central laboratories spend about 53% of their time directly on R&D compared with 39% for US engineers may be another strength of the Japanese innovation management system. (Westney and Sakakibara 1986*b*). See also Doeringer, Evans-Klock, and Terkla (1998) with respect to merit pay and promotions for workers (primarily blue-collar) in Japanese manufacturing transplants in the USA.

227. This is changing, but probably gradually, as manufacturing companies begin to adopt merit pay systems (see note 275 below and accompanying text).

228. Some companies offer an option to receive (a portion of) pension payments as an annuity (either until death or over a fixed number of years following retirement) in lieu of a single lump sum payment on retirement.

    Persons who have reached up to manager's (buchou) status can usually keep working in management positions longer. As Japan's population ages, persons below this rank are also being offered opportunities to work past 60, but usually in different jobs and with salaries reduced 40–50% (Michio Sugai, 'Boomers get Option to Retire Older', *Nikkei Weekly*, March 21, 2005: 21).

229. Defined contribution pension plans (401-k type plans), although becoming more common, probably still constitute a small percentage of total pension payments by large manufacturing companies.

230. Perhaps this is changing. The *Nikkei Weekly* (May 1, 2006) reported that 'Toray Industries group intends to bring on some 300 midcareer and other workers— the most among manufacturers—to accelerate development of materials used for digital electronics and aircraft', and that 'Mitsubishi Electric plans to take on 160 mid-career engineers and other workers, up 45.5% amid brisk sales of machine tools'. It is not clear, however, to what extent these mid-career hires involve R&D personnel from non-affiliated companies. Recent discussions with R&D personnel in major manufacturing companies suggest that mid-career hires of R&D personnel from unaffiliated companies are still rare.

231. I have confirmed this is the case through interviews with officials in the personnel department of one major electronics company, through published media reports about another, and through discussions with R&D employees and with an academic researcher studying a third company. These sources suggest that such reforms have been adopted by most of the major electronics companies. I am not sure about the extent of similar reforms in other industries.

232. They must make this choice at the time of employment. If they choose to manage pension funds on their own, they receive company contributions toward their pensions as supplements to their semiannual cash bonuses.

233. Interview in April 2006 with personnel officials in a major electronics company.

234. These plans are in effect *defined benefit plans* where the benefits accrue linearly. They are not *defined contribution (401-k type) plans* (although their effect may be somewhat similar) because the employer company manages its pension funds and because benefits are fixed as a proportion of salary. Unlike US companies that have switched to defined contribution plans, the Japanese companies still assume the market risks associated with guaranteeing a particular level of benefits to their employees.

235. The company that described its pension reforms indicated that it did not believe that employees hired since April 1999 were leaving at higher rates than young employees hired just prior to 1999.

236. Aoki (1990: 22). The extent to which these attitudes still persist is an important question. Despite the pension reforms, my sense is that reorientation of research teams to seeking out and absorbing outside technologies has not proceeded to a large extent except that more corporate researchers are being sent to universities. In other words, there is more emphasis on absorbing early stage academic discoveries, but much less on partnering with independent new companies. The former requires a much smaller readjustment of corporate research styles and focus than does the latter.

237. Lynn, Piehler, and Kieler (1993).

238. See the cases in Chapter 4 and also the case of the OPSR protein science consortium presented earlier in this chapter. See also Kneller (2003) summarizing discussions with large pharmaceutical companies.

One of Japan's largest pharmaceutical companies boasted that with twenty persons engaged in finding and maintaining collaborations it had entered the first rank of companies engaged in partnering (Osigo and Matsuzaki 2006). However, these efforts are mainly directed toward overseas partnerships. Also, US and European pharmaceutical companies generally have many more employees engaged in seeking out and managing alliance partnerships. Roche, for example, has approximately 100, according to a 2005 discussion with Roche officials responsible for alliances.

239. Recalling the importance of manufacturing keiretsu noted in Chapter 6, Japanese automobile companies sit atop a high technology chain consisting of many large and small high technology companies, and directly and indirectly account for 60–70% of sales by Japan's machine tool industry ('Auto industry sits atop tech chain'. *Nikkei Weekly*, Jan. 16, 2006: 14). These include not only close long-term cooperation with small to mid-size subcontractors but also with other large industrial companies. Recalling the close

cooperation and demand articulation between suppliers and users described in Chapter 2 (and also by Kodama 1991), Laage-Hellman (1997) describes successful long-term cooperation between Nippon Steel and several auto manufacturers. He also describes a successful collaboration between Toshiba and Cummins to develop ceramic diesel engine components. Myers (1999) provides an in-depth history of the successful collaboration between IBM and Toshiba in the 1980s to jointly develop and manufacture flat-panel displays for notebook computers. Both companies felt they learned from each other and that the project benefited from complementary strengths.

Not all collaborations are smooth. Recall from Chapter 6 the formation of Epilda and the squabbles between its parents.

As for collaborations with overseas ventures, the comments I have heard from US and European biotechs engaged in collaborations with Japanese pharmaceutical companies are generally positive and depict stable, long-term relationships. In the field of siRNA, a number of Japan's largest pharmaceutical companies have collaborations with a single US biotech company dating to the 1990s. But at least four Japanese biotechs whose core business is drug target identification using siRNA have had a hard time obtaining customers from among Japan's pharmaceutical companies. Might their relative late-comer status be a factor? Or might they be better off as one or two united companies? Absolutely. But I suspect that there is a chicken-and-egg dynamic, where one component is the tendency of large Japanese companies to ignore domestic ventures.

240. e.g. who will assume new posts related to the collaboration or whether there are persons in house who might do the same work as the collaborator is proposing.

241. e.g. the need to work through various levels of the corporate bureaucracy to develop a consensus how to handle relations with the outside partner.

242. Such a startup, founded by a professor in a major national university, actually exists.

243. See the case of Big Crystal in Chapter 4.

244. As if to confirm what I had long suspected, in early 2006 the head of merger and acquisitions in a major Japanese bank who tracks the interest of Japanese pharmaceutical companies in equity investments in biotechs observed that the pharmaceutical companies usually do not want a controlling influence over the overseas companies in which they invest. However, they would usually expect to control any Japanese venture in which they invest. The pharmaceutical company that invested in Gene Angiogenesis and the parent of Internal Search Engine (Chapter 4) are exceptions that prove the general rule.

245. See the spin-off cases in Chapter 6 and the case of Phoenix Wireless in Chapter 4.

246. *Organizational relationship management overload* may also be a barrier to collaboration in with ventures. As noted in Chapter 2, Japanese companies emphasize a smooth technical and business interface between collaborating companies and they devote much energy to making these relationships work. It may be that the number of relationships they can handle is limited. Thus, venture companies that pop up with new technologies have a high threshold to overcome to be taken seriously by large Japanese companies, because being taken seriously has traditionally entailed the commitment of substantial resources.

247. Appendix 7.4, Figures 7A4.1 and 7A4.2 show age-specific mid-career departure rates for firms with at least 1000 employees (large firms) and those with 30–99 employees (small firms), respectively. Graphs for other firm size groups, as well as graphs for specific manufacturing industries, are not shown. In the latter cases, exceptions to the pattern shown in Figure 7A4.1 for firms with at least 1,000 employees are (*a*) plastic products manufacturing, which had higher mid-career retirement rates in 1995 than during any year 2000–4 (this industry is unusual in that total employment in large firms increased from 1995–2004) and (*b*) metal products manufacturing, where mid-career departures seem to oscillate in a high-low pattern year by year. I limited my analysis to male employees because female employment is much less than male employment (see Fig. 7A4.4), and because a large proportion of women leave the work force in their 20s and 30s to raise families.

248. Press reports in 2005 mentioned sizable layoffs by Matsushita, Sankyo, and Sony. Discussions with R&D scientists and managers in major electronics companies indicate that shedding of workers was widespread in all electronics companies, and that even as late as 2005 most still wanted to reduce payrolls. Nevertheless, even in Sankyo and Sony, reductions in their Japanese workforces were achieved mainly through transfers to subsidiaries, early retirement, etc., not by layoffs.

249. According to an executive familiar with several of the seller and buyer companies.

250. This company tried (apparently more or less successfully) to ensure that its good engineers kept working. In contrast, when IBM instituted its second Career Path in 1987 as a means to ease out unwanted senior personnel (by offering a two years terminal leave of absence with full salary and US$10,000 for educational expenses) the persons whom the company hoped would stay left while those whom it hoped would take advantage of the program chose to remain.

251. According to a person familiar with the industry, about 30% of the 1990 workforce in large electronics companies were contract workers from outside corporations, but by 2005 most companies had raised (or were trying to raise) this to about 40%.

252. On this last point, see: 'New Lives for Corporate Dropouts', *Nikkei Weekly*, March 6, 2006, 7.
253. The largest percentage declines in female manufacturing employment were in the largest and smallest firms, while the largest absolute decline was in the smallest firms (see Appendix 7.4, Figure 7A4.4).
254. Also known as Keidanren.
255. Also Chairman of the Board of Toyota.
256. Keynote address to the 2006 Keidanren Employer–Employee Forum by Chairman Okuda, held on January 12, 2006 in Tokyo. *Italics* indicate speaker's emphasis as indicated on the Japanese transcript available at www.keidanren.or.jp. Masahiko Aoki, Professor Emeritus at Stanford and one of the most respected economists concerning Japan's economic system, echoed the same perspective in an English language interview (*Nikkei Weekly*, Jan. 9, 2006, 5):

...I think the lifetime employment and seniority systems have been at the heart of Japanese institutions. Under those systems, people have been able to secure stability in their livelihoods in exchange for considering an occupation at one place as their vocation and working there diligently. Because the systems are solid, it will take about a generation for them to collapse. After that, lifetime employment elements and a flexible labor mobility will likely coexist.

257. About 5% consider work in such a venture desirable, and over half consider such work unacceptable. Rates are similar for MS and Ph.D. students.
     However, in a university that has been the training ground for government and big business elites, such interest might be lower than in most other universities.
258. See, e.g. JPD-SED (2005), reporting results of a survey of new company recruits participating in two-day training courses held by a government-affiliated organization, suggesting that about 26–38% of new recruits expect to spend their entire career in one company, compared with 16–22% of new recruits in 1999. (The lower estimates within each range are for respondents who had been on the job for half a year, while the higher estimates are for respondents who had just joined corporations.)
     Household-based interview surveys by the Japan Institute for Labor Policy and Training (Nihon ryoudou kenkyuu kikou, available at www.jil.go.jp) show that overall 77% of regular employees (and 79% of part-time employees) supported lifetime employment compared with 72% in 1999. Among men in their 20s, the support rate was 64% in both years (rates for women in their 20s were slightly higher). Overall 43% of employees hoped to work in the same company all their life in 2004 compared with 40.5% in 1999. Only 26% expected to work for more than one company compared with 24% in 1999. Only 13% expected to start their own business compared with 15% in 1999.

259. See new Section 21(4) of the Law to Prevent Unfair Competition [Fusei Kyousou Yobou Hou]. This section authorizes criminal as well as civil penalties against persons who disclose trade secrets of their former employers to their new employers, if they do so with the purpose of gaining profit or aiding competitors of their former employer. Criminal penalties can include up to five years' imprisonment. The new employer can also face criminal and civil penalties, including injunctions. Prior to 2005 it was less clear that criminal penalties would apply, except in cases of acquisition of trade secrets by theft, fraud, or assault. The fact that criminal penalties now can apply even to acquisition of trade secrets in the normal course of work might expose to liability employees who move to ventures where they will continue work begun with their former employers.

260. Prior to 2005, plaintiff companies would have had to publicly disclose in court the trade secrets that were allegedly misappropriated by departing employees. See: Examples of Regulations Governing so-called In Camera Procedings [Iwayuru in-ka-me-ra shinri ni kansure kitei no rei] at http://www.soumu.go.jp/gyoukan/kanri/jyohokokai/pdf/050125_sanko2.pdf

261. And also in nondisclosure agreements that many employees are expected to sign when they leave employment.

262. Indeed, it might make sense for Japanese courts to consider carefully the likely proximate damages that would occur to the former employer's investments in deciding whether to enjoin an employee from moving to another company. In other words, following Hyde's suggestion (2003), a rational condition for upholding unfair competition (including trade secret) suits by employers against departing employees might be that the employer show that the change of jobs would likely result in significant harm to a specific project the employer is undertaking.

263. A 2003 METI White Paper referred to four district level court decisions between 1970 and 1998 that upheld covenants not to compete for periods ranging from one to five years (Trade Secret Management Guidelines [Eigyou Himitsu Kanri Shishin] issued January 30, 2003).

264. I have searched for IP-related decisions by the Tokyo District Court under key words such as unfair competition and trade secrets. Of approximately 300 cases, a research assistant scanned over 100 and identified 2 that addressed this issue indirectly.

    In addition, I was granted access to the lists and disposition of all cases brought before the Tokyo District Court dealing with patent or unfair competition issues from 1994 through 2004, no matter whether they resulted in a final court ruling. Selecting all such actions brought in 1994, 1999, and 2004 (844 in total), I selected a 30% random sample (145) and then, on the basis of the type of action (infringement, injunction, etc.), the main technology at issue (an item in the database) and publicly available information on the plaintiffs and defendants, I tried to determine which cases in the sample might have involved unfair competition charges against R&D employees who had moved to high technology ventures. I identified only one such case (out of 145). In other words, the number of such cases indeed is probably quite small.

Japanese patent attorneys, whom I have queried do not know of specific court cases dealing with this issue.

265. The threat was reportedly made by the main VC backer of a venture company in response to suggestions that the venture's lead manager might be recruited to form another startup.

266. The challenges in balancing work and family obligations are intense in most modern societies. However, these challenges are particularly severe in Japan and are felt most intensely by women (See, e.g. Tolbert 2000; Otake 2006). For women with minimal family obligations, career opportunities in 2005 are greater and closer to men's than they ever were before. Nevertheless, most Japanese organizations have evolved less than those in the USA to allow both men and women to accommodate the needs of family as well as work. For professionals, obligations to be present at work (or at meetings or social events with work colleagues) in the evening and weekends are frequent. Also flexible work patterns to accommodate care for young children that allow the mother (and/or father) to remain professionally involved in the work of her organization and subsequently to resume a full-time career (with promotion potential not markedly below that of colleagues that did not have to care for children) are rare. In my interviews with the eight largest Japanese pharmaceutical companies in 2002, only one said it hoped that its female researchers would remain employed after marriage, or return to work after taking time off to care for young children and had policies in effect to encourage their return (Kneller 2003). It is common for a married professional woman either to have no children or to have one child and then no more, saying that she does not want to impose further on her work colleagues the inconvenience of her having to take time from work to care for a child.

267. I am surprised at the amount of time children even in elementary grades four to six spend cramming to prepare for examinations to reach preferred schools in the next rung of the educational ladder. Probably about 20% of students in these grades in Tokyo (roughly half the students who will attend some sort of university or junior college after high school) are involved in cramming programs that, for most serious students, leave little unstructured time to explore and develop their own interests. The way these programs impart knowledge (in cram schools or using self-study books) aims at accumulating a lot of factual knowledge, without opportunities to develop skills in self-expression (written or oral) and critical analysis, and also without opportunities to observe nature and to develop interests and skills in experimentation. This phenomenon is also present in other large Japanese cities. Parents say that a generation ago, children spent less time cramming. Cramming to enter elite secondary schools is motivated by parents' belief (largely self-fulfilling) that public junior high schools do not provide a good education that enables their children to be competitive in entrance examinations for senior high schools and universities,

and that bullying and other disruptive behavior pervades public junior high schools. Also by entering an elite private secondary school, students are usually spared an otherwise mandatory round of cramming and test taking to enter senior high school (10th–12th grades)—even public senior high schools admitting students only by examination set by the individual schools. Tokyo metropolitan area statistics show the percentage of sixth graders taking entrance exams for private junior high schools rising from 15% to 17% from 2003 to 2006, but percentages within Tokyo itself are higher (*Asahi Daily* (English edition, combined with the *International Herald Tribune*), Feb. 11–12, 2006, 35).

The public elementary schools, which occasionally organize field trips and have some inquiry-based science classes, scale back homework and the other demands on their students, knowing that a large proportion of students are heavily burdened with cram school assignments. *In other words, the public elementary schools are ceding responsibility for educating children to the cram schools whose avowed goal is to impart knowledge necessary to pass junior high school, high school and university entrance exams.* Of course children from families that value education do not spend sixteen hours per day with their noses in books. A few take part in organized sports (but much less than in the USA). Some take music lessons, although it is disheartening to hear parents frequently say that it is important for their children to take music or art lessons in order to be admitted to elite junior high schools. There is widespread fatigue, angst, and cynicism among children, and many take long breaks to watch television and play computer games. But nevertheless, for children whose parents hope they will attend good universities and have professional careers, from the latter years of elementary school on, most time is structured, and the need to prepare for entrance exams to keep climbing the academic ladder is ever present.

I say this largely out of personal experience observing our eldest daughter progress through the Japanese public elementary school system and conversations with her classmates and parents. (And, yes, we have sent our daughter to two different cram schools over the past four years, and she still attends a cram school to keep up her Japanese.) In fifth grade, we switched her to an international English-language school in Tokyo. We found the level of mathematics, in terms of concepts, equivalent to that taught in the most competitive cram schools (quite high compared to what I studied in US public schools in the 1960s). Moreover, the level of mathematics and Japanese language taught in the cram schools is generally several months ahead of that in the public schools. However, reinforcement of mathematical concepts and problem solving speed is higher among the public school students who attend cram schools compared to my daughter's international school. Both the public and international schools give students time for hands on experimentation and discovery in science classes—the public school devoting more time to science instruction generally, the international school going more deeply into particular topics. But the single glaring contrast between the curricula of the two school systems is the amount of time spent in the international school to develop *skills of self-expression and critical analysis*—in writing, in oral and computer-based presentations, even in drawing—and also the degree to which students are encouraged to do projects or to explore subjects that interest them. The Japanese public school tries to do this to some degree through occasional student projects. However, there is *absolutely none of this in the two cram schools* my daughter has attended, which are both large with dense neighborhood branches nationwide and where students often direct more of their energy than the public schools.

268. i.e. high examination scores and admission to elite schools.

269. For persons interested in R&D careers, they reach this point usually on graduation from a good university.

270. Conversations with friends from well-to-do neighborhoods around Washington, DC suggest convergence towards the Japanese model.

271. See Trends in International Science and Mathematics Study (TIMSS) scores for 2003, where Japanese eighth graders ranked fifth in both mathematics and science while US eighth graders ranked fourteenth in mathematics and eighth–ninth in science (tied with Australia). (Some major developed countries such as France, Germany, and England do not appear in these rankings.) See also the OECD's Program for International Student Assessment (PISA), which compares reading, mathematics and science abilities among students in most OECD countries. These rank Japanese 15-year-olds among the top four while US 15-year-olds score close to the OECD average (data available from the US National Center for Education Statistics).

272. See Westney and Sakakibara (1986*a*, 1986*b*).

273. So far, my discussions have mainly been with University of Tokyo students and faculty. One professor who works closely with many electronics manufacturers put the matter this way.

'It would be difficult for me to try to assign my students to various companies. If I recommended one of my best students go to Company A, Companies B and C might feel I was not treating them fairly. It is better for me to leave the initiative with my students, and to let this be a matter between the students and the companies.'

The situation may be different in less prestigious universities, where professors have to go to bat for their students in order for them to find good jobs in top-tier companies.

274. This is not a question I have so far raised directly with many entrepreneurs and corporate officials, so reality may differ from what I have surmised in the text above. The Chapter 4 case studies of Chip Detect and Internal Search Engine nevertheless seem consistent and show variations in the possible responses of parent companies. In Chip Detect's case, relations with the parent remain distant, although the terms of separation were favorable for Chip Detect. In Internal Search Engine's case, the separation was amicable and both sides

continue to make the relationship mutually advantageous. However, Chip Detect's technology is closer to its parent's core business than is Internal Search Engine's, and more key employees left the parent in the former than the latter's case. Also Internal Search Engine's parent is unique in encouraging its employees to form independent spin-offs. Therefore, not surprisingly, whether departures occur amicably or are regarded as betrayal depends on the number and importance of the employees who leave and how close their work is to the parent's core business.

Those who leave are also sensitive about the notion of *betrayal*. A senior manager in a major electronics company used just this term to describe his initial reaction to overtures that he join a US venture, and also the frequency with which employees moved between US ventures. His new boss, the founder of the venture that recruited him, tried to assuage his guilt and surprise by saying, 'Just think of all these ventures [in this field of technology] as one company.'

275. Because most Japanese merit pay systems have been in effect only a few years, it probably is too early to judge their effectiveness in improving productivity within companies, much less in providing a signal to some employees that they should leave in mid career. (To my understanding, the latter is not a goal of most merit pay systems.) A 2004 survey of both company officials and employees found general agreement that introduction of merit pay increased work effort and output, as well as cooperation among co-workers. But employees also felt that their extra effort was not rewarded in their salaries and bonuses, evaluations often were not fair, and the evaluation criteria did not match work goals (JILPT 2004).

276. Lincoln and Gerlach (2004: 329).

277. For example,

- sustained, widespread interest among university researchers in forming bioventures,
- the ability of patents to prevent encroachment into their core business by large companies that may have a large presence in university laboratories,
- the relative success of bioventures in pursuing niche technologies with commercial potential that are not being pursued by pharmaceutical or other large companies,
- the existence of specialized biomedical VC companies that are able to help ventures grow not only by financial investments, and
- access to university discoveries not enjoyed by ventures in other S&E fields.

278. The decline of large basic-research-oriented corporate laboratories such as Bell and PARC and Sarnoff, and the shift to contract research in the central laboratories of large Japanese corporations, all are indications that most corporate research now is application oriented. Fundamental research that will give rise to the next generation of technologies is now almost the exclusive purview of universities and GRIs.

279. Chesbrough (2003) and IBM's Emerging Business Opportunities Program (which supports ventures and other SMEs developing hardware and software related to IBM's business) indicate that the flow of technology from ventures to large US corporations is substantial and quite important, even outside bio-medicine. This is substantiated by discussions with officials of IBM, Johnson & Johnson (with respect to medical devices) and other companies, as well as data on university licensing presented in Part II above.

280. See notes 85, 178–179 and 202 above, as well as Chapter 2.

281. i.e. research to bridge the gap between basic research findings and development aimed specifically at commercialization. Examples of translational research might include development of a working prototype, research into methods to scale up production, and in the case of drugs, studies with live animals showing proof of concept.

282. This is probably a valid assumption in view of the sharp rise in industry-sponsored joint research as well as the rise in government-sponsored commissioned research, which includes government-sponsored consortium research.

283. See references cited under notes 3–5. What about the lower level engineers, scientists, programmers, technicians, and secretaries in ventures, those who do not have great say in management and may not even have large stock options? To what extent does a venture's success depend on motivating all its employees? Assuming that such motivation is important (but that stock options are not distributed equally) how do ventures sustain a sense of shared commitment in the face of potential job cuts during economic downturns or in the event of mergers/acquisitions? These are questions about which surprisingly little seems to be written—yet the answers bear on the lives and careers of many Americans, as well as the sustainability of what appears to be a unique, and so far successful, innovation system.

284. The ever present danger of earthquakes may have contributed to the preference for individual homes. However, apartment buildings are now built to withstand sizable quakes and most families still say they prefer a house of their own.

285. Perhaps characteristic Japanese qualities such as politeness, reluctance to confront others, and reticence except among close associates are, at their most basic level, means to maintain social harmony while at the same time preserving and respecting a core sphere of individual privacy. However, these same qualities may not be well suited for ventures which thrive on rapid communication and ability to make deals rapidly, sometimes with previously unknown entities.

In this regard, see Yamagishi, Cook, and Watabe (1998) and other works by Yamagishi suggesting that general trust (the tendency to trust another person regardless of whether he or she is bound by the same stable social relations, i.e. is a member of the same family or work group) is lower in Japan than America, and this is due largely to the closed nature of key social groups in Japan, particularly work-related groups.

286. Or, in the case of universities, let them have expression in semi-autonomous fiefdoms (*kouzas*) subject to only weak departmental and university-wide coordination.
287. See also Whittaker (1997) and Friedman (1988).
288. e.g. Sony and a few of the machine tool companies.
289. At least on the shop floor and within work units.
290. Aoki (1990). The typical western corporation Professor Aoki has in mind is one that relies on hierarchical authority and explicit written instructions and procedures (the opposite of tacit knowledge) for control and coordination. Whether this is an accurate characterization of successful US manufacturing corporations today may merit clarification.
291. Musha (2006).
292. This is, of course, a risky prediction offered not as much out of confidence that it will come to pass, but to catalyze further discussion. One reason it may be proved wrong is that large Japanese manufacturers may devise a uniquely effective system of close cooperation with universities. This would go beyond simply appropriating the bulk of university IP, to involve students and professors helping companies to make strategic decisions as to which new technologies to pursue, even though commercialization may be far in the future. In other words, universities would not only play a role akin to that of Bell Labs, PARC or Sarnoff Research Center, but they would also help companies to conduct strategically targeted translational research to demonstrate the commercial feasibility of some of their key discoveries. Some companies are establishing research centers in major universities, which suggests steps in this direction.

Nevertheless, I do not think this is a scenario that many professors, corporate managers, or government officials consider feasible or desirable. Fundamental culture and goals are too distinct between universities and corporate laboratories. The resources that both sides would have to commit to carry out such a transformation of university research would be too great, and universities would be loath to lose so completely their neutrality vis-à-vis competing companies. Nevertheless, there are some concrete steps in this direction, such as the contractual right, noted in Chapter 3, of companies sponsoring joint research to censor academic publications—a right which some researchers indicate is frequently exercised, although usually in the form of requests for changes. If it did come about in some sort of creeping manner (perhaps in response to reduction of government support for university R&D, or, conversely the government increasing funding for such centers in an explicit attempt to aid Japanese industry), it would be to the detriment of Japanese science. It would also be a further blow to startups, whose access to university discoveries and the energy of university researchers would be even more curtailed. This scenario may be less far fetched, however, in the case of other countries where university R&D is more dependent on corporate joint research funding, e.g. China and Korea.

293. 'Biotech, then and now' (interview with George Rathmann), *Business Week Online*, Sept. 19, 2005.
294. VC investment in the USA was US$21.7 billion and US$21.6 billion in 2005 and 2004, respectively, compared with US$18.9 billion in 2003 and US$104.8 billion in 2000, the recent nadirs and peaks (National Venture Capital Association (NVCA) at www.nvca.org).
295. In 2003, 21% of total industry R&D in the USA was conducted by companies with less than 500 employees, compared with 6% in 1984 (Global Insight 2004).
296. OECD (2004*b*). Between 1999 and 2002, 0.4% of US GDP was accounted for by VC investment in high technology sectors, compared with 0.64% for Israel and about 0.36% for Canada, the next closest country. At the other end of the spectrum of OECD countries, high-tech VC investments accounted for about 0.01% of Japan's GDP, 0.05% of Italy's, 0.06% of Australia's, 0.08% of Germany's, 0.09% of Denmark's, 0.11% of France's, 0.12% of Finland's, 0.14% of the Netherlands's, 0.17% of Korea's, 0.21% of the UK's, and 0.26% of Sweden's. (Sweden is the fourth ranked country after Canada.) (Data for Australia, Japan, and Korea are from 1998–2001.) The percentage for Japan has probably increased since 2001 (see Chapter 5).
297. Most ventures tell me they try to match salaries in large industry leaders to within 5–10%, but they cannot match on pensions and other benefits, except through stock options.
298. See Thelen and Kume (2003), n. 19, with respect to Germany.
299. See Jaeger and Stevens (1999) which shows, e.g., that about 52% of US male heads of households in 1996 had been employed in the same organization less than ten years compared with about 42% in 1982. Higher education was associated with greater likelihood of job change. Kambourov and Manovskii (2004) also found that mobility between and within US industries increased for all age and education groups between 1968 and 1993, which they attributed partly to 'the increasing variability of occupational demand shocks over time'.
300. See OECD (2004*a* & *b*) and n. 296. The OECD has established a scale to measure employment protection legislation (EPL) that takes into account protections against individual dismissal, protections against collective layoffs, and regulations on temporary employment. The USA scores lowest on this scale, followed by the UK, Canada, New Zealand, Ireland, Australia, Switzerland, Hungary, Japan, and so forth. According to this scale, EPL protection in Japan is about two to three times stronger than in the USA, while in Italy, Germany, Belgium, Norway, Sweden and France, it is about four times stronger than in the USA.

    The Pearson correlation coefficient for a simple comparison of percentage of GDP invested in high-tech VCs and strength of EPL is −0.52. Japan's level of VC investment is lower than expected on the basis of its moderate level of EPL, while Sweden's level of VC investment is higher than expected on the basis of its high EPL. Remove Japan and the correlation coefficient becomes

−0.55, remove Sweden and it becomes −0.61, remove both and it becomes −0.65.

Of course, there are other determinants of high technology entrepreneurship than the level of employment protection. (In the case of Japan, these are discussed above.) Nevertheless, this simple analysis suggests a causal relationship between the ease with which companies can dismiss employees and the suitability of that environment for ventures.

301. Casper (2000).
302. Leave Development to Us (Chapter 4) is one of the earliest ventures founded on this strategy, but subsequently others have been formed.
303. Thelen and Kume (2003).
304. See, e.g. Begley (2002) and Hyde (2003). The senior Japanese R&D manager with experience in both Japanese and US IT companies (see n. 274 above) expressed shock at how scientists and executives in a US company reported to work one morning to find notices that they had to clean out their offices by the end of the day and would be given only two weeks' severance pay.
305. See Hyde (2003) and notes 3–5.
306. Factors that make such pressures higher than in large Japanese or German companies may include pressures to return value to stockholders and the importance of stock incentives in executive compensation packages.
307. The varieties of capitalism literature draw a number of contrasts between liberal market economies (LMEs, of which the USA is the archetypal example, but which also include the UK, Canada, Australia, and New Zealand) and coordinated market economies (CMEs, among which Germany is probably the archetype, but among which Japan and Korea are often included, along with many other countries of Continental, especially northern, Europe). These include:

- Organization of production by arm's-length market transactions vs. coordination among the principal actors (employers' associations, unions, and government),
- financing and control via equity investors vs. loans or equity investments by banks or other corporations,
- at-will hiring and firing of employees vs. legal or cultural restrictions on the same,
- low vs. high levels of in-house or employer-funded training, and
- technology transfer via labor mobility vs. inter-company associations.

(Hall and Soskice 2001)

Because these distinctions are inter-related and based on established social and institutional norms, shifting from one type of system to the other is not easy. (See also Whitley 2000.)

308. Hall and Soskice (2001: 42–3) present graphical data showing that the relative innovativeness by industry of US companies (as measured by the proportion of European patents they hold) is almost a mirror image of the relative innovativeness of German companies. For example, a higher than average proportion

of total European patents in biotechnology, pharmaceuticals, medical engineering, optics, semiconductors, and IT were issued to US companies, while for all these same industries except pharmaceuticals, the proportion issued to German companies was less than expected. But on the other hand, in transportation, agricultural machines, mechanical elements, engines, and thermal processes, the proportion of European patents issued to German companies was higher than average, while the proportion issued to US companies was lower than average. Without much explanation, Hall and Soskice assert that the former group of industries is characterized by radical innovation while the latter group is characterized by incremental innovation.

309. See previous note and the figures in Chapter 1.

310. This may be the primary rationale for not only a liberal market type of capitalism but also for strong easy to use patent protection, and a system of ownership of university IP that encourages university, faculty and student entrepreneurship. Shift from a liberal to a coordinated market economy, undermine IP protection, shift to a Japanese style of university–industry cooperation, and large companies will survive, but innovation by new companies would be crippled.

311. This is especially so if inventions are assumed to be radical or incremental primarily on the basis of the industry in which they arise (i.e. if industrial classification becomes a surrogate for innovativeness). For example, was the introduction of computerized numerical controllers (pioneered largely by Fanuc) in the machine tool industry radical or incremental? Hall and Soskice (2001: 39) indicate the machine tool industry is characterized by incremental innovation, but that particular innovation (and some of those that it enabled) would probably best be characterized as radical. Streb's (2003) history of the German chemical cartel, I.G. Farben, and its core progeny, BASF, Bayer and Agfa, describes how these leading companies (in an industry, polymers, that seems to fit neatly into neither classification) produced a mixture of incremental and radical innovations, with radical innovations being more prevalent in its interwar and early postwar years as opposed to the late postwar years.

In addition, attributing leadership to the USA on the basis of patent numbers in industries supposedly characterized by radical innovation should be done with great caution in view of the tendency for some companies in these industries to amass large patent portfolios to be used as strategic arsenals in negotiations with other companies (see Part II above).

312. Preferably on a patent-by-patent or product-by-product, rather than an industry-by-industry, basis.

313. I have shown this in the case of pharmaceuticals. Goodman and Myers (2005) have shown this in the case of *essential* third generation mobile telecommunication patents, although the dominance of new companies is attributed almost entirely to Qualcomm, with Nokia, Ericsson, Motorola, and NTT DoCoMo trailing by significant margins. (Nortel, Lucent, and IBM did not declare their patents to the *International Third Generation Partnership Projects* to establish standards to aid component parts manufacturers and to ensure a degree of systems interoperability. Thus their contributions are not assessed.)

314. In Germany in early 2006, the Christian Democrats and the Social Democrats agreed that the trial period for newly hired employees, during which they could be dismissed relatively easily, should be extended from six to twenty-four months. Similar draft legislation in France sparked riots in early 2006 leading to suspension of the reform efforts.

315. For descriptions of employer–employee relations in Germany, see Hall and Soskice (2001) and Thelen and Kume (2003).

316. Prior to this law, profits from such sales were taxed at around 40%. Dougherty Carter (2005). 'Less "Germany Inc."—more outward', *International Herald Tribune* (*New York Times*), Sept. 3–4, 9.

317. Vogel (2003).

318. At least, however, there are the beginnings of political debate in Germany concerning job security. See Survey of Germany: 'Waiting for a Wunder and Squaring the Circle', *Economist*, February 9, 2006.

319. In other words, joint and commissioned research contracts between companies and universities must contain an irrevocable override of article 73 of Japan's Patent Law, lest coinventing companies be able to do as they wish, for free, with university coinventions, while denying universities the option of licensing to third parties.

320. In particular, when the only information at issue is in the moving person's head, courts should probably require the plaintiff company to show a high likelihood of specific, substantial injury meaning that (*a*) it would have to halt a particular project from which it expected to gain significant revenue or (*b*) if it continued with the project, sales revenue would likely be substantially reduced. These criteria are similar to those proposed by Hyde (2003). Courts should require that trade secrets covered under no-compete clauses and the LPUC be specifically and narrowly defined by companies, and not simply cover almost every facet of the work of researchers and R&D managers.

321. For example, occasional windfall profits for universities should not be viewed as a sign the system of university–industry technology transfer is failing, in view of the importance of universities and biotechs in discovering and developing innovative drugs. However, teaching and research should remain the paramount missions of universities. Conflicts of interest must be managed so as to ensure these goals are upheld and that the safety of research subjects is not jeopardized. The basic principle underlying NIH guidelines on licensing of research tools and genome inventions, *that publicly funded inventions should be licensed nonexclusively unless exclusive licenses are necessary to promote their development* (e.g. by venture companies) should probably be applied to most publicly funded university inventions. Fortunately, it appears that many leading universities are complying with the spirit of these guidelines trying to promote wide access to university discoveries, while at the same time providing venture companies with the exclusive rights they need if it appears that the ventures are the best candidates to develop the inventions (Pressman, Cook-Deegan, McCormack et al. 2006).

322. See Jaffe and Lerner (2004).

REFERENCES

Acs, Zoltan J. and Audretsch, David B. (1987). 'Innovation, Market Structure, and Firm Size', *Review of Economics and Statistics*, 69(no. 4, Nov.): 567–74.

————— (1988). 'Innovation in Large and Small Firms: An Empirical Analysis', *American Economic Review*, 78(no. 4, Sept.): 678–90.

————— (1991). 'Innovation And Technological Change: An Overview' (ch. 1), and 'R&D, Firm Size and Innovative Activity' (ch. 3), in Acs and Audretsch (eds.), *Innovation and Technological Change: An International Comparison*. Ann Arbor, MI: University of Michigan Press, pp. 1–23, 39–59.

Anderson, Stuart (1996). 'Foreign-born Engineers and Scientists Don't Undercut Wages: They Earn More', online report from the Cato Institute, Washington, DC at www.cato.org

Aoki, Masahiko (1990). 'Toward an Economic Model of the Japanese Firm'. *Journal of Economic Literature*, 28(March): 1–27.

Arora, Ashish and Merges, Robert P. (2004). 'Specialized Supply Firms, Property Rights and Firm Boundaries', *Industrial and Corporate Change*, 13(3): 451–75.

Association of University Technology Managers (AUTM). 'Licensing Surveys and Survey Summaries for Designated Years', available through www.autm.org

Baba, Yasunori, Yarime, Masaru, Shichijo, Naohiro, and Nagahara, Yuichi (2004). 'The Role of University–Industry Collaboration in New Materials Innovation: Evolving Networks of Joint Patent Applications', paper presented at the International Schumpeter Society Conference 'Innovation, Industrial Dynamics and Structural Transformation: Schumpeterian Legacies', Milan, June 9–12.

Barley, Stephen R. and Kunda, Gideon (2004). *Gurus, Hired Guns, and Warm Bodies: Itinerant Experts in a Knowledge Economy*. Princeton, New Jersey: Princeton U. Press.

Barnett, Jonathan M. (2003). 'Private Protection of Patentable Goods', Research Paper 28, Fordham University School of Law, available http://ssrn.com/abstract-434585

Begley, Sharon (2002). 'Angry Engineers Blame Shortage on Low Pay, Layoffs and Age Bias', *Wall Street Journal*, July 5, A9.

Block, Z. and Ornati, R. (1987). 'Compensating Corporate Venture Managers', *Journal of Business Venturing*, 2: 41–51.

Branchflower, David and Oswald, Andrew (1992). 'Entrepreneurship, Happiness and Supernormal Returns: Evidence from Britain and the US', NBER Working Paper No. 4228. Washington, DC: National Bureau of Economic Research.

————— (1998). 'What Makes an Entrepreneur?' *Journal of Labor Economics*, 16:26–60.

Branchflower, David and Oswald, Andrew and Stutzer, Alois (2001). 'Latent Entrepreneurship Across Nations', *European Economic Review*, 45(May): 680–91.

Branstetter, Lee G. and Sakakibara, Mariko (2002). 'When Do Research Consortia Work Well and Why? Evidence from Japanese Panel Data', *American Economic Review*, 92(1): 143–59.

Callon, Scott (1995). *Divided Sun: MITI and the Breakdown of Japanese High-Tech Industrial Policy, 1975–1993*. Stanford, CA: Stanford University Press.

Calomiris, Charles W. and Himmelberg, Charles P. (March 1995). 'Government Credit Policy and Industrial Performance (Japanese Machine Tool Producers, 1963–91)', World Bank Policy Research Working Paper No. 1434.

Carlson, Bo (1990). 'Small-Scale Industry at a Crossroads: U.S. Machine Tools in Global Perspective', in Acs and Audretsch (eds.), *The Economics of Small Firms*. Dordrecht, Netherlands: Kluwer, pp. 171–95.

Casper, Steven (2000). 'Institutional Adaptiveness, Technology Policy, and the Diffusion of New Business Models: The Case of German Biotechnology', *Organization Studies*, 21(5): 887–914.

Chandler, Alfred D. (1990). *Scale and Scope: The Dynamics of Industrial Capitalism*. Cambridge, MA: Belknap (Harvard University Press).

—— (2001). *Inventing the Electronic Century: The Epic Story of the Consumer Electronics and Computer Industries*. New York: Free Press.

Chesbrough, Henry W. (1999). 'The Organizational Impact of Technological Change: a Comparative Theory of National Institutional Factors', *Industrial and Corporate Change*, 8(3): 447–85.

—— (2003). *Open Innovation: The New Imperative for Creating and Profiting from Technology*. Boston, MA: Harvard Business School Press.

Christensen, Clayton M. (1993). 'The Rigid Disk Drive Industry: A History of Commercial and Technological Turbulence', *Business History Review*, 67: 531–88.

—— and Raynor, Michael E. (2003). *The Innovator's Solution*. Boston, MA: Harvard Business School Press.

Chuma, Hiroyuki (2001). 'Sources of Machine-Tool Industry Leadership in the 1990s: Overlooked Intrafirm Factors', Discussion Paper No. 837. Yale University Economic Growth Center.

Cockburn, Iain and Henderson, Rebecca (1997). 'Public–Private Interaction and the Productivity of Pharmaceutical Research', Working Paper No. 6018. National Bureau of Economic Research.

Cohen, Wesley M. and Klepper, Steven (1991). 'Firm Size Versus Diversity in the Achievement of Technological Advance', in Acs and Audretsch (eds.), *Innovation and Technological Change: An International Comparison*. Ann Arbor, MI: University of Michigan Press, pp. 183–203.

—— Nelson, Richard R., and Walsh, John P. (2000). 'Protecting Their Intellectual Assets: Appropriability Conditions and Why U.S. Manufacturing Firms Patent (or Not)', NBER Working Paper No. 7552. Cambridge, MA: National Bureau of Economic Research.

Coleman, Samuel (1999). *Japanese Science: From the Inside*. London: Routledge.

Corey, E. Raymond (1997). *Technology Fountainheads: The Management Challenge of R&D Consortia*. Boston, MA: Harvard Business School Press.

Crabtree, Penni. 2006. New drug price tag: $1.2 billion—Biotechs criticized for rising treatment costs. San Diego Union-Tribune. 10 Nov. 2006.

Cukier, Kenneth (2005). 'Survey: Patents and Technology', *The Economist*, Oct. 20.

Dahms, A. Stephen and Trow, Stephen C. (May 2005). 'US Biotechnology Companies and Foreign Nationals: The Changing Dynamics of Access to H-1B Visas', *Nature Biotechnology*, 23(5): 629–30.

Dibner, Mark D. (1988). *Biotechnology Guide U.S.A.*, 1st edn. New York: Stockton Press.

—— (1995). *Biotechnology Guide U.S.A.*, 3rd edn. Research Triangle Park: Institute for Biotechnology Information.

—— (ed.). (1999). *Biotechnology Guide U.S.A.* 5th edn. London: Macmillan Reference.

Doeringer, Peter B., Evans-Klock, Christine, and Terkla, David G. (January 1998). 'Hybrids or Hodgepodges? Workplace Practices of Japanese and Domestic Startups in the United States', *Industrial and Labor Relations Review*, 51(2): 171–86.

Etzkowitz, Henry (2002). *MIT and the Rise of Entrepreneurial Science*. London: Routledge.

Feldman, Maryann, Feller, Irwin, Bercovitz, Janet, and Burton, Richard (2002). 'Equity and the Technology Transfer Strategies of American Research Universities', *Management Science*, 49(1): 105–21.

FitzRoy, Felix R. and Kraft, Kornelius (1991). 'Firm Size, Growth and Innovation: Some Evidence from West Germany', in Acs and Audretsch (eds.), *Innovation and Technological Change: An International Comparison*. Ann Arbor, MI: University of Michigan Press, pp. 152–9.

Flanigan, James (2005). 'Why High-Tech Jobs Fly Off to India: Foreign Brainpower Lets Research-Oriented U.S. Firm Cut Costs', *International Herald Tribune* (*New York Times*), Nov. 18, 13.

Florida, Richard L. and Kenney, Martin (1988). 'Venture Capital-Financed Innovation and Technological Change in the USA', *Research Policy*, 17: 119–37.

Fransman, Martin (1995). *Japan's Computer and Communications Industry: The Evolution of Industrial Giants and Global Competitiveness*. Oxford: Oxford University Press.

Friedman, David (1988). *The Misunderstood Miracle: Industrial Development and Political Change in Japan*. Ithaca, NY: Cornell University Press.

Ganguli, Ishani (2006). 'Complement', *The Scientist*, 20(7): 34.

Gans, Joshu S., Hsu, David H., and Stern, Scott (2000). 'When Does Start-up Innovation Spur the Gale of Creative Destruction', NBER Working Paper No. 7851. Cambridge, MA: National Bureau of Economic Research.

Gilson, Ronald J. (1999). 'The Legal Infrastructure of High Technology Industrial Districts: Silicon Valley, Route 128, and Covenants Not to Compete', *New York University Law Review*, 74(June): 575–627.

Global Insight (2004). 'Venture Impact 2004: Venture Capital Benefits to the U.S. Economy' (available at www.nvca.org).

Goodman, David J. and Myers, Robert A. (2005). 3G Cellular Standards and Patents. Proceedings of IEEE WirelessCom 2005, June 13–16. Available at http://www.frlicense.com/wireless2005-b.pdf

Gorman, Robert A. and Ginsburg, Jane C. (1993). *Copyright for the Nineties: Cases and Materials*. Charlottesville, VA: Michie.

Hall, Bronwyn H. and Ziedonis, Rosemarie Ham (2003). 'The Patent Paradox Revisited: An Empirical Study of Patenting in the U.S. Semiconductor Industry', 1979–1995', *RAND Journal of Economics*, 32(1): 101–28.

Hall, Peter A. and Soskice, David (2001). 'An Introduction to Varieties of Capitalism', in Hall and Soskice (eds.), *Varieties of Capitalism: The Institutional Foundations of Comparative Advantage*. Oxford: Oxford University Press.

Hara, Takuji (2003). *Innovation in the Pharmaceutical Industry: The Process of Drug Discovery and Development*. Cheltenham, UK: Elgar.

Hayashi, Takayuki (2003). 'Effect of R&D Programmes on the Formation of University–Industry–Government Networks: Comparative Analysis of Japanese R&D Programmes', *Research Policy*, 32: 1421–42.

Hecker, Daniel (2005). 'High Technology Employment: A NAICS-Based Update', *Monthly Labor Review* (July): 57–72.

Hutchinson, Jamie (2002). 'Plasma Display Panels: The Colorful History of an Illinois Technology', Electrical and Computer Engineering Alumni News, University of Illinois at Urbana-Champaign 36 (no. 1). Available at http://www.ece.uiuc.edu/pubs/plasma/plasma1.html

Hyde, Alan (2003). *Working in Silicon Valley*. Armonk, NY: M.E. Sharpe.

Jaeger, David A. and Stevens, Ann Huff (1999). 'Is Job Stability in the United States Falling?' Discussion Paper Series No. 35. IZA, Bonn, Germany: Institute for the Study of Labor Law.

Jaffe, Adam B. and Lerner, Josh (2004). *Innovation and Its Discontents: How Our Broken Patent System Is Endangering Innovation and Progress, and What to Do About It*. Princeton, NJ: Princeton University Press.

Japan Institute for Labour Policy and Training (JILPT) (2004). 'Survey of Employee Work Motivations and Employment Management' [Roudousha no hataraku iyoku to koyou kanri no arikata ni kansuru chousa], available at www.jil.go.jp

Japan Productivity Center for Socio-Economic Development (JPC-SED) (2005). 'Survey of Attitude Changes after Half a Year Among FY 2005 New Company Employees' (Summary), [2005 Nendo shinniu sha-in hantoshikan no ishiki henka chousa].

Jensen, Richard and Thursby, Marie (2001). 'Proofs and Prototypes for Sale: The Licensing of University Inventions', *American Economic Review*, March: 240–59.

Johnson, Chalmers (1983). *MITI and the Japanese Miracle: The Growth of Industrial Policy, 1925–75*. Stanford, CA: Stanford University Press.

Kalafsky, Ronald V. and MacPherson, Alan D. (2002). 'The Competitive Characteristics of U.S. Manufacturers in the Machine Tool Industry', *Small Business Economics*, 19: 355–69.

Kambourov, Gueorgui and Manovskii, Iourii (2004). 'Rising Occupational and Industry Mobility in the United States: 1963–1993'. Discussion Paper Series No. 1110. IZA, Bonn, Germany: Institute for the Study of Labor Law.

Kneller, Robert W. (2003). 'Autarkic Drug Discovery in Japanese Pharmaceutical Companies: Insights into National Differences in Industrial Innovation', *Research Policy*, 32: 1805–27.

—— (2005a). 'Correspondence: The Origin of New Drugs', *Nature Biotechnology*, 23(5) (May): 529–30.

—— (2005*b*). 'Correspondence: The National Origins of New Drugs', *Nature Biotechnology*, 23(6): 655–6.

Kodama, Fumio (1991). *Emerging Patterns of Innovation: Sources of Japan's Technological Edge*. Boston, MA: Harvard Business School Press.

Kortum, Samuel and Lerner, Josh (1999). 'What Is Behind the Recent Surge in Patenting?' *Research Policy*, 28: 1–22.

Kuzunoki, Ken and Numagami, Tsuyoshi (August 1998). 'Interfunctional Transfers of Engineers in Japan; Empirical Findings and Implications for Cross-Functional Integration', *IEEE Transactions on Engineering Management*, 45(3): 250–62.

Laage-Hellman, Jens (1997). *Business Networks in Japan: Supplier–Customer Interaction in Product Development*. London: Routledge.

Leaf, Clifford (2005). 'Law of Unintended Consequences', *Fortune*, Sept. 7.

Lerner, Josh (1995). 'Patenting in the Shadow of Competitors', *Journal of Law and Economics*, 38(Oct.): 463–95.

Lincoln, James R. and Gerlach, Michael L. (2004). *Japan's Network Economy: Structure, Persistence, and Change*. Cambridge, MA: Cambridge University Press.

Lynn, Leonard H., Piehler, Henry R., and Kieler, Mark (1993). 'Engineering Careers, Job Rotation, and Gatekeepers in Japan and the United States', *Journal of Engineering and Technology Management*, 10: 53–72.

Markoff, John (2005). 'Silicon Valley Loves—and Fears—China: Venture Capitalists See Boomerang Risk', *International Herald Tribune* (*New York Times*), Nov. 5–6, 1.

McKelvey, Maureen (1996). *Evolutionary Innovations: The Business of Biotechnology*. Oxford: Oxford University Press.

Merges, Robert P. (1995). 'Intellectual Property and the Costs of Commercial Exchange: A Review Essay', *Michigan Law Review*, 93(May): 1570–615.

—— (2004). 'A New Dynamism in the Public Domain', *University of Chicago Law Review*, 71: 183–203.

Murray, Fiona (2002). 'Innovation as Co-Evolution of Scientific and Technological Networks: Exploring Tissue Engineering', *Research Policy*, 31: 1389–403.

Musha, Ryoji (2006, Vice Chairman Deutsche Securities). Interview in '*Weighing in on the Japanese Economy*', *International Herald Tribune* (*New York Times*), May 27–8, 16.

Myers, Robert A. (1999). 'The IBM Origins of Display Technologies, Incorporated', Discussion Paper MITJP 99-01 of the MIT Japan Program, Science, Technology, Management. Cambridge, MA: Center for International Studies, Massachusetts Institute of Technology.

National Science Board (2004). *Scientific and Engineering Indicators 2004* (two volumes). Arlington, VA: National Science Foundation.

Organization for Economic Cooperation and Development (OECD) (2004*a*). *OECD Employment Outlook 2004*, ch. 2: Employment Protection Regulation and Labour Market Performance. Paris: OECD.

—— (2004*b*). 'Venture Capital: Trends and Policy Recommendations'. Report Prepared by Baygan, Gunseli of the OECD Secretariat, available at http://www.oecd.org/dataoecd/4/11/28881195.pdf

Okada, Yosuke and Kushi, Takahito (2004). 'Government-Sponsored Consortium Research in Japan. A Case Study of the OPSR Program', OPIR Research Paper Series No. 22. Available at http://www.jpma.or.jp/opir/research/paper_22.pdf

O'Neil, Maryadele J. (ed.) (2006). *The Merck Index: an Encyclopedia of Chemicals, Drugs and Biologicals, 14th Edition.* Whitehouse Station, New Jersey: Merck Research Laboratories.

Osigo, Yoshinori and Matsuzaki, Yusuke (2006). '___ Aims to Be All-Around Player', *Nikkei Weekly*, March 6, 12.

Otake, Tomoko (2006). 'Women in Japan: Equality Still Has a Long Way to Go', *Japan Times*, March 12.

Paral, Rob and Johnson, Benjamin (2004). 'Maintaining a Competitive Edge: The Role of the Foreign-Born and U.S. Immigration Policies in Science and Engineering', *Immigration Policy in Focus* (Immigration Policy Center) (August): 1–15.

Pavitt, Keith, Robson, Michael, and Townsend, Joe (1987). 'Size Distribution of Innovating Firms in the UK: 1945–1983', *Journal of Industrial Economics*, 55(March): 291–316.

Pressman, Lori, Burgess, Richard, Cook-Deegan, Robert M., McCormack, Stephen J., Nami-Wolk, Io, Soucy, Melissa, and Walters, Leroy (2006). 'The Licensing of DNA Patents by US Academic Institutions', *Nature Biotechnology*, 24: 31–9.

'Protein Engineering in Japan: Report of a U.S. Protein Engineering Study Team on its Visit to Japan' (1992). US Government limited circulation report.

Rader, Ronald A. (2006). *Biopharmaceutical Products in the U.S. and European Market, 5th edition.* Rockville, Maryland: Bioplan Associates.

Rai, Saritha (2005). 'India Gains in Role of Designing Chips'. *International Herald Tribune* (*New York Times*), Nov. 19–20, 13.

Rathmann, George B. (1991). 'Biotechnology Startups', in Vivian Moses, Ronald E. Cape (eds. of 1st edn. 1991) and Derek G. Springham (ed. of 2nd edn. 1999), *Biotechnology: The Science and the Business.* Amsterdam: Harwood Academic, pp. 47–58.

Roessner, David, Bozeman, Barry, Feller, Irwin, Hill, Christopher, and Newman, Nils (1997). 'The Role of NSF's Support of Engineering in Enabling Technological Innovation', Final Report Prepared for the National Science Foundation by SRI International. Arlington, VA: SRI International.

Rosenberg, Nathan (1998). 'Technological Change in Chemicals: The Role of University–Industry Relations', In Ashish Arora, Ralph Landau, and Nathan Rosenberg (eds.), *Chemicals and Long-Term Economic Growth: Insights from the Chemical Industry.* New York: John Wiley & Sons, pp. 193–230.

Rothwell, Roy and Dodgson, Mark (1994). 'Innovation and Size of Firm', in M. Dodgson and R. Rothwell (eds.), *Handbook of Industrial Innovation.* Cheltenham, UK: Elgar.

Sakakibara, Mariko (1997). 'Evaluating Government-Sponsored R&D Consortia in Japan: Who Benefits and How?', *Research Policy*, 26: 447–73.

—— (2001). 'Cooperative Research and Development: Who Participates and In Which Industries Do Projects Take Place?', *Research Policy*, 30: 993–1018.

Saxenian, AnnaLee (1994). *Regional Advantage: Culture and Competition in Silicon Valley and Route 128*. Cambridge, MA: Harvard University Press.

—— (1999). *Silicon Valley's New Immigrant Entrepreneurs*. San Francisco, CA: Public Policy Institute of California.

—— and Hsu Jinn-Yuh (2001). 'The Silicon Valley–Hsinchu Connection: Technical Communities and Industrial Upgrading', *Industrial and Corporate Change* 10(no. 4): 893–920.

Scherer, F. M. (1991). 'Changing Perspectives on the Firm Size Problem', in Zoltan J. Acs and Joseph Schumpeter (eds.), *1942. Capitalism, Socialism and Democracy*. New York: Harper Perennial.

Shane, Scott (2001). 'Technological Opportunities and New Firm Creation', *Management Science*, 47:(Feb.): 205–20.

—— (2004). *Academic Entrepreneurship: Academic Spinoffs and Wealth Creation*. Cheltenham, UK: Elgar.

Silicon Valley Network (2005). *Index of Silicon Valley* (at www.jointventure.org).

Streb, Jochen (2003). 'Shaping the National System of Inter-Industry Knowledge Exchange: Vertical Integration, Licensing and Repeated Knowledge Transfer in the German Plastics Industry', *Research Policy*, 32: 1125–40.

Takahashi, Hiroshi (2007). *Jouhou tsuushin kakumei: inobaeshion ni taisuru seifu no yakuwari* [The politics of the telecommunications revolution: the role of government in innovation]. (Doctoral dissertation in Japanese submitted to RCAST, University of Tokyo, and approved in Feb.)

Thelen, Kathleen and Kume, Ikuo (2003). 'The Future of Nationally Embedded Capitalism: Industrial Relations in Germany and Japan', in Kozo Yamakura and Wolfgang Streeck (eds.), *The End of Diversity? Prospects for German and Japanese Capitalism*. Ithaca, NY: Cornell University Press.

Thursby, Jerry and Thursby, Marie (2003). 'Industry/University Licensing: Characteristics, Concerns and Issues from the Perspective of the Buyer', *Journal of Technology Transfer*, 28: 207–13.

Tolbert, Kathryn (2000). 'Women Opt for Jobs Instead of Children', *Japan Times*, Aug. 22, 3 (reprinted from the *Washington Post*).

Uchitelle, Louis (2005). 'If You Can Make It Here', *New York Times*, Sept. 4, sec. 3 p. 1.

Vogel, Steven K. (2003). 'The Re-Organization of Organized Capitalism: How the German and Japanese Models are Shaping Their Own Transformations', in Kozo Yamakura and Wolfgang Streeck (eds.), *The End of Diversity? Prospects for German and Japanese Capitalism*. Ithaca, NY: Cornell University Press.

Westney, D. Eleanor and Sakakibara, Kiyonori (1986a). 'The Role of Japan-Based R&D in Global Technology Strategy', in Mel Horwitch (ed.), *Technology and the Modern Corporation*. New York: Pergamon. 217–32.

———— (1986b). 'Designing the Designers: Computer R&D in the United States and Japan', *Technology Review*, April: 24–69.

Whitley, Richard (2000). *Divergent Capitalisms: The Social Structuring and Change of Business Systems*. Oxford: Oxford University Press.

Whittaker, D. H. (1997). *Small Firms in the Japanese Economy*. Cambridge, MA: Cambridge University Press.

Willoughby, Kelvin W. (1997). *New York's Evolving Bioscience Industries: Managing Knowledge, Production and Services for Economic Development*. New York: Center for Biotechnology & New York Biotechnology Association.

—— (1998). 'Utah's Bioscience Technology Industry Complex: A Census and Analysis of the Biotechnology, Medical Technology and Life-systems Technology Industries in the State of Utah'. Salt Lake City: Utah Life Science Association & State of Utah Division of Business and Economic Development.

—— (2005*a*). 'Intellectual Property Management in Entrepreneurial Technology Firms in the United States', Presentation at RCAST, University of Tokyo, 25 August 2005. Also additional analyses related to these data conducted July, 2005.

—— (2005*b*). 'How Do Entrepreneurial Technology Firms Really Get Financed and What Difference Does it Make?', Paper for the U.S. Association for Small Business and Entrepreneurship, Technology Entrepreneurship Division, Annual Conference, Tucson, AZ, January 12–15, 2006.

Wilson, John W. (1985*a*). *The New Venturers: Inside the High Stakes World of Venturing*. Reading, MA: Addison-Wesley.

—— (1985*b*). 'America's High-Tech Crisis: Why Silicon Valley Is Losing Its Edge', *Business Week*, March 11.

Wulf, William A. (President of the National Academy of Engineering) (2005). The importance of foreign-born scientists and engineers to the security of the United States. Statement before the Subcommittee on Immigration, Border Security, and Claims of the Committee on the Judiciary of the US House of Representatives (Sept. 15).

Yamagishi, Toshio, Cook, Karen S., and Watabe, Motoki (1998). 'Uncertainty, Trust, and Commitment Formation in the United States and Japan', *American Journal of Sociology*, 104(1): 165–94.

Zucker, Lynne G. and Darby, Michael R. (1997). 'Present at the Biotechnology Revolution: Transformation of Technological Identity for a Large Incumbent Pharmaceutical Firm', *Research Policy*, 26: 429–46.

Glossary

AIST: National Institute of Advanced Industrial Science and Technology (Japanese: *Sangyou gijutsu sougou kenkyuu jo*). MITI/METI's main intramural research organization. Prior to 2001 AIST and its laboratories were directly under MITI. Since 2001, AIST has been an independent administrative entity under METI. It comprises twenty-two research institutions, twenty-four more narrowly focused research centers, an advanced computing center, a depository for patented organisms, and various regional collaborative centers, research initiatives, research teams and administrative offices. Its 1999 budget was about US$450 million. For comparison, this was about one-fourth of NIH's intramural budget.

Angels: Individual investors in pre-IPO companies. In the USA and Japan such individuals often are the main sources of initial funding for new technology-based companies. They are often friends and family members, but in the USA, former entrepreneurs and other successful business persons who put their own money at risk often are the most important sources for such investments.

AUTM: Association of University Technology Managers. The main organization representing Canadian and US TLOs and other university technology transfer professionals. Through its annual surveys, AUTM is also the main source of data on university technology transfer activities.

CFR: Code of Federal Regulations, the official compilation of US government regulations.

Commissioned Research: In Japanese, *jutaku kenkyuu* or *itaku kenkyuu*. Contract research that does not involve exchange of researchers. The legal framework governing research commissioned to universities in Japan is described in Kneller (2003*a*), cited under Chapter 3 References.

CREST* (Core Research for Evaluational Science and Technology, in Japanese: *senryaku teki souzou kenkyuu suishin jigyou*): A major JST program to support primarily university research. Applications must be for collaborative research involving several laboratories, and they should be targeted on one of the approximately twelve new research themes that JST announces each year.

DDS: Drug delivery system(s), for example encasing a drug in a lipid envelope so that it is more likely to reach its target organ intact.

DHEW and DHHS: The former stands for the US Department of Health Education and Welfare. Created in 1953, its name was changed to the Department of Health and Human Services (DHHS) in 1980 when a separate Department of Education was formed. DHEW and DHHS are important in this book because the Public Health

Service (PHS) operates within these departments, and under PHS are the USA's main health and biomedical R&D agencies. The largest of these is NIH, but also included are the Centers for Disease Control and Prevention (CDC)and the FDA.

Donations: In Japanese, *kifukin* or *inin keirikin*. Officially classified as charitable gifts, these are the most common form of industry support for research in Japanese universities (see Chapter 3).

ERATO* (Exploratory Research for Advanced Technology, in Japanese: *souzou kagaku gijutsu suishin jiggyou*): A JST program to fund about four new large scale research projects each year to conduct pioneering research in areas where Japan needs to boost its S&T capabilities. Some of the projects have received considerable praise in Japan and overseas.

Extramural research: Research funded by an organization (typically a government S&T agency) that is carried out outside that organization's own laboratories. Often extramural research is synonymous with grant or contract research. However, it is useful to distinguish external research funding by organizations such as NIH or METI that have their own laboratories. Research these organizations fund that is conducted in their own laboratories is called 'intramural' research.

FDA: US Food and Drug Administration. The US government agency responsible for ensuring the safety and efficacy of drugs as well as some food products. It, like NIH, is one of the agencies within the US Public Health Service.

GRI: Government research institute. Research institutes owned by or affiliated with government agencies (national or local) or other public organizations, but not including universities or academic medical centers. Examples are NIH's intramural laboratories, the US Department of Energy's laboratories (including contractor-operated laboratories so long as they are not on university campuses), UK Medical Research Council institutes and units (not within universities), French CNRS laboratories (excluding university partnership laboratories on university campuses), Max Plank and Fraunhofer Institute laboratories in Germany, and AIST and Riken in Japan.

IC: Integrated circuit, as used in computers.

IIS: Institute for Industrial Science, a major applied science and applied engineering research institute in the University of Tokyo. IIS and RCAST are the principal occupants of one of the university's main research campuses. This campus is located about 45 minutes from the main engineering, basic science, and medical campus.

Imperial Universities: The former Japanese Imperial Universities are the University of Tokyo and Kyoto, Osaka, Tohoku, Nagoya, Hokkaido, and Kyushu Universities. These universities together with TIT constitute the generally acknowledged eight leading national universities in Japan.

Innovation: Inventive activity with commercial utility or potential to benefit health, *or* improvements in existing processes and products. (Thus, my use of this term is broader than a more limited definition that refers mainly to improvements to or derivations on existing processes and products.)

IP: Intellectual property—principally patents, copyrights, trademarks, trade secrets, and in Europe, some data bases.

IPO: Initial public offering. This refers to the first sale of a company's (e.g. a venture company's) stock on public stock markets. Stock markets are subject to regulations to protect investors and ensure the integrity of the market for the general investing public. In order to qualify for sale of their stock on public stock markets, companies have to make various disclosures about their financial status and management. Depending on the particular market, they also need to meet minimum requirements regarding investment capital, sales, etc.

IPR: Intellectual property rights: the rights associated with IP or ownership of IP.

IT: Information technology, including both software and hardware, computers, integrated circuits, communications and information storage devices, etc.

Joint Research: In Japanese, *kyoudou kenkyuu*. Contract collaborative research that permits (or can consist exclusively of) researcher exchange. The legal framework governing joint research with Japanese universities is described in Kneller (2003*a*), cited under Chapter 3 References.

JPMA: Japan Pharmaceutical Manufacturers' Association, the main trade association of the Japanese pharmaceutical industry.

JPO: Japan Patent Office.

JSPS: Japan Society for the Promotion of Science, a legal corporation under Monbusho and later MEXT that funds external research and scientific exchanges.

JST: Japan Science and Technology Corporation, in Japanese *kagaku gijutsu shinkou kikou*. The main extramural research funding organization under STA, prior to its merger with Monbusho. Now a semi-autonomous corporation under MEXT, it continues to manage large extramural funding programs that can cover various areas of S&T. It has had responsibility for managing national inventions arising from national universities and certain other government laboratories—a responsibility that has been diminished since national universities became independent corporations.

Kansai: The Kansai plain in which is located the cities of Osaka, Kyoto, Kobe and Nara, and which constitutes Japan's second major center of population, industry, and education after Kanto.

Kanto: The Kanto plain which is synonymous with the habitable (non-mountainous) portions of the Tokyo metropolitan region. It includes the major cities of Tokyo and Yokohama as well as smaller cities such as Chiba and Saitama. It constitutes Japan's center of government and its largest center of population, industry, and education. With a population of 33–5 million, it constituted the world's largest metropolitan region by far in 2005.

Keiretsu: A **bank** or **financial keiretsu** is a group of companies usually linked by cross shareholdings and reliant on a particular major bank for loans. In the period between approximately 1950 and 2000, there usually were considered to be six such keiretsus, those centered on the Mitsubishi (later Tokyo-Mitsubishi and most recently Mitsubishi-UFJ) Bank, Mitsui Bank, Sumitomo Bank, Fuji Bank (successor to the Yamada Bank), Sanwa Bank, and First Industrial Bank. However, only the first three were considered to have sufficient strength and organizational identity to significantly influence behavior among their members. As banks have merged and companies have sold cross-held shares, the strength and importance of these keiretsu affiliations has diminished. Nevertheless, such affiliations still exist at least for the Tokyo-Mitsubishi Bank (now merged with UFJ holdings which in turn was created by the merger of Sanwa Bank with two other financial institutions) and the now merged Sumitomo and Mitsui Banks and the concept remains useful in some circumstances.

A **manufacturing keiretsu** is a group of supplier companies linked to a large manufacturing company (or main suppliers to that manufacturer), not only by sales but also by consultative relationships. The supplier companies might include independent companies that could even sell to competitors of the main manufacturer (although rarely would they do so against the wishes of the main manufacturer), independent companies that depend on the main manufacturer for most of their sales, and partially owned subsidiaries of the main manufacturer. The term **shita-uke** often applies to companies in the latter two categories that take their cues largely from the main manufacturer or one of its main suppliers.

MAFF: Japanese Ministry of Agriculture, Forestry and Fisheries.

METI: Japanese Ministry of Economy, Trade and Industry. Until 2000, the Ministry of International Trade and Industry (MITI).

MEXT: Japanese Ministry of Education, Culture, Sports, Science and Technology (in Japanese, *Monbu kagaku shou*). Created in 2000 by the merger of Monbusho and STA.

MEXT grants-in-aid:* In Japanese, *kagaku kenkyuuhi hijoukin*, abbreviated as *kakenhi*. MEXT's main program for funding project-specific university research. There are many categories, ranging from support for projects by young researchers to large scale projects in special priority areas. The total 2002 budget was 170 billion yen. Just over half of this was for projects proposed by individual researchers (categories S, A, B, and C). These S, A, B, and C category grants are the mainstay of support for individually initiated university *basic* research projects. They are roughly equivalent to R01 grants, the mainstay form of support by NIH for individual investigator-initiated projects.

MHLW: Japanese Ministry of Health, Labor and Welfare. Formed in 2000 from the merger of MHW and Ministry of Labor.

MHW: Japanese Ministry of Health and Welfare. Merged in 2000 with the Ministry of Labor to form MHLW.

MITI: Japanese Ministry of International Trade and Industry. Name changed to Ministry of Economy, Trade and Industry in 2000.

MOF: Japanese Ministry of Finance, also known by its former Japanese name, *ookurashou*.

Monbusho: Japanese Ministry of Education, Science, Sports and Culture (sometimes also referred to as MESSC or simply the Ministry of Education or MOE). Monbusho merged with STA in 2000 to form MEXT.

National universities: Universities under the jurisdiction of Monbusho/MEXT. In 2006 they numbered 87 plus four separate research institutions. Prior to 2004, these were simply branches within MEXT. That year, they were incorporated as semi-independent **national university corporations** with nominal control over key matters such as finance and personnel. However, they remain dependent on the national government for most of their financing. National universities account for the vast majority (approximately 75%) of university R&D in Japan. The most prestigious and also the best-funded national universities are the seven former Imperial Universities (the University of Tokyo, and Kyoto, Osaka, Tohoku, Nagoya, Hokkaido, and Kyushu Universities) and the Tokyo Institute of Technology. (Listed in order of MEXT Grants-in-aid received in 2003.) A few private universities, such as Keio and Waseda, and a few private medical schools are also major academic R&D centers, but their R&D funding is less than that of any of the eight national universities listed here.

NEDO:* New Energy Development Organization, an independent administrative agency under MITI/METI that funds most of MITI/METI's extramural competitive R&D projects—but not including SBIR type projects funded by another organization under METI.

NIH: US National Institutes of Health, the US government's main health research agency and part of the US Public Health Service (along with the FDA, Centers for Disease Control and Prevention (CDC) and several other agencies) under the Department of Health and Human Services. Not only is NIH is the world's largest funder of extramural biomedical research (its annual 2004 budget was nearly US$30 billion), it also has the world's largest intramural biomedical research laboratory—located for the most part in the Maryland suburbs of Washington, DC.

Nikkei Shimbun: Japan's leading financial and business daily newspaper, equivalent to the *Wall Street Journal* or *Financial Times* (translated as Japan Financial Times, or Japan Economic Times).

Nikkei Weekly: A weekly compilation of English language articles, most of which are translations of recent articles in the *Nikkei Shimbun* and other newspapers published by the same publisher (e.g. *Japan Industrial Times*).

NME: New molecular entity, i.e. new small molecule drugs, that typically function by blocking or enhancing the action of an enzyme in the body, usually by fitting into the active site of a receptor on that enzyme. Compare **NTB**.

NSF: US National Science Foundation, the main US agency for funding basic, non-biomedical, and S&T research. Most of its funds go to engineering, physics, chemistry, biology, computational science, earth and planetary science, and social science departments in US universities. In addition, however, it also funds the USA's main polar and Antarctic research programs.

NTB: New therapeutic biologic. New drugs that are human-made copies or derivatives of naturally occurring enzymes, hormones, antibodies or other proteins, or protein conjugates. However, this term does not include vaccines, blood products, agents for cell or gene therapy, or tissues. These latter products are reviewed by the FDA's Center for Biologics Evaluation and Research (CBER) while NTBs and NMEs are reviewed by FDA's Center for Drug Evaluation and Research (CDER).

NTT: Nippon Telephone and Telegraph Corporation. Until 1985, NTT was wholly-owned by the Ministry of Posts and Telecommunications. It had a legal monopoly over telephone and telegraph communications and was responsible for building and maintaining Japan's telecommunications network. Partially privatized in 1985, it has since lost some of its monopoly control. NTT owns several R&D laboratories. At least in the 1970s and 1980s, some of these were among the strongest laboratories for basic and applied telecommunications research in Japan.

OECD: Organization for Economic Cooperation and Development, an intergovernmental organization headquartered in Paris whose members include most of the world's industrialized countries. Its primary function is to compile economic and S&T data and to prepare analytical or advisory reports based on such data.

OPSR: Organization for Pharmaceutical Safety and Research, an agency under the MHW/MHLW that funded medical research and also provides support for persons suffering from adverse drug reactions. In 2004 it was merged with two pharmaceutical and device regulatory agencies under MHLW to create the Pharmaceutical and Medical Devices Agency (PMDA) and in 2005 its medical research support functions were taken over by the National Institute of Biomedical Innovation (NIBI). Basically, its role in supporting biomedical research was analogous to the research support roles of JSPS (for Monbusho) and NEDO (for MITI/METI).

Peer review: Shorthand for the process for reviewing and choosing between competing applications for R&D funding, so long as a significant component of the selection process involves review of applications by a panel of persons who are supposed to have specialized knowledge of the subject of the applications. Typically, these are applications from university researchers to government funding agencies.

Postdocs: Postdoctoral researchers. In the USA, this refers to persons who have recently obtained doctoral degrees but are pursuing semi-independent research in a university to increase their research experience and publication record. In the USA, being a postdoc is part of the normal career track to obtaining a tenure track position (assistant professorship) and ultimately tenure (associate professorship). But as described in the text, a postdoc position in Japan is still off the main career track. Also in Japan, *postdoc* often has a looser meaning that includes Ph.D. candidates, particularly those who have completed their coursework and are pursuing thesis research.

PRESTO* (Precursory Research for Embryonic Science and Technology, in Japanese: *sakigake kenkyuu*): Another major JST program to support primarily university research. It is like CREST except that awards are usually to individuals and a special subprogram supports research by Ph.D. candidates and postdoctoral researchers.

PRI: Public research institutes: a term used to refer to GRIs and universities, private as well as public. (Even though, strictly speaking, it would not include private universities.)

PTO: US Patent and Trademark Office.

R&D: research and development 'Research' is often conceptually broken down into 'basic, fundamental, or curiosity motivated' research and 'applied' research that has particular industrial, commercial, or health applications in mind.

RCAST: Research Center for Advanced Science and Technology. One of several research and education centers in the University of Tokyo, RCAST was established in 1987 to promote interdisciplinary research in an environment open to collaboration with outside organizations. RCAST shares the Komaba Research Campus of the University of Tokyo with IIS.

S&E: science and engineering—a term often coincident with S&T.

S&E Indicators: Science and Engineering Indicators, published every two years by the National Science Board of NSF. The 2004 edition is available at http://www.nsf.gov/sbe/srs/seind04/start.htm

S&T: science and technology. This term refers to engineering, computer science, the various fields of natural science, and non-clinical fields of medical science.

SBIR: US Small Business Innovation Research, a Congressionally mandated program that requires US S&T agencies to set aside approximately 3 percent of their total R&D budgets for research grants to US small businesses. It is a significant source of funds to promote research in small, mostly new companies and to bridge the gap between academic research and research that has progressed sufficiently to attract private venture capital or corporate investment.

SEC: US Securities and Exchange Commission which regulates the sales of stock and related reporting requirements for all companies whose stock is traded on US public exchanges. Information reported to the SEC, particularly the 10-K annual reports, are important sources of information about companies.

Shita-uke company: (literally *sub-contractor*). This term usually implies a supplier in a long-term relationship with a larger manufacturing customer—a relationship that embodies a significant degree of (although not necessarily exclusive) dependence on orders from that main customer. This dependence often also requires adherence to the wishes and business plans of the main customer, but it may also involve obligations on the part of the main customer to treat the subcontractor fairly and to continue to give it its business.

SME: Small- and medium-sized enterprise. In the US context, I generally use this term to coincide with the official US definition of a 'small business', i.e., an independent company, with its principal place of business in the USA, with no more than 500 employees, and at least half of whose voting stock is owned by US citizens or permanent residents (13 CFR §121.4). In the Japanese context, I generally use this tem to coincide with the official Japanese definition of SME (chushou kigyou), i.e., an independent company with no more than 300 employees and less than 3 billion yen (~US$28 million) in paid-in capital.

SMRJ: Organization for Small and Medium Enterprises and Regional Innovation. Its formation in 2004 by the merger of the Japan Small and Medium Enterprise Corporation (JASMEC), Japan Regional Development Corporation, and Industrial Structure Improvement Fund, was intended to rationalize many of the METI-affiliated programs to promote SMEs and regional business development. Government VC companies, such as the Tokyo Metropolitan Small and Medium Enterprise Investment Corporation, are now under the SMRJ umbrella.

Spin-off: A new company based on technology from an existing company. Formation can be either at the direction of the existing company (usually as a means of diversification) or it can be against the wishes of the existing company, as when engineers who are frustrated that their projects are not being developed leave to form a new company that will be the vehicle to continue development of their projects.

Sponsored research: Research funded by an outside organization, typically a private company or government-affiliated agency. As used in this book, the research is usually project or theme specific, as opposed, for example, to an award or an endowment that can be used freely for purposes decided primarily by the recipient. *In the Japanese context*, when I refer to *industry sponsored research*, I mean research funded (or largely funded) by private companies under **Commissioned Research** or **Joint Research** or **donations.** However, when I use the term *formal* or *contractual* industry sponsored research, I refer only to **Commissioned** or **Joint Research**. When I refer to *government sponsored research*, I mean *contractual* **Commissioned** or **Joint Research** funded by government affiliated organizations, or research funded by **MEXT Grants-in-aid.** These mechanisms are discussed in Chapter 3.

STA: Japan Science and Technology Agency. Merged with Monbusho in 2000 to form MEXT.

Startup: An independent new company based on university discoveries.

TIT: Tokyo Institute of Technology. Sometimes considered to be Japan's equivalent of MIT, TIT and the seven former Imperial Universities are generally considered to be the eight leading national universities in Japan.

TLO: Technology licensing office or technology licensing organization. This is the general term used in Japan to refer to a university licensing, technology transfer, or technology management organization. It is also widely used in the USA as a generic term with the same meaning.

USC: US Code, the official compilation of US laws.

VC: venture capital.

Venture Companies (or ventures): New, entrepreneurial, technology-based companies. Although a traditional definition would limit these to companies financed primarily by equity investments by venture capital, angels, etc., in this book, I also include new companies whose operations are largely self-financed through revenue or even loans. In this book, two distinct subcategories are ventures based on university discoveries (*startups*) and ventures based on discoveries in an existing company (*spin-offs*).

---

* See Chapter 3, Appendix, for the approximate size of these programs.

Index

Abbott 160, 242
ABI 201
academic qualifications 60, 81, 82, 336, 352–3
academic recruitment and promotion 63, 86, 289, 303
Acs, Zoltan J. 314, 316, 317
Adachi, S. 223
Adtron 331
Advanced Telecommunications Research Institute (ATR) 345
Affymetrix 158, 201
Agility 203, 221
air purification 108–9
Ajinomoto 196, 204, 217
Akiyoshi Wada 200
Alexion 321
Allen, Paul 332
alliances 168–9, 174, 181, 187–8, 303
Amdahl, Gene 252, 333
Amersham 199, 200–1
Amgen 195, 242, 295, 318, 320
Anderson, Stuart 335
angel investment 121, 124, 169–73, 182–3, 303
Angel Tax Incentive 135
Anges MG 146, 177, 186
Angiogene (UK) 187
Aoki, Hatsuo 35
Aoki, Masahiko 87, 215, 334, 352, 353, 354, 357, 364
Apple 251, 255, 332, 338
Applied Biosystems Incorporated (ABI) 199
Arora, Ashish 328
Asahi Glass 198–9
Asanuma, Banri 225, 340
Assay and Systems 166
Astra Zeneca 315
AT&T 23, 38
  Bell Laboratory 211, 256
ATT 326
Audretsch, David B. 314, 316, 317
autarkic innovation ch. 3 generally 113, 128–31, 168, 174, 276, 280–2
automobile industry 206–7, 266, 354–5
Aventis 196

Baba, Yasunori 36, 74
Barley, Stephen R. 315
Barnett, Jonathan M. 71, 326
basic research 56, 61–2, 249, 289, 302, 330–1
Bayer 34, 160
Beard, Tom 339
Begley, Sharon 334, 366
Biomolecular Engineering Research Institute (BERI) 348
biomolecules 102–3
Biotechnology Guide USA 160
bioventures/biotechnology companies 20, 22, 35, 43, 50, 72
  definition 32, 318
  government support 132, 133–7
  main investors 177–80
  and pharmaceutical companies 128–31
  revenue and investment data 162
  satellite 208
  skills shortage 132
  and universities 131, 176
  US 42, 168, 226
  venture capital 132–3
  see also ventures/venture companies
Blackburn, Robert 155
Block, Z. 315
Boyer, Herbert 72, 331
Branchflower, David 314, 316
brand name value 158
Branstetter, Lee G. 70, 347
bureaucracy, interventionist 268, 340–1
Burroughs 253
business incubators 134, 145

Callon, Scott 340, 341, 342
Calne, Roy 322
Calomiris, Charles W. 340
Canada 121, 243
cancer, genetic basis 123–6
cancer drugs 118–20
Canon 192, 199
careers 351–2
Carlson, Bo 340
cartilage regeneration 250
Casio 24
Casper, Steven 322, 366
Celera 157, 201

Centers of Excellence (COE) 64–5, 69
Central Commercial and Industrial
    Bank 266
central personnel office 278
Cetus Corp. 72
Chabrow, E. 78
Chandler, Alfred D. 316, 331, 333, 334
chemical industry 52–3, 227
Chesbrough, Henry W. 35, 214, 215, 223,
    227, 315, 316, 334, 363
China 225, 251, 253, 256–7, 264, 292, 301
Chinese 111
Chinese universities 101
Christensen, Clayton M. 35, 220, 314, 331,
    333
Chugai 161–2
Chuma, Hiroyuki 87, 352
Cisco Systems 202, 243–4, 331
Clare, Jeffrey J. 219
Clark, Kim B. 214, 215
clinical trials 241, 244
Cockburn, Iain 320
Cohen, Stanley 331
Cohen, Wesley M. 155, 314, 324, 325, 326,
    328, 363
Cole, Robert E. 214
Coleman, Samuel 86
collaboration 101, 107, 110, 112, 355–6
    barriers to 280–1
    benefits of 156
    small-large company 151–2
    see also consortia; partnerships
Coller Capital 212
Commercial Code 135, 136
commissioned researcher agreements 78,
    86
Commodore 338
communications 66–7, 87–8
companies:
    large: advantages of 236–7; dominant
        role 271, 272; source of talent 114;
        need for new 291–2
    relationship with universities 11–12, 21,
        28–30, 45, 57, 233, 289–91
    small: advantages of 235–6; disadvantages
        of 267
    SMEs 10–11, 23
company growth 237–8, 255
company size:
    definitions 77, 314, 317
    and innovation 1, 235–43, 299–300
company training 278
Compaq 250, 252, 333

competition 256, 262
computer hardware manufacture 24–8
consortia:
    government support 268–77, 343–4
    inclusion of ventures 274–5
    and innovation 272–3
    negative effect 274, 276
    patents 349–50
    projects 341
    and universities 275
    value of 347–8
consulting 47
contract research organizations (CROs) 244
Control Data Corporation 250, 252, 333
Cook, Karen S. 88, 363
coordinated market economies 299, 366
copyright law 324
Corning Glass 269, 328
Cotrell, Thomas 36
Court of Appeals for the Federal Circuit
    (CAFC) 247
Crabtree/Carbtree, Penni 319
crystals 103–5
Cukier, Kenneth 334
customized antibodies 121

Dahms, A. Stephen 335
Daiichi Pharmaceuticals 195
Daiichi Sunory Pharma Co. Ltd 195
Daiken Chemical Co. 16
Daisy Systems 333
Damadian, Raymond 38
Darby, Michael R. 33, 70, 319
developing countries 304
Development Bank of Japan 99, 135
development cycles 324
Dibner, Mark D. 70, 72, 147, 160, 180
DNA extractor 122
DoCoMo 23, 37
Dodgson, Mark 314, 317
Doi, Takeo 87
donepezil 21
Dore, Ronald 215
Dougherty Carter 368
downsizing 282–3, 299
    see also layoffs
Dragon Genomics 197
DRAM chips 269
drugs:
    clinical trials 241, 320, 330
    development 117–20
    discovery 239–43
    in-licensed 20

drugs: (*cont.*)
  innovative 22, 33
  methodology 30–1
  molecular targets (receptors) 127–8
  output and market size 261–2
  pipeline 31
  rational discovery and design 126–7
  US approvals 306–8
Dyer, Jeffrey H. 206, 223, 224, 340

economic downturn 258, 261, 265, 267
economy, national 256
education 60, 234, 359–61
  gateway to success 285–6
Edwards, Mark G. 217
Effector Cell Institute 146, 187
Eisai 21, 33
Eisenberg, Rebecca S. 71
Elpida 213
Elysium 164
EMI 39
employees:
  incentives 136, 315, 362
  morale 298
  see also labor
employment:
  female 283, 357
  lifetime 12, 114, 263, 279, 284, 297, 357
  trends 310–13
employment protection legislation 365
engineering 52–3, 143
engineers 37, 334–5
environment and energy 143
erythropoietin (EPO) 195, 216–17
established newcomers:
  advantages 193–4
  in biomedicine 194–202
  overseas contacts 196–7
Etzkowitz, Henry 78
EUV LLC 342–3
Evolvable Systems Research 164
Exploratory Research for Advanced
    Technology (ERATO) 61, 84, 347
Extreme Ultraviolet Association
    (EUVA) 270, 350
Exxon 211

Fairchild Semiconductors 250, 252
Fanuc 213, 264, 265, 266, 339
Fast, Norman 227
fiber optic networks 202–3
finance 143–4, 168, 180–1
Fishmann, Mark 321

Flanigan, James 336
Florida, Richard L. 321
Ford 206
Foremski, Tom 36
France 241, 243, 300
Fransman, Martin 214, 215, 326, 328, 334,
    340, 341
Frederick, Jim 215
Freear, John 182
Friedman, David 215, 225, 314, 340, 364
Fujikura 269
Fujisawa 34
Fujitsu 25, 36, 199, 212–13, 222, 264, 331,
    332, 333, 341
Fujitsu Research Institute (FRI) 147–8
Fukugawa 77
Fukuoka 177
funding:
  by government 57–8, 61–3, 68–9, 102–3,
    170, 202, 268–77, 343–4, 350–1
  university 46, 53, 63–6, 74, 77, 87, 302

Ganguli, Ishani 321
Gans, Joshua S. 328
Gates, Bill 332
GE 251
Gellman Research Associates 316
Gemmill, Trent R. 219
gender imbalances 60, 283, 285
Gene Angionenesis 182
Gene Logic 201
gene sequencing 199–201, 204, 219, 250
Genentech 45, 72, 146, 161–2, 242, 295,
    317
General Electric-Yodogawa Medical Systems
    (GE-Y) 29
genetic engineering 240, 331
Gerlach, Michael L. 223, 226, 362
Germany 78, 240–1, 243, 299, 339
  cross-shareholding 300–1
  innovation 366
  labor contracts 297–8
  MT companies 264
  universities 322
Gershwiler, James 182
Giddings & Lewis 339
Gilson, Ronald J. 223, 333
Glaser, Donald 72
Glaxo 33
Glober, Garry 39
glutamic acid 196
GM 206
Goodman, David J. 322, 367

Google 243
Goto, Akira 155, 215, 222
government:
  budgets 221
  funding 57–8, 61–3, 68–9, 102–3, 170, 202,
    268–77, 343–4, 350–1
  grants and contracts 183
  role in Internet development 28
  support 234; for bioventures 116, 132,
    133–7; for consortia 343–4
government research institutes (GRI):
  drug discovery 241–2
  patents 7–9
  startups 102–5, 108–9, 114
  ventures from 114
granulocyte colony stimulating factor
    (G-CSF) 195, 216–17
group 67, 293, keiretsu definition 222

Haas Automation 265, 339
Hall, Bronwyn H. 324, 326, 327, 331
Hall, Peter A. 299, 366, 367, 368
Hara, Takuzi 33
Harada, Tsutomu 225, 227, 228
hard disc drives 250–1, 253, 255
Hashimoto, Takehiko 73
Hayakawa Electric 9
Hayashi, Takayuki 347
health insurance 258–9, 337
Hecker, Daniel 337
Henderson, Rebecca M. 33, 214, 215, 320
hepatocyte growth factor (HGF) 117–18
Herbert, Peter 152
Hewlett Packard 113, 255, 331
Hicks, Diana 323
hierarchical relationships 66–7, 87, 292–3
high efficiency fiber optic cable 328–9
High Voltage, Inc. 71–2
Hikino, Takashi 225, 227, 228
Himmelberg, Charles P. 340
Hitachi 24–5, 29, 39, 199–202, 204–5, 213,
    255, 331–2, 341
Hitachi Life Science 220
Hitachi Metals 103
Hokkaido University 166
Honda 9, 192
Hong Kong 292
Hood, Leroy 219
Hosono 77
Hsu, David H. 328
Hsu, Jinn-Yuh 336
Hudson, Marianne 182
Human Genome Sciences 157

Hunkapillar, Dr 199, 219
Hyde, Alan 314–15, 327, 331–3, 335, 337,
    358, 366

IBM 23–7, 211, 250–3, 267, 269, 315, 326,
    329–30, 333, 338, 342
IBM Japan 36
IC chips, flaw detection 99–100
I. G. Farben 367
Ikeda, Masayoshi 225
immigration 257, 304, 335–6
in-licensing 120, 130
  late-stage 20
  see also licensing
incentive packages 136, 315, 362
Incyte 157
independence, personal need for 292–3
India 253, 256, 257, 301
industrial policy 263, 266, 268
industry concentration 238
information gatekeepers 280
Ingersoll Milling Machine Co. 339
initial public offering (IPO) 59, 80, 95,
    143–4, 177
  by company age 176, 183
  by company characteristics 175–6
  first wave of 173–5
  listing requirements 170, 174
innovation:
  autarkic 19–39
  capabilities 233
  and company size 1, 235–43, 299–300
  and consortia 272–3
  data availability 238
  early stage 1
  R&D costs 238–9
  reasons for strength of 262–82
  sources of 1
  sustainable 233, 254–61
innovation systems 299–300
institutional patent agreements (IPAs) 44
insularity 280
insurance companies 175, 188
Intec Web and Genome Informatics
    Corp. 177
integrated circuits 96–7
Intel 250, 252, 254, 255, 270, 333
intellectual property (IP) 136, 233, 305
  importance for ventures 114–16
  national ownership 46
  pre-1998 system 45–8
  university ownership 43–4
interluken 196, 217

International Patent Classification (IPC) 2, 12
Internet 23, 28–9, 39, 250, 331
INVENT 342–3
inventions, industry sponsored 55
investment:
  angel 121, 124, 169–73, 182–3, 303
  pre-IPO 170
Iolon 203
iRNA type drugs 22, 35
Israel 296
Italy 264
Ito, Kiyhiko 222–4, 225, 227–8
Itochu 155

J-Phone 23
Jacobs, Irwin 332
Jaeger, David A. 365
JAFCO 177, 184, 205, 208, 222
Jaffe, Adam B. 369
Japan Asia Investment Corp. 188
Japan Business Federation 284
Japan Chamber of Commerce 165
Japan Machine Tool Builders Association 339
Japan Patent Office (JPO) 2
Japan Science and Technology Agency (STA) 201
Japan Science and Technology Corporation (JST) 50, 61, 170, 274
  Core Research for Evaluational Science and Technology (CREST) 61, 272, 346
Japan Tobacco 197–8, 204
Jasdaq 146, 121
Jensen, Richard 322
JMNet 177
job searching 286–7
job security 296–7
Jobs, Steven 315, 332
Johnson, Chalmers 340
Johnson, Lanny 16
Johnstone, Bob 23–4, 35, 215
Joint Industry, University and Government Cooperative Research Program 272
joint research 51–6
Juniper Networks 202

Kalafsky, Ronald V. 339
Kambara, Hideki 200–1, 219
Kaneka 180
Keio University 137–45, 163, 197, 207, 95
keiretsu 205–14, 354–5
  definition 222–3

Kenney, Martin 321
Key Technology Center 271, 345, 348
Kieler/Keilen, Mark 354
Kirin 194–6, 204, 215, 216
Kishi, Yoshinubo 219
Kissei, Kyorin and Mitsubishi Pharma (Japan) 187
Kiyonari 340
Klepper, Steven 314
Kneller, Robert W. 31–2, 35, 72–3, 85–6, 160–1, 218, 317, 319, 338, 347, 352
knockout mice 122–3
Kodak 247
Kodama, Fumio 37, 215
Koike 340
Komag 253
Korean 111
Korput, K. W. 70
Kortum, Samuel 326
kouza system 60–1, 63, 82, 303
Kume, Ikuo 365, 366, 368
Kunda, Gideon 315
Kurtzman, Joel 38
Kushi, Takahito 346, 347
Kuzunoki, Ken 352
Kyocera 9
Kyoto University 39, 64, 166, 278

Laage-Hellman, Jens 355
labor:
  mobility 28–9, 136, 233, 251–7, 279, 282–8; limiting factors 300–1
  redeployment 338
  see also employees
Lam, Alice 35
Latker, Norman J. 71
Lattice Technology 182
Law to Prevent Unfair Competition (LPUC) 303
laws:
  Civil Rehabilitation Law 135
  Employment Security Law 136
  Law for Facilitating the Creation of New Businesses 134
  Law of Special Measures to Revive Industry 48
  Law on Temporary Measures to Facilitate Specific New Businesses 133
  Law to Promote the Transfer of University Technologies 48
  Law to Strengthen Industrial Technology 48–9, 50, 51, 136
  Law on Trust Companies 136

Limited Partnership Act for Venture
   Capital Investment 134–5
National University Incorporation Law 49,
   136
Newly Incorporated Business Law 133
Patent Law 136, 368
Public Pension and Insurance Law 184
Small- and Medium-Size Business Creation
   Activity Promotion 134
SME Challenge Law 135
TLO law 136
Worker Dispatching Law 136
layoffs 356
   see also downsizing
Leigh, Beatrice 35
Lequio Pharma 182
Lerner, Josh 187, 326, 330, 369
liberal market economies 366
licensing 321
   exclusive 49, 51–2
   see also in-licensing; patents
licensing departments 161
life sciences 52–3, 137–42
light emitting diodes (LEDs) 36
light filtering films 100–2
Lincoln, James R. 226, 362
Linux 26
LSI Logic 255
LTT Bio-Pharma 159, 182
Lucent Technology 211–12
Lycos 243
Lynn, Leonard H. 354

machine tool industry 263–8
machine tools 264–5, 309, 338–41
McKelvey, Maureen 70, 72, 319, 321, 330
MacMillan, I. 227
McPherson, Alan D. 339
magnetic resonance imaging (MRI) 28–30,
   38
Makino 264
Mallard, John 38
management 144–5
managers 131, 148
manufacturers, relationship with customers
   27–8
manufacturing:
   competence 259
   high quality 27
   Good Practice 157
Markoff, John 337
Marubeni 155
materials 143

materials-chemistry technology 275
Matsushita 222, 251, 332
Matsushita Electric 205
Matsushita (Panasonic) 25
Matsuura 340
Maxtor 251, 253, 331
May, John 182
MBV 157, 160, 296
Medibic 186
MediciNova 187
MediNet 146, 180, 186
Merck 33, 34–5
mergers 213, 324
Merges, Robert P. 36, 316, 324, 326, 328
merit pay systems 362
mevastatin 21, 33
microprocessors 328
Microsoft 24, 26, 251, 255, 332
middleware 26, 36
Millennium Project 158
Ministry of Agriculture 166
Ministry of Economy, Trade and Industry
   (METI) 76, 99, 102, 135, 149, 163,
   183, 269
   New Energy Development Organization
      (NEDO) 62, 149, 170, 184, 270–1,
      274
Ministry of Education, Culture, Sports,
   Science and Technology (MEXT)
   48–9, 59–60, 62, 64, 82, 83–5, 170,
   272
Ministry of Health and Welfare (MHW) 34
Ministry of Internal Affairs and
   Communications (MIC) 271, 345
Ministry of International Trade and Industry
   (MITI) 263, 269, 341, 342
Ministry of Post and
   Telecommunications 341, 345
Mitsubishi 341
Mitsubishi Electric 24, 205, 222
Mitsubishi Kasei 213
Mitsubishi Motors 205
Mitsubishi Petrochemical 213
Mitsubishi Pharma 159, 182
Mitsubishi Shouji 155
Mitsui Busan 155
Mizuno, Yuji 227–8
Mizuta, Yuji 36
Molecular Dynamics 199, 201
Monbusho 74
Monsanto 315
Mori Seki 340
Motorola 326

moulded plastic 106–8
MRI scanners 251
MSEs 245
Murray, Fiona 70, 217, 322, 331
Musha, Ryoji 364
Myers, Robert A. 37, 227, 322, 355, 367

Nagahara 74
Nagata, Aliya 155
Nagoya stock exchange 177, 187
Nakagawa, Katsuhiro 162, 185
Nakane, Chie 87
Nakayama 77
Nanosys 328, 331
nanotechnology 6–7, 15, 204–5, 245, 248, 345
Nasdaq Japan 133, 170, 177
National Centre for Genomics Research 220
National Institute of Advanced Industrial Science and Technology (AIST) 77, 95, 137–45, 147, 163–4, 166, 207
Nayak, Krishna 39
NEC 24, 25, 192, 199, 213, 255, 328, 341
Nelson, Richard R. 314, 324, 325, 326, 328, 363
Netscape 243
networks 226–7, 233, 265
New Energy Development Organization (NEDO) 62, 149, 170, 184, 270–1, 274
new molecular entities (NMEs) 240, 243, 317–18, 319
new therapeutic biologics (NTBs) 240–1, 243, 319–20
New Venture Partners 212
New Ventures Group (NVG) 211–12
Nichia 24, 36
Nikkei Financial Daily 184
Nikkei Weekly 17, 157
Nippon Telephone and Telegraph Corporation (NTT) 23, 37, 269, 290, 330, 341, 342, 345
Nishiguchi, Toshihiro 225
Nissan 206, 225
Nixon, President Richard 71
no-wet-lab venture 297
Nokia 329
Nomura Securities 205, 222
Nonaka, Ijujiro 215
Norvatis 196
Novartis 321
Noyce, Robert 252

Numagami, Tsuyoshi 352

Odagiri, Hiroyuki 73, 215, 222–3, 224, 225, 227, 340
Ogura, Yoshiaki 70
Okada, Yosuke 346, 347
Oki 25
Okuda, Chairman 357
Okuda, Hiroshi 284
OncoTherapy 146, 180, 186
O'Neil, Maryadele J. 318
OPSR 271–2, 274, 346, 349
Oracle 26
Organization for Small and Medium Enterprises and Regional Innovation (SMRJ) 104, 134, 135, 163, 184
Ornati, R. 315
Orsenigo, Luigi 33
Osaka Securities Finance Co 177
Osaka Stock Exchange, Hercules 146, 170, 187
Osaka University 105
Oswald, Andrew 314, 316
outsourcing 251, 257–8

Pallarito, Karen 315
partnerships 104–5, 161, 277, 281
   see also collaboration; consortia
patent attorneys 115, 153
patent law 52, 136, 368
patent pools 326
patent rights, pharmaceuticals 44
patents 4, 254, 275, 303–4
   by industry 2–10
   by organization type 3–10
   cost of 14–15, 155, 324–5
   effectiveness 155, 246–9, 325–6
   European 366–7
   importance for bioventures 133
   infringement 115, 153–4
   IPAs 44
   issued and applied for 14
   joint applications 54
   methodology 12–14
   process 149
   rights assigned 333
   awarded to individuals 7–8
   US/Japan comparison 3–10
   see also licensing
Paull, Robert 152
PC components 23
peer review 62, 85, 302, 304

pension benefits 279
pension funds 175
    private 134, 171
    public 184
pension reform 114, 279–80, 287, 354
pension system 110, 150, 184
Perkin Elmer 199, 201
Perseus Proteomics 182
personal handy phone system (PHS) 151
personnel:
    training and movement 137, 242
    see also employees; labor
personnel management policies 263, 277–81
Pfizer 30
pharmaceutical industry, protection 34
pharmaceuticals/pharmaceutical
        companies 19
    alliances 174, 187–8
    autarkic innovation 20–2
    and bioventures 128–31
    definition 32
    European and American 129–30
    exclusive patent rights 44
    and venture companies 260
Phoenix Wireless 275
Piehler, Henry R. 354
pilsicainide 195
pipelining 78
Pisano, Gary P. 33
Pixar 315
plasma display panel high definition
        television (PDP HDTV) 251
Plasmaco 332
Polaroid 247, 326
Pollack, Andrew 219
Porter, Eduardo 38
Powell, Walter W. 70
Precision Systems Science 177
Precursory Research for Embryonic Science
        and Technology (PRESTO) 61
printed circuit boards (PCB) 97–9, 106
Proctor and Gamble 196, 217
promotions 278–9
Protein Engineering Research Institute
        (PERI) 216, 345, 348–9
public listing 135

Qualcomm 244, 251, 255, 332, 338

Rader, Ronald A. 318
Rai, Saritha 337
Rathmann, George B. 242, 295, 321
Raynor, Michael E. 35

RCA, Sarnoff Laboratories 256
Read-Rite 253
Red Hat 329
research and development (R&D):
    alliances 130
    efficiency ratio 316
    innovation costs 238–9
    SMEs 23
Research Institute for Innovative Technology
        for the Earth (RITE) 345
research partnerships 161, 277
researchers:
    mobility 28–9
    supply of 233
retirement, early 152
Rifkin, Glenn 38
Riken 220
Rind, Kenneth 227
Robinson, Robert J. 182
Roche 161–2, 198
Roche Diagnostics 201
Roe, Mark J. 223
Roessner, David 331, 332
Romanos, Michael A. 219
Rose, Elizabeth L. 222
Rothwell, Roy 314, 317
Rowen, Henry S. 162
royalties 47, 49, 76
Rtischev, Dimitry 214
Russia 264

Sakakibara, Mariko 341–3, 346–8, 351–3, 361
salaries 278
Samsung 331
Sankyo 21, 33, 34
Santur 203, 220
Sanyo 9, 24, 36
Saxenian, AnnaLee 332, 333, 335, 336
SBI Holdings 184
Scherer, F. M. 314, 316, 317
Schering-Plough 20–1, 30, 160
Schumpeter, Joseph 237–8
Science and Technology Agency (STA) 200
Scorer, Carol A. 219
Seagate 251, 253, 255, 331
search engine, internal 109–11
Seiko Instruments 200
self-financing 168
SEMATECH 351
semiconductors 23, 248, 255, 269–70
Shane, Scott 71, 76, 165, 316, 321, 328
shareholding, cross 287
Sharp 9, 24, 25, 36, 192

Shibuya, Takahiro 227
Shichijo 74
Shinohara, Kazuko 86
Shonan Fujisawa 166
Siegel, E. 227
Siegel, R. 227
Siemens 29, 30, 39, 339
Silicon Valley model 233
Singapore 292
Sinkula, Michael 152
sleeping university inventions 47, 74
Small Business Investment Corporations
    (SBICs) 133, 135
Small and Medium Enterprise Agency 149
Small and Medium Enterprise
    Corporations 266
Smith, Loyd 219
Smith, W. Novis 71
Smith-Doerr, L. 70
SmithKline Beecham 30
social pressures 285, 359
Socolof, Stephen J. 227
software 52–3, 142, 175–6, 260
    open source 37
    prepackaged 36
    startups 94, 95
software industries 24–8
    independent 27
    price competition 26
Sohl, Jeffrey E. 182
Soiken 146, 180, 186
Sony 9, 24, 209
sorivudine 218
Sosei 188
Soskice, David 299, 366, 367, 368
Sperry Rand 250, 252
spin-offs 211–12
    definition 14
    see also tethered spin-offs
Stanley Electronics 24, 36
startups 42, 245
    average size 50
    biomedical 94
    biotechnology 72
    by industry 94
    by technology fields 137–43
    conflicts of interest 58–60, 80
    core 76
    definition 14
    disadvantaged 47, 56
    employment 94
    founding professors 131
    growth of 50–1

number of 79, 93
rate of 49–50
software 94, 95
technologically innovative 145
    see also ventures
statin drugs 21, 33
Stern, Scott 328
Stevens, Ann Huff 365
stock options 315
stock ownership 149
Streb, Jochen 367
Stutzer, Alois 314
subcontracting 265–6
Sugimoto, Dr 33
Sumitomo Electric Industries (SEI) 208, 226,
    269
Sumitomo Pharmaceuticals 208
Sumitomo Shouji 155
Sun Microsystems 37, 243, 255, 329, 331
Suntory 194–6, 204, 216, 224
Suntory Biomedical Research, Ltd. (SBR) 194
Suzuki, Hiroto 137, 162
Suzuki, Jun 215
Sweden 296
Switzerland 241, 243

Taiwan 251, 264
Taiyo Industrial Co. 189
Takahashi, Hiroshi 341
Takahashi, Satoshi 219
Takara Biotechnology 197, 201, 218
Takara Shuzo 197
Takeda 33, 161
Takia, Shinji 36
Tanabe Pharmaceuticals 187
Tandy 338
Tatsura, T. 223
taxation 169, 328
technology licensing office (TLO) 48, 49, 53,
    54–6, 75, 76
technology transfer 43–58, 136, 249, 291–2
Tekeuchi, Hirotaka 215
terrorism 336
tethered spin-offs:
    advantages 193–4
    control versus flexibility 207–14
    engines of innovation 205–7
Texas Instruments 247, 250, 252, 342
Thayer, Ann M. 152
Thelen, Kathleen 365, 366, 368
thin-film disks 253
Thursby, Jerry 322
Thursby, Marie 322

Thyssen 339
Tobisha 341
Tohoku University 80
Tokuhisa, Yoshio 225, 227, 228
Tokushima University 80
Tokyo District Court 115
Tokyo Fire and Marine Insurance 188
Tokyo Metropolitan Government 128
Tokyo Municipal Government 104
Tokyo Stock Exchange, Mothers 133, 146,
    170, 177, 180, 187
Torii Pharmaceutical Company 198
Toshiba 24, 25, 26, 27, 29, 39, 199, 255,
    278
Toyoda, A. Maria 162
Toyoda, Sakichi 222
Toyota 182, 192, 206, 225
Toyota Motors 221
trade secrets 252, 254, 284–5, 332, 358–9
trading companies 116
training 137, 278
TransGenic 146, 177, 186, 188
translation research 290
Trimble, Robert B. 219
Trow, Stephen C. 335
trust 363
Tsai, Alexander 187
Tsukuba University 146, 183
Tushman, Michael 314

Uchitelle, Louis 339
unemployment 284
United Kingdom 240, 243
    data 238–9
    SMEs 155
    University of Sussex, SPRU 316
United States 243, 299
    academic qualifications 81, 82, 336
    alliances 181
    angels 182
    bioscience finance 180–1
    biotechnology companies 42, 168, 226
    conflicts of interest 80–1
    consortia 351
    copyright law 324
    corporate pensions 184
    data 238–9
    Department of Health and Education
        (DHEW) 44
    Department of Labor 184, 327
    drug approvals 306–8
    drug innovation 22
    educational system 286

engineers 334–5
entrepreneurship 294–9
ERISA 184
faculty-company relations 57
Food and Drug Administration
    (FDA) 239–40, 243
government funding 57
Government Patent Policy 71
immigration 257, 304, 335
intellectual property 305
job stability 296–7
machine tool companies 264–5
manufacturing 259–60
MIT 57
Nasdaq 174
National Institutes of Health (NIH) 44, 61,
    62, 83–4, 295, 320, 321
National Science Foundation (NSF) 28,
    61, 62, 85
Patent and Trademark Office (PTO) 12–13
patents 1–2, 14–15, 44–5, 56, 247, 303–4,
    323–5
pharmaceutical companies 129–30
science funding 323
Silicon Valley 251, 254, 255–6, 258, 259,
    266, 334
software developers 27
spin-offs 211–12
Stanford University 103–4, 109, 113, 117
startup employment 94
tax law 328
technology transfer 291
therapeutic bioventures 146, 147
trade secrets 332
universities 65, 121, 196, 242–3
University of Illinois 251
university startups 245
venture capital 171, 338, 364–5
venture companies 19, 43, 168, 221
universities:
    access to patients 244
    and bioventures 131, 176
    collaboration with ventures 116–17
    and consortia 275
    donations to 46, 56
    drug discovery 241–2
    elite 63–4
    funding: activity-specific 53, 77;
        project-specific 46, 74, 302;
        unequal 63–6, 87
    joint research 78
    licences 321
    life sciences 245

universities: (*cont.*)
  ownership of intellectual property
    (IP) 43–4
  patent applications 4–5, 7–9
  preemption of discoveries 51–6
  relationships with companies 11–12, 21,
    28–30, 45, 57, 34, 289–91
  research 38
  role of 288–91
  room for reform 301–3
  spin-offs 292
  US 65, 121, 196, 242–3
  and venture companies 260–1
  windfall profits 368
University of Tokyo 7, 64, 79–80, 95, 137–45,
    147, 152, 163–4, 207, 220, 289, 295
university venture, definition 14
Unix 26
UNOVA 339
UP Science 208, 211

Van de Graff 72
Van Osnabrugge, Mark 182
venture capital 93, 132–3, 171, 338, 364–5
venture capital (VC) companies:
  angels 171–3
  hands-on 172, 186
  importance of 116
  investment levels 171
  nature of 171
  size of investment 184–5
Venture Enterprise Center 133
ventures/venture companies:
  alliances 144–5
  biomedical case studies 117–33
  collaboration with universities 116–17
  dearth of 261–2
  definition 15
  demographics of 258, 295
  employment 143–4
  family bias against 97
  finance 143–4
  foreign sales 113
  from GRIs 114
  government support 116
  importance of IP 114–16
  management 144–5

no-wet-lab 297
non-life science case studies 95–117
non-pharma 243–51
overseas experience 113–14
and pharmaceuticals 260
recruitment 152
role in drug discovery 239–43
and universities 260–1
US 19, 43, 168, 224
vitality of 232
see also bioventures; startups
Very Large Scale Integration (VSLI)
  project 269, 290, 342, 350
Vogel, Steven K. 368
von Hippel, Eric 227

Wada, Dr 201
Walsh, John P. 324, 325, 326, 328, 363
Watabe, Motoki 88, 363
Western Digital 251, 331
Westney, D. Eleanor 351, 352, 353, 361
Wetzel, William 182
Whittaker, D. Hugh 17, 225, 314, 340, 364
Willoughby, Kelvin W. 180, 182, 314, 329
Wilson, John W. 314, 321, 332–3, 337
wireless communication 111–12, 150–1
Wolfe, Josh 152
women's employment 283, 357
Wozniak, Steven 332
Wulf, William A. 335
Wyeth 322

Xerox 253, 256

Yamagishi, Toshio 88, 363
Yasunori 74
Yoshida, James A. 225, 227, 228
Yoshikawa, Akikiro 219
Yoshikazu, Nakayama 16
Yu, Robert 217
Yukio Sakamoto 213

Zahra, S. 226
Ziedonis, Rosemarie Ham 324, 326, 327,
  331
Ziegler, Frederick D. 219
Zucker, Lynne G. 33, 70, 319